www.wadsworth.com

wadsworth.com is the World Wide Web site for Wadsworth and is your direct source to dozens of online resources.

At *wadsworth.com* you can find out about supplements, demonstration software, and student resources. You can also send email to many of our authors and preview new publications and exciting new technologies.

wadsworth.com
Changing the way the world learns®

Environment, Energy, and Society

A New Synthesis

CRAIG R. HUMPHREY
The Pennsylvania State University

TAMMY L. LEWIS
Muhlenberg College

FREDERICK H. BUTTEL
University of Wisconsin

Australia • Canada • Mexico • Singapore • Spain • United Kingdom • United States

Sociology Editor: Lin Marshall
Assistant Editor: Analie Barnett
Editorial Assistant: Reilly O'Neal
Marketing Manager: Matthew Wright
Project Manager, Editorial Production: Jerilyn Emori
Print/Media Buyer: Robert King
Permissions Editor: Stephanie Keough-Hedges
Production Service: Peggy Francomb/ Shepherd, Inc.

Printed in the United States of America
1 2 3 4 5 6 7 05 04 03 02 01

Library of Congress Cataloging-in-Publication Data
Humphrey, Craig R.
Environment, energy, and society : a new synthesis / Craig R. Humphrey, Tammy L. Lewis, Frederick H. Buttel.— 1st ed.
p. cm.
Includes bibliographical references and index.
ISBN 0-534-57955-8
1. Environmentalism—Social aspects. 2. Environmental policy. 3. Social ecology. 4. Human ecology. I. Lewis, Tammy L. II. Buttel, Frederick H. III. Title.

GE195 .H86 2001
304.2—dc21 2001045316

Copy Editor: Julie Kennedy
Cover Designer: Janet Wood
Cover Images: leaf © PhotoDisc 2001, children © Bridget McCoy 2001, globe © Janet Wood 1995, all other images © Index Stock Imagery 2001
Text/Cover Printer: The Maple-Vail Book Manufacturing Group
Compositor: Shepherd, Inc.

Wadsworth/Thomson Learning
10 Davis Drive
Belmont, CA 94002-3098
USA

For more information about our products, contact us:
Thomson Learning Academic Resource Center
1-800-423-0563
http://www.wadsworth.com

International Headquarters
Thomson Learning
International Division
290 Harbor Drive, 2nd Floor
Stamford, CT 06902-7477
USA

UK/Europe/Middle East/South Africa
Thomson Learning
Berkshire House
168-173 High Holborn
London WC1V 7AA
United Kingdom

Asia
Thomson Learning
60 Albert Street, #15-01
Albert Complex
Singapore 189969

Canada
Nelson Thomson Learning
1120 Birchmount Road
Toronto, Ontario M1K 5G4
Canada

To Our Children

Gwen and Michelle (C.H.)
Anna and Isabel (T.L.)
and Allison (F.B.)

Brief Contents

Contents

Preface

A growing number of sociologists throughout the world are concerned about the impact of the biophysical environment on the processes by which societies change and how that social change, in turn, affects the environment. In this book we attempt to integrate *environmental sociology* in a comprehensive way. Although the book is appropriate for environmental sciences courses, our intended audience is advanced undergraduates and beginning graduate students in sociology. The book is also relevant to students in other sociological subdisciplines: demography, social movements, gender, development, and the sociology of energy, for instance.

Our primary goal is to update Humphrey and Buttel's *Environment, Energy, and Society* (Wadsworth, 1982), the first comprehensive synthesis of the field for students and colleagues. Substantial theoretical advances have been made in environmental sociology since 1982. Some of these advances include the greening of Marxist theory, incorporating gender into environmental analyses, and new work on ecological modernization. We have incorporated these new ideas into the text.

The world has also undergone significant environmental changes in the last twenty-five years. One of the new emphases throughout the book is on how the continuing process of globalization affects social-environmental changes. The world is increasingly connected through transnational relationships between actors from more-developed and less-developed countries. Transnational interactions around environmental issues take a number of forms. We analyze international conferences on environmental issues, development policies related to population and hunger, the world petroleum trade, the World Bank's environmental lending practices, and environmental movement oranizations' attempts to improve the environment not just in their home countries, but across borders. In addition to looking at these global connections, we include in each chapter examples from both the more-developed and less-developed nations. Some of the examples we analyze include the tragic death of the Brazilian rubber tappers' leader, Chico Mendes; the toxic waste incident in the Love Canal neighborhood; the role of American petroleum dependence in the Persian Gulf War; the anti-environmental movement; the role of corporations such as 3M and Ben and Jerry's in shaping a more sustainable world; and the future of international environmental agreements.

To help students draw connections between these issues and their lives, five of the chapters feature boxes that focus on the United States. These boxes illustrate how students make a difference in the environment, for example, through their consumption habits and activism. Because these changes are so substantial, we consider this work to be more than an update. We consider *Environment, Energy, and Society: A New Synthesis* to be a first edition.

Much of this book elaborates on the assumption that social structure and change can be best understood in terms of conceptual models or paradigms. We use three sociological paradigms—conservative, managerial, and radical—that are rooted in the work of the three most prominent classical sociological theorists—Emile Durkheim, Max Weber, and Karl Marx. Because *Environment, Energy, and Society: A New Synthesis* works with these scholarly legacies, the book has a structural orientation. These classical scholars' explanatory frameworks and empirical interests are clearly diverse; nonetheless, all employ broadly similar variables—culture, bureaucratic power, and social stratification. They also focus on the common problematic of understanding the origins and implications of capitalist industrialization. Drawing upon these classical works, we reshape them into policy-oriented analytical paradigms and show how these paradigms can be used sociologically to explain major changes in the biophysical environment.

The "pluralist-functionalist perspective," which we refer to with the shorthand of the "conservative paradigm," tends to assume that cultural values, beliefs, and attitudes are the key social forces in human societies because shared values are the basis of social integration and, hence, social organization. The conservative paradigm, consequently, seeks to explain social problems—including environmental pollution and natural resource scarcity—by identifying what social values or other cultural elements impel institutions into a state of disharmony in relation to the natural environment. The "power elite" or "bureaucratic elitist" perspective, referred to in this book as the "managerial paradigm," views political power and domination as the key force in society. Sociologists who hold this view, therefore, explain environmental problems in terms of the control that dominant elites have over the economy, the state, the mass media, and other social institutions, and stress the role of power and domination exercised through state bureaucracies. Finally, proponents of the radical or "social class or conflict" paradigm view environmental problems as the result of the action of the property-owning capitalist class and the inherent tendency of capitalism to be expansionist, a tendency that perpetuates the ever-increasing consumption of natural resources and the build-up of waste in the environment.

ORGANIZATION

Throughout this book we examine a variety of major societal-environmental interrelationships using the explanatory power of the three sociological paradigms. Chapter 1 traces the history and development of environmental sociology in the United States. We also discuss two exemplary studies in environmental sociology—one from the less-developed countries and one from the United States. Chapter 2 introduces the three sociological paradigms used throughout the book and explicates their analytical and explanatory styles by showing how each perspective has been applied to the issue of tropical deforestation. Chapter 3 discusses world population growth, population policy, and the relationship between population growth and contemporary environmental problems. Chapter 4 considers the development of modern agriculture, the environmental constraints of this development, and debates about the causes of world hunger. Chapter 5 discusses changing world fuel consumption patterns, the growth of American petroleum dependency, the link between that dependency and the Persian Gulf War, and alternative energy futures. Chapter 6 explores the contemporary environmental movement, featuring macro-, meso-, and micro-level sociological analyses of this enduring social movement. Chapter 7 focuses on the concept of sustainable development, the history of sustainable development as a concept, and conservative, managerial, and radical interpretations of sustainable development. Chapter 8 envisions the future of both environmental sociology and the environmental movement in the early twenty-first century.

ACKNOWLEDGMENTS

Our graduate students provided generous assistance in formulating and developing many ideas used in the book. Lucie K. Ozanne, now a lecturer at Lincoln University, Canterbury, New Zealand, helped us include the work on feminist theory and the environment. Anne Rademacher, now a doctoral candidate at Yale, helped in updating the chapter on food and world hunger. Erik Johnson, now a doctoral candidate in sociology at Penn State, researched energy and world population growth for multiple chapters.

Many undergraduates also played an essential role in this undertaking, and we thank the following students for their research and assistance: Sara Bauer, Allison Bentz, Kari Hernquist, Lindsay Hoffman, Lindsay O'Leary, Martin

Piotrowski, Hannah Smith, Lindsay Stenback, and Mark Stevens. Special thanks go to Elisabeth Workman for her outstanding preliminary editing.

We, Fred Buttel and Craig Humphrey, had wanted to revise *Environment, Energy, and Society* for many years but found ourselves inextricably involved in competing professional commitments. We tried, but the effort just would not get off the ground. That problem was solved when Professor Tammy Lewis joined us in 1996 to revise the text. Professor Lewis was and is the spark, the catalyst that allowed us to achieve a dream that had lingered with us for years.

The editors and consultants of Wadsworth have played a decisive role in all aspects of our endeavor. Eve L. Howard, Wadsworth's publisher, supported the project from the beginning. Lin Marshall, our Wadsworth sociology editor, and her able assistant, Reilly O'Neal, provided warm guidance, humor, and gentle prodding to keep us on time with the publication schedule. Wadsworth also led us to Shepherd, Inc. and Julie Kennedy, whose copyediting improved our writing style and streamlined the voluminous references and citations throughout the text.

Finally, we thank Wadsworth for commissioning members of the Section on Environment and Technology to review our first draft: Penelope Canan, University of Denver; Charles E. Cipolla, Salisbury State University; David J. Frank, Harvard University; Robert Gramling, University of Southwestern Louisiana; Ed Knop, Colorado State University; Harry R. Potter, Purdue University; Thomas K. Rudel, Rutgers University; Rik Scarce, Michigan State University; and Adam S. Weinberg, Colgate University. We take full responsibility for any errors, but these reviewers' constructive criticism and suggestions were essential to the quality of the book. Our sincere hope is that colleagues will reach beyond this work to define and explore further the continually developing field of environmental sociology.

1

Exploring Environmental Sociology

For the residents of Xapuri, on the western edge of Brazil's Amazon frontier, Christmas 1988 was not the same. Instead of celebration, Xapuri's mood was deeply mournful. Chico Mendes (Figure 1.1), a rubber tapper who extracted wild rubber, Brazil nuts, and other forest products for his income, a populist left-wing politician, and the organizer for the Xapuri Rural Workers' Union, was dead. Chico was a victim of a shotgun assassination at his own home on a muggy, moonless December 23 night, by a local cattle rancher, Darly Alves de Silva and Darly's son, Darci. More than a thousand people surrounded the Xapuri church singing praise, mourning, and bearing final witness to their martyr's burial. As Mendes's biographer, Revkin, later writes, "The funeral brought together the two sides of Chico's life—people from the forest and those from the outside (worldwide), who had found in this simple rubber tapper an indispensable ally."[1] Rubber tappers and friends of Chico's cause traveled hours by foot, and labor leaders, celebrities, and leftist politicians from around the world flew into Acre, Xapuri's Brazilian state, to participate in the procession of Chico's coffin to the site where he would finally rest, next to Ivair Higino de Almeida, another local activist assassinated in the same year as Chico.

At the time of his death, there were disturbing environmental anomalies occurring internationally. The United States suffered from an enduring heat wave destroying crops and starting forest fires in the West. As Yellowstone National Park suffered devastating losses of forests and wildlife from ravaging fires, national leaders were beginning to take a closer look at global warming policies

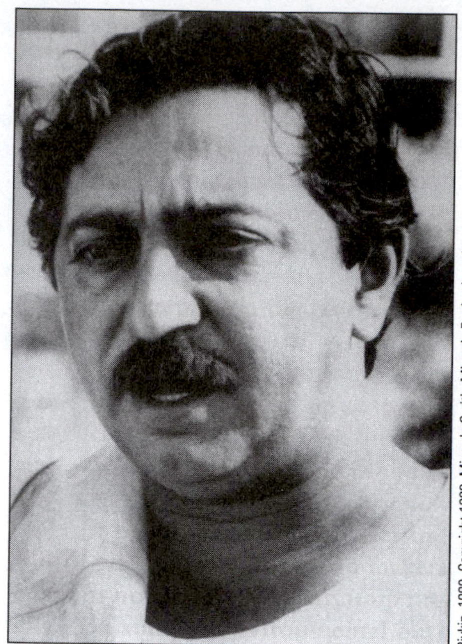

FIGURE 1.1 Francisco "Chico" Alves Mendes Filho

and the implications of the greenhouse effect. Newspapers and television showed picture after picture of dead forests, making it look like the world's forests were simultaneously burning and dying. Thus, the world was ready for—indeed, desperately needed—a human face to attach to the rain-forest cause.

So it was not surprising that international media transformed Chico Mendes into an icon. He symbolized grassroots environmentalism; several national news media labeled him the Ghandi of the rain forest. But the media's message was incomplete. Mendes was, in essence, a rubber tapper, a leftist politician, and a union organizer who was schooled as a child by a Communist political refugee. While the media portrayed Mendes as the leader protecting rain forests, he also extracted resources from the forest to make money to survive and live off the land.

What allowed Mendes to be different from other rain-forest entrepreneurs was his effort to replenish and revitalize the forest and to stop the destruction of slash-and-burn cultivation encouraged by the Brazilian government and cattle ranchers. As Mendes once said, "I'm not protecting the forest because I'm worried that in 20 years the world will be affected. I'm worried about it because thousands of people living here depend on the forest, and their lives are in danger everyday."[2] Thus, Mendes's cause and fight for the rain forest represented not just an environmental issue but one for social justice as well in the face of social inequality so pronounced in Brazilian society.

Mendes's struggle to protect the rubber tappers' natural resources began when he realized that Brazilian society was changing in ways that could make his way of life obsolete. By the 1970s Brazil was one of the world's most indebted countries and owed the United States over 120 billion dollars. To repay this debt Brazil launched a rural development plan to increase exports through mining, logging, and cattle ranching. The government provided roads, low-interest loans, and tax breaks to rural entrepreneurs, including the powerful cattle ranchers.

To obtain the loans ranchers had to demonstrate that they had a legal title to their land. The surest way to establish title was to clear and fence the land, selling valuable hardwoods to logging companies in the process. The land that the ranchers claimed, however, was land that the rubber tappers controlled. Only fraud and intimidation created the opportunities for state-subsidized ranching in the Amazon. Such tactics, as Warren notes, are not a problem on the Brazilian frontier, "home to thousands of highly professional claim jumpers and gunslingers, where courts, police, and politicians find bribery irresistible and investigations exhausting and bewildering."[3]

To protect the rubber tappers from the ranchers and negligent government officials, Mendes became politically active, first with the Xapuri Rural Workers Union and later through the National Workers' Party and a national organization of the rubber tappers. Chico taught his coworkers to resist the bribery of the ranchers and practice *empates,* Portuguese for stalemate. By physically surrounding land about to be logged for ranching, the tappers created a human barrier to the loggers' chainsaws. In 1985 the First National Rubber Tappers' Congress convened, attended by workers, representatives from ministries of industry and commerce, education, health, and agriculture as well as national congressional delegates. The Congress produced a list of demands focusing official attention to agrarian reforms: rubber and food policies and the need for improved health and education programs in rural Brazil.

As the ideas and proposals from the Congress captured Brazilian political leaders' attention, Mendes turned his political efforts internationally. In 1987 he traveled to the United States, meeting with members of Congress and, on a visit to Miami, meeting with those attending the Governors' Meeting with the Inter-American Development Bank. Mendes, on both occasions, urged officials to approve loans for rural development, such as public road building in the Amazon, contingent upon Brazilian forest land conservation programs. He also traveled overseas, making his cause known in Great Britain, the number one importer of Brazilian lumber and rubber. In 1987 Mendes received the prestigious Global 500 Award from the United Nations Environmental Programme, an award for outstanding contributions to environmental protection. Thus, for the first time, Mendes's efforts on behalf of the rubber tappers, other rural minorities, and the rain forest received worldwide attention at the highest levels of political power. Chico Mendes played no small role in wedding environmental protection and environmental justice.

By 1988 Brazil began to establish forest reserves in response to pressure from lending agencies such as the Inter-American Development Bank. One of the reserves was the Cachoeira, a closed canopy (undisturbed) rain forest near Xapuri. The Reserve was near the vast land holdings that Darly Alves de Silva and his brothers had assembled through threats and intimidation of the rubber tappers and Indians who struggled to survive on the frontier. Chico and his coworkers successfully stymied Darly and his family's efforts to expand their claims to the forest. Establishing the Cachoeira, from the Alves' viewpoint, suggested officials in Acre sympathized with Chico and the rubber tappers' cause. Darly, already wanted for murder in the Brazilian state of Parana, resorted to his most treacherous tactic. On the night of December 23, 1988, the social tension between the struggle to improve human rights through rain-forest conservation and frontier entrepreneurship reached a violent, tragic moment.

The cascading events in the Mendes tragedy exemplify the social forces at work that are of central interest to environmental sociology. Central to the tragedy are different groups struggling with social change. Mendes and the rubber tappers struggle to preserve their way of life as ranchers expand their control of forest land. A debt-ridden government, however, encouraged the ranchers' entrepreneurship. Brazil encumbered the debt in an effort to transform the Brazilian economy from one dependent on commerce and agriculture to one revolving around commerce and industry. The Inter-American Development Bank, a post–World War II development agency, financed by the world's most economically developed countries, created the opportunity for Brazil to do so.

While the entire sociological story encoded in the Mendes tragedy is hardly new—the story could be traced as far back as Columbus's quest for the New World or Marco Polo's expedition into the Orient—it does have a qualitatively new dimension. The rubber tappers' plight is now of concern to international environmental organizations who see the rubber tappers' interests as parallel with their own concerns about the protection of worldwide biodiversity. The politics of protecting human rights join with a newer social movement for the protection of the rights of nature. This new, more complex social movement, the movement's origins, and how the movement deals with issues of economic development and social change in countries, rich and poor, is an important intellectual focal point in the field of environmental sociology.

Given the increasing prevalence of stories such as the Chico Mendes tragedy, it comes as no surprise that leading scholars in environmental sociology find the field being revitalized in North America and quickly being institutionalized as an area of inquiry throughout the world.[4]

Environmental sociology is an emerging field with boundaries and an agenda continually being discussed and refined.[5] The initial character of environmental sociology, in our view, is decisively shaped by the intellectual climate or context within which the field developed during the late 1960s and 1970s. In this chapter, we will define environment and environmental sociology, we will present examples of exemplary work, we will discuss the historical roots of the field, and we will outline key issues in the field.

DEFINING TERMS

The concept of environment stems from the French word *viron,* meaning a circle, a round, or the country around.[6] The environment thus refers to "external conditions and influences affecting the life of an organism"[7] or entire societies, or, "the physical and biotic infrastructure" supporting populations of all kinds.[8] The environment is the physical and material bases of all life, including land, air, water, and the vital material resources and energy in which societies are embedded.

While scientists have been studying the environment for centuries, the field of ecology is relatively new. Ernest Haeckel, a German botanist, first used the term ecology in the *History of Creation,* published in 1868. Haeckel derived the concept of ecology from the Greek word *oikos*—a house or one's surroundings. The word ecology also provides the root for economics. Some writers descriptively refer to ecology as the study of nature's economy.[9] Hawley notes that, while the term came into use after Haeckel's publication, the science of ecology did not become an enterprise until after texts began appearing early in the twentieth century.[10]

The broadest ecosystem commonly envisioned for the earth would be the biosphere or the "thin film of life covering the earth's surface,"[11] embedded between the planet's crust and its unique atmosphere.[12] Within the biosphere, living species from human populations to the tiniest of amoebae constantly consume and recycle carbon, hydrogen, sulfur, oxygen, and other chemicals. Green plants consume carbon dioxide and sunlight, manufacture organic compounds and oxygen in the process of photosynthesis, and provide the converted biomass as food and energy for the animal world. Herbivores consume the biomass in plants, serve as food for carnivorous species including humans, and restore nutrients to the soil and water as they die and undergo biomass decomposition.

The action of producing and consuming biomass does not occur in the biosphere per se; the action occurs in the vast diversity of interconnected ecosystems that form the biosphere's building blocks. Each ecosystem has a dynamic structure formed by the diversity of resident populations genetically capable of surviving in the unique climatic, seismic, and botanical conditions present at any given biospheric location. Hierarchical structures of interdependent food chains are a common way of envisioning the division of labor within ecosystems. There is an inverse relationship between the biomass requirements of a species in a given system and the species' population size.

The complexity and size of populations within a given ecosystem are fixed by the matter and energy within. Thus, ecosystem growth theoretically changes from a rapid early growth phase when the indigenous populations are few and small in number to a later steady state. However, ecologists now recognize that the steadiness of this last stage of ecosystem development is turbulent.[13] Additional population growth beyond the carrying capacity of an ecosystem, a limit set by the available energy and nutrients necessary for biomass production, is not uncommon. Thus, dynamic population decline and rejuvenation are common, even within the so-called steady state of ecosystem development.[14]

Environmental Sociology

Buttel describes environmental sociology as an effort to understand the "material embeddedness of social life."[15] Environmental sociologists are interested in specifying the ways that society connects with the material or biophysical environment, the cultural values and beliefs that prompt people to use the environment in particular ways, and the eventual implications of these interrelationships for social consensus and conflict.

A good way to define the field involves a look at how members of the Section on Environment and Technology in the American Sociological Association describe their work to prospective members. Their brochure, "Let us tell you about the Section. . . ," reads:

> Many of society's most pressing problems are no longer just "social." From the maintenance of genetic diversity to the disposal of radioactive wastes, from toxics in the groundwater below us to global warming of the atmosphere above, the challenge of the 20th and 21st centuries are increasingly coming to involve society's relationships with the environment and technologies upon which we all depend. These interrelationships involve a range of complexities and implications that could not have been imagined in the early days of sociology. They are of sociological interest not just because of their significance to society, but also because of their significance to sociology. The Section on Environment and Technology was founded to provide a home for sociologists interested in exploring these issues.

Environmental sociology, then, focuses on interrelationships between society and the material world; how they act upon each other.

Michael Bell uses different words to convey the same idea about the scope and focus of the field.[16] Environmental sociology, for Bell, is the study of the *community* in the broadest sense. Environmental sociologists study the sometimes cooperative and sometimes deeply conflicting interrelations between people, the atmosphere, land, water, plants, and animals. As Bell writes, "Environmental sociology studies this largest of communities with an eye to understanding the origins of, and proposing solutions to, these all-too-real social and biophysical conflicts."[17]

ENVIRONMENTAL SOCIOLOGY: TOPICS AND TYPES

Four major research topics predominate in environmental sociology: (1) analyses of natural settings that are relatively unchanged, or at least not intentionally altered, by human activity, such as research on organized efforts to protect biodiverse ecological communities; (2) analyses of modified, often rural, environments affected either by unwanted physical intrusions such as the contamination of groundwater by toxic chemical pollution or by planned modifications such as reservoir impoundments and offshore oil drilling; (3) the examination of built environments (not entirely separate from modified environments) including the impacts of accidents involving the chemical contamination of buildings and accidents in nuclear power plants; and (4) analyses of issues such as the environmental movement as a reaction to environmental problems; the analysis of the women's health movement and family planning in relationship to population growth; or, the merging of struggles by grassroots groups such as the Brazilian rubber tappers and international environmental organizations concerned about protecting tropical rain forests.

The specific focal points of environmental sociology are diverse because humans relate to the environment in both symbolic and nonsymbolic ways. In strictly symbolic analyses environmental sociologists question how different groups compare and contrast in their views about problems such as the prospect of global warming, irrespective of whether global warming really is a biophysical problem. Nonsymbolic work may also be symbolic, involving cultural values and attitudes, but this kind of environmental sociology also examines the material or biophysical consequences of behavior. For example, there is work on human migration into rain forests and consequences of migration for the forest;[18] or, the impact of commercial offshore oil drilling on community growth in nearby coastal areas.[19]

Table 1.1 identifies these major interests of environmental sociologists. Studies of the overall growth of the U.S. environmental movement since the late 1960s or the movement's current structure involve the symbolic aspect of environmental issues.[20] Analyses of social inequalities between neighborhoods, racial composition, and exposure to toxic chemicals exemplify a kind of environmental sociology examining the objective consequences of connections between the material environment and groups of people.[21] Some of this research does not involve specific environmental contexts. We include work on trends in public opinion about environmental protection, nationally or internationally, in this category.[22] This work focuses on people's reaction to general environmental conditions, even though these conditions may not directly impinge on their own personal lives.

An important recent development in purely symbolic environmental sociology is work on the social construction of environmental problems. In this development "the process of claims making is treated as more important than the task of assessing whether these claims are truly valid or not."[23] The controversy over the prospect of global warming, including analysis of the scientists and institutes supporting research on both sides of the issue, is exemplary.[24] For scholars with an interest in the significance of different paradigms within scientific communities, and how these paradigms shape the development of environmental science and policy debates, the social construction approach is a welcome addition to the field.

An equally important example of the social constructionism is the globalization of official interest and concern for environmental issues and problems

Table 1.1 A Typology of Research Areas in Environmental Sociology

Type of Environment	Examples	LEVEL OF INTERACTION	
		Symbolic	**Nonsymbolic**
Natural	Atmosphere; forests; lakes and oceans; mineral deposits	Northern Spotted Owl controversy; growth of ecotourism; controvery over resource extraction in areas designated as wilderness	Effects of energy policies on global warming; impact of national economic development policy on rain forests; plant genetics and the changing structure of agriculture
Modified	Community air and water pollution, toxic waste depository siting, soil erosion, land use controversies	Social inequality in exposure to toxic waste; Third World women and natural resource conservation; growth controls in communities with scenic beauty	Soil erosion and poverty; capital intensive extraction technology and ecological simplification, oil spill controversies
Built	Technological accidents, impact of highways and airports, housing, commercial building design	Anti-high rise movements; nuclear power controversies; rent control efforts; equity problems in gentrification	Impact of increased capital mobility on urban landscape; recycling and resource conservation; inequality in exposure to industrial waste products
Social	Family planning policies; population redistribution policies; controversy over local development efforts	Feminist critique of family planning; ecofeminism; politics of abortion policies; environmental attitudes; the environmental movement	Behavioral consequences of population density; impact of noise pollution on communities

SOURCE: Adapted from Dunlap and Catton (1975b:77) and Humphrey and Buttel (1982:5).

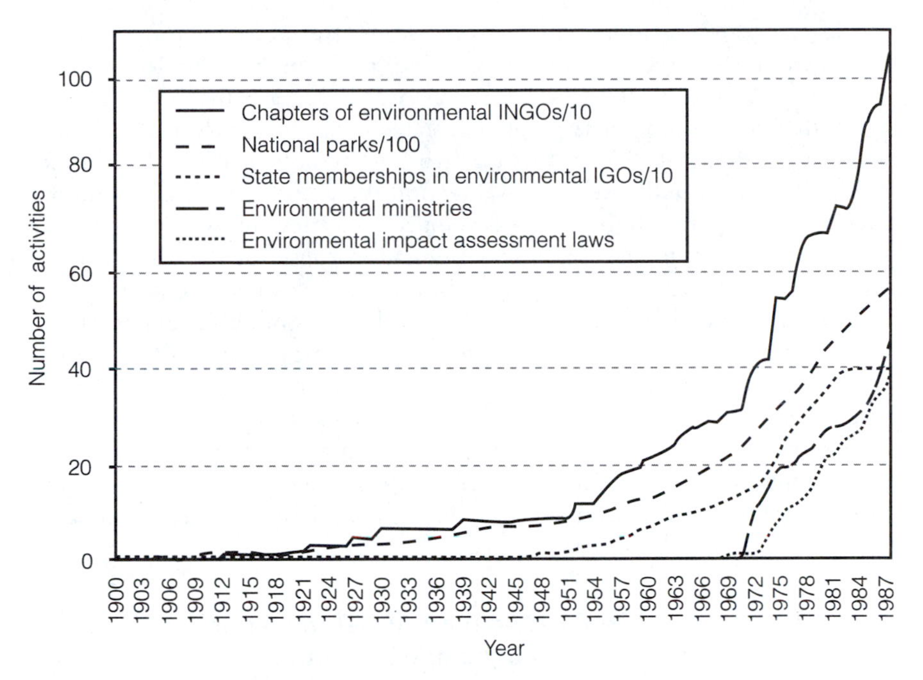

FIGURE 1.2 The Cumulative Numbers of Five National Activities, Worldwide, 1900–1988

NOTE: INGOs are international nongovernmental organizations; IGOs are intergovernmental organizations.

SOURCE: Frank, Hironak, and Schofer, 2000.

among nation-states. Political action such as the making of international environmental conventions: the Vienna Convention for the Protection of the Ozone Layer (1985) or the Montreal Protocol on Substances that Deplete the Ozone Layer (1987) exemplify this globalization process.

Other examples of the globalization of environmentalism include the passage of new environmental legislation in many countries and the formation of nongovernmental international environmental organizations such as Greenpeace International and the World Wide Fund for Nature. Figure 1.2 dramatically indicates the worldwide growth of chapters of international nongovernmental organizations, of public officials' efforts to set aside land for national parks, of the establishment of national environmental ministries, and nationally mandated environmental assessment prior to initiating public works such as roads, bridges, and flood control projects. The steep growth in environmentalism is especially pronounced since the formation of the United Nations in 1945; since the first UN Conference on the Environment in Stockholm in 1972; and, since the formation of the UN Environmental Programme—a product of the 1972 Stockholm Conference. As the world became more politically integrated, and as scientists became more fully aware of human impacts on the biosphere, environmental regulation and policymaking became a routine task in more and more nations worldwide.[25]

Several excellent studies illustrate the nature of work in environmental sociology. Bina Agarwal applies feminist theory to the problem of deforestation in the less-developed countries.[26] Agarwal's work illustrates the growing interest in the relationship between gender, state economic development efforts, and the environment. While Agarwal's work is symbolic because she analyzes culturally transmitted patriarchal values and the pro-development commitments of public officials in the less-developed countries, her work also is nonsymbolic. Agarwal focuses on how deforestation physically changes the daily work routines of rural women villagers in the less-developed countries.

Work by Adeline Levine traces the symbolic roots of public opposition to toxic waste sites and the environmental justice movement in the United States.[27] Levine's work is also nonsymbolic because she focuses on a built environment where tons of underground toxic chemical waste seeped into the basements, playgrounds, and swimming pools in a working-class residential community. Women, in both of the following cases, take the lead in social action to deal with an environmental problem, something that will occur again and again throughout this book.

An Ecofeminist Account of Deforestation in the Less-Developed Countries

Francoise d'Eaubonne coined the term, ecofeminism, in 1974 to identify theoretical work on the potential for women to bring about an ecological revolution and to ensure survival on the planet. Ecofeminists' imperative argument, in essence, is that the domination of women parallels the domination of nature. This mutual domination of women and nature leads to environmental destruction and gender inequality by the controlling patriarchal society.[28] Ecofeminists take as their principal theoretical concern the identification of the connections between the twin oppressions of women and nature. These principal ecofeminist concerns include a critique of patriarchal thinking that sanctions and legitimates this oppression of women and nature.

Agarwal's *Cold Hearths and Barren Slopes* is truly exemplary ecofeminism. *Cold Hearths and Barren Slopes* shows that, for a large proportion of people in the less-developed countries, firewood constitutes the main and, for some, the sole source of inanimate energy. Given that in the majority of these countries 70 percent or more of the population resides in rural communities, wood may be the most important source of domestic energy, especially for cooking, in use by people in the less-developed countries. Because women are the prime movers in the daily care and feeding of their families in these countries, deforestation is directly connected to their ability to play their family caretaking roles.

Not widely recognized until the past decade, the shortage of wood fuel in the less-developed countries now is a problem of crisis proportion. The shortages are evident in women's longer hours spent gathering fuel and fodder; in families going hungry because they do not have enough firewood to cook their food; in the rising prices of firewood in cities; and, in the shift to cattle dung and other substitute fuels for domestic use. The wood fuel crisis is evident in the growing confrontations between forest communities and forestry officials; and, in the growing hardship of people's lives with increasing ecological degradation.

Although the wood fuel shortage is of a crisis proportion, this energy-related hardship is not experienced equally throughout the less-developed countries. The crisis most severely affects the poorest households, those who depend on the wood fuel gathered free on village public commons and state forests. While more affluent households will either purchase wood fuel or substitute other fuels, the poor will forego the purchase of other goods to purchase their fuel. Along with the poor, women also bear additional burdens, as mentioned earlier, because they are the traditional wood gatherers.

Agarwal insists that the wood fuel crisis is not a result of population growth and poverty that drives more and more poor women to gather wood fuel. The crisis results "from the particular uses to which forest land and wood resources have been put over the years by specific classes of people, and cannot be traced in any straightforward way to the gathering of firewood by the poor for their domestic use."[29] The depletion of forests can be traced to social conditions created by state officials. As in the Brazilian Amazon, economic development-oriented state officials encourage the privatization of large tracts of forest land for export crop cultivation, pastures, plantations, and industrial development.

State privatization of land, according to Agarwal, perpetuates a major aspect of the wood fuel crisis—the maldistribution of material resources between different classes, ethnic groups, and the sexes, which affects both the absolute availability of wood and its relative distribution between users. The use of land and the effects land uses have on deforestation reflect the unequal distribution in the ownership of and access to land among people, a land tenure pattern. The persistence of illegal cutting of government-owned forests by local timber merchants, who make large profits with these state-controlled resources, is closely linked to the economic and political power that these merchants command.

The solution to the wood fuel crisis in less-developed countries such as India, according to Agarwal, is state-based agrarian reforms. Crucial to this reform is a more egalitarian land tenure pattern both with respect to land ownership and with respect to access to land. Ensuring alternative, viable means of livelihood for the poor, according to Agarwal, is equally crucial in finding a solution to the wood fuel crisis. The now prevalent grassroots social movements in less-developed countries, often led by women, can serve as the catalysts for these reforms. The Chipko Movement, revived in 1972 within the Indian Himalayan foothills to protest the commercial harvesting of trees, is one such catalyst. Another catalyst for agrarian reform are the protests against potential deforestation and land loss from proposed hydroelectric projects in India's Narmada Valley and in the state of Kerala. Women's militancy in these instances reflects their family survival values. Women organize and work for "an existence that is based on equality, not dominance over people, and on cooperation and not dominance over nature."[30]

Because of these grassroots social movements, state-based technical assistance programs to alleviate the wood fuel crisis are being closely scrutinized. Critics say efforts in state programs to promote improved wood-burning stoves, for example, have a gender bias. Although women are the potential users of the innovation and are in the best position to evaluate the merits of new wood-burning stoves, men usually handle the little household cash available and make decisions on how they spend their money. Rural women usually have no direct access to institutional credit or to disposable cash to purchase new innovations. Also, women seldom have access to information on innovations such as more efficient stoves. The state extension workers who provide information about household energy production usually are men, and there often is an ideological bias in extension services that works against the direct involvement of or consultation with village women. For ecofeminists such as Agarwal, equal training and opportunities for women to participate in the development and implementation of these kinds of state-based technical assistance programs and equality in access to credit are keys to programs' success.

In addition to the problems associated with the diffusion of wood stove technology, state-based reforestation programs or social forestry to expand the supply of wood fuel have achieved only limited success. Social forestry has not been successful, even in the sense of ensuring the planting and maturing of trees, let alone providing for the daily needs of fuel, fodder, and food among poor rural households. Grassroots groups, in many instances, actively oppose social forestry programs. In Niger, Senegal, Ethiopia, the Philippines, and several Indian states, villagers pull down the planted trees, uproot eucalyptus saplings, plant trees upside down, and start forest fires. In other instances, a passive resistance to schemes appears in the poor success of most government-promoted, community self-help projects, reflecting half-hearted efforts by local villagers.

Agarwal finds the reasons for such hostility and limited success of the projects embedded in a range of interlinked social forces that begin with all the levels of the state. Reforestation programs largely fail because of inadequate involvement of local people, especially women, in the design and implementation of programs. They also fail because of social inequalities in village class structure. Owners of large blocks of land can use part of a block for reforestation programs, and the profits forest land owners make from the sale of wood fuel exacerbates village inequality. The antagonistic attitudes of state forest officials toward local people also contributes to program failure. Finally, the programs do little to satisfy the purported aims of social forestry, namely the provision of wood fuel and other forest products for rural people's domestic needs.

Agarwal identifies successful social forestry programs in South Korea and China. She attributes their success to the relatively egalitarian distribution of land and wealth in these countries. Effective village leadership also is instrumental, but the effectiveness of governance and leadership is greater where the village class structure is relatively homogeneous. For ecofeminists such as Agarwal, state-based development programs that work to eliminate gender biases in the development and implementation of programs to deal with the wood fuel crisis and the reform of access to productive land and capital are essential policies to achieve social justice and ecologically sound land uses.

The Movement against Locally Unwanted Land Uses

A second example of work in the field, one involving the sociology of the built environment and also one where women play a leading role, is the growing literature on public opposition to locally undesirable land uses such as toxic waste facilities. We draw upon the writing of three environmental sociologists, Andrew Szasz, Adeline Levine, and Robert Bullard, to illustrate this example.

Szasz's *Ecopopulism: Toxic Waste and the Movement for Environmental Justice* empirically demonstrates the growing "public dread" of hazardous waste sites near homes and residential neighborhoods across the United States beginning in the late 1970s. Figure 1.3 shows the growing public opposition to toxic waste facilities. A national survey conducted by the U.S. Environmental Protection Agency (EPA) in 1970 indicates that a substantial majority of adults, at the time, were willing to live within five miles of a hazardous waste facility. A survey by the President's Council on Environmental Quality ten years later indicates that the majority of Americans wanted no part of living near a hazardous waste site even fifty miles from their homes. By 1980 hazardous waste sites were nearly as welcome as living within fifty miles of a nuclear power plant, another dreaded land use.

According to Szasz's *Ecopopulism,* American popular opposition to toxic waste facilities reached a turning point in 1978. At that time the national news media carried vigilant coverage of what became a two-year episode with a serious toxic waste incident in the working-class neighborhood of Niagara Falls, New York, known as the Love Canal. The frightening episode in the Love Canal neighborhood began thirty-five years earlier. Between 1942 and 1954 Hooker Occidental Chemical Company dumped 21,000 tons of waste products in an abandoned canal ten feet deep, sixty feet wide, and one thousand feet long. Figures 1.4 and 1.5 show the location of the community, the neighborhood, and the canal. W. T. Love built the canal in the late nineteenth century as part of an unsuccessful industrial development effort along the Niagara River. With few regulations for any form of waste management at the time, Hooker readily obtained the right to use the clay-lined canal as a chemical waste dump in 1942, nearly fifty years after Love abandoned his effort. Between 1942 and 1952 Hooker filled the canal with benzene hexachloride, mercaptans, sulfides, and quantities of the very toxic dioxin, among many other chemical wastes. In 1953 Hooker sold the now earth-covered site and surrounding land to the Niagara Falls School Board for one dollar, exonerating the company of any liability from future uses of the site.

In 1954, during the zenith of American suburbanization, the Niagara Falls School Board opened the 99th Street Elementary School at the edge of the canal, shown in Figure 1.5. The board also sold portions of the parcel once owned by Hooker to the city of Niagara Falls and to a private investor. The

Cumulative Percentages of Americans' Willingness to Accept a Waste Disposal Site at Various Distances from their Homes, 1970

	Willing	Unsure	Unwilling	Don't Know
Within 1 mile	36.8	27.9	35.0	0.3
Within 5 miles	57.9	21.4	20.4	0.3
Within 10 miles	67.4	18.8	13.5	0.3
Within 25 miles	70.1	17.4	12.2	0.3

SOURCE: U.S. Environmental Protection Agency, 1973.

Cumulative Percentages of People Willing to Accept New Industrial Installations at Various Distances from their Homes, 1980

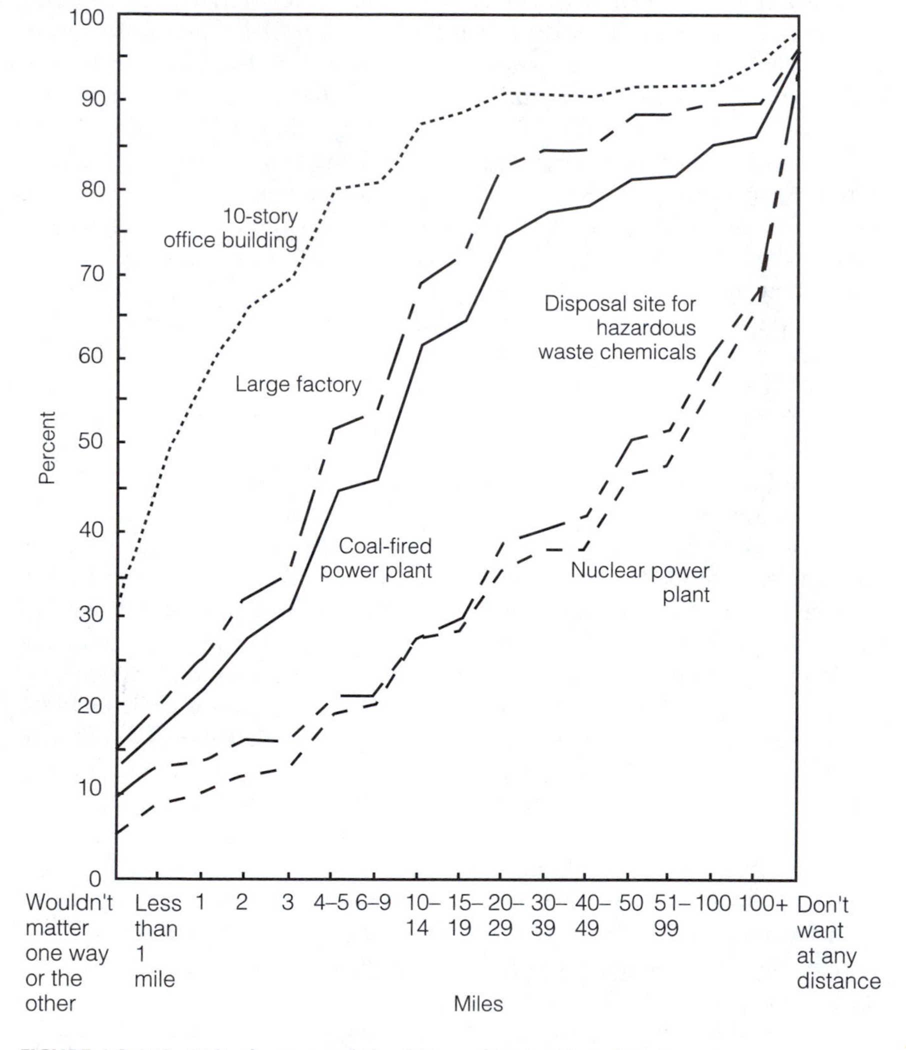

FIGURE 1.3 U.S. Attitudes Toward the Siting of Toxic Waste Treatment Facilities, 1980

SOURCE: U.S. Council on Environmental Quality, 1980.

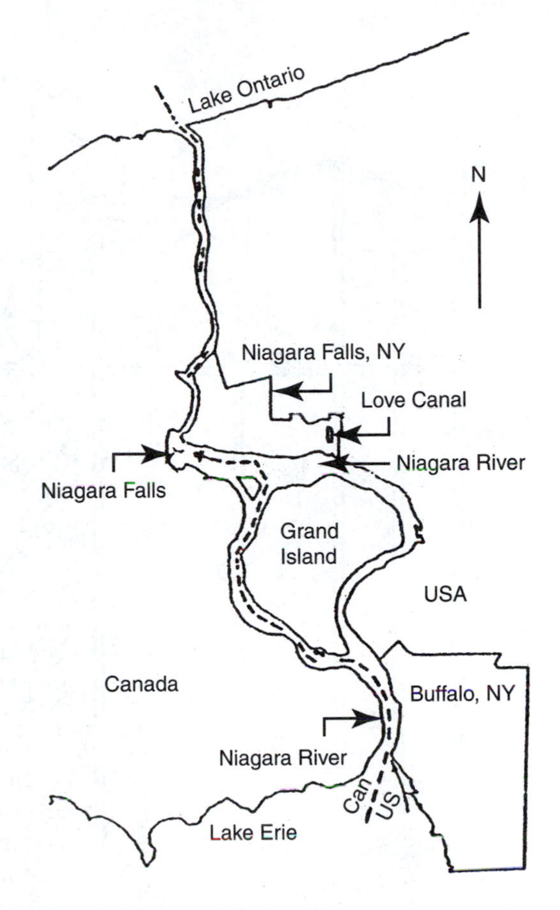

FIGURE 1.4 Niagara Falls, New York, and the Love Canal

SOURCE: Levine, Adeline. 1982. *Love Canal: Science, Politics, and People.* Lexington, MA: D. C. Heath, and Co.

road construction, water and sewer lines, and other environmental modifications that followed shortly thereafter not only increased the accessibility and desirability of the area for residential subdivision, but also broke the clay lining of the canal—something no one knew about. Thus, the subdivision activities created the potential for the subsurface migration of chemical leachates throughout a new and growing working-class suburb.

Nearly twenty-five years passed before, in 1976, an unprecedented spring snowmelt pushed toxic chemicals from below the surface of the Love Canal neighborhood into the homes and yards of the mostly young families near the school. Malodorous fumes in people's basements became annoying; house foundations cracked; and underground swimming pools heaved with the upward pressure of groundwater and waste chemicals. Chemical residue began appearing inside basements and pools, and water-pumping equipment soon corroded or clogged from the harsh chemical substances.[31]

During the discovery phase of this toxic waste incident, according to Adeline Levine's *Love Canal: Science, Politics, and People,* the local residents encountered an important social transformation. At first they experienced their

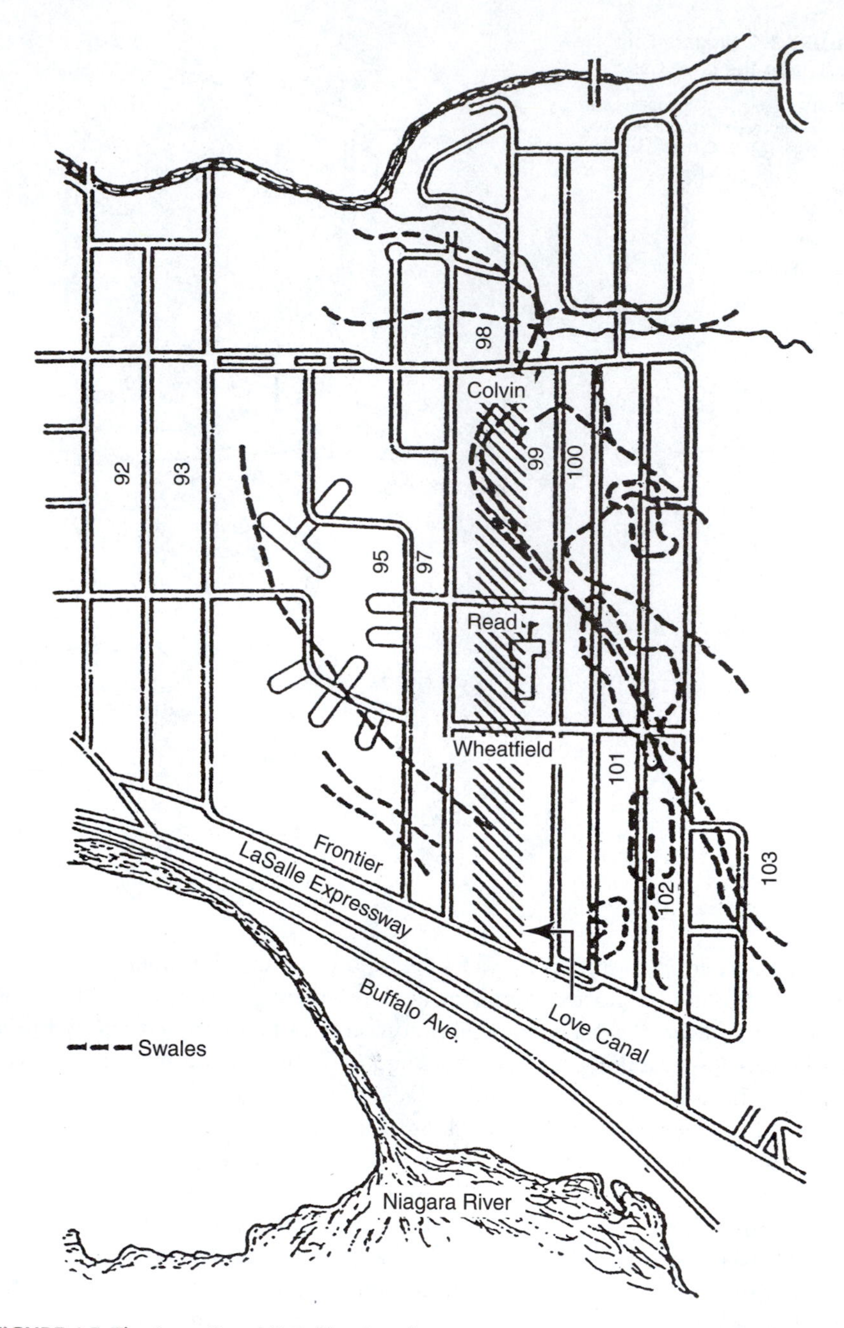

FIGURE 1.5 The Love Canal Neighborhood

NOTE: The 99th Street School building is located between Read and Wheatfield. The LaSalle Housing Project (Griffon Manor) is between 93rd and 95th Streets.

SOURCE: Levine, Adeline. 1982. *Love Canal: Science, Politics, and People.* Lexington, MA: D. C. Heath, and Co.

annoyance and fear as individuals concerned about property values, their own personal health, and, of course, the health of their children. But as public officials at the local, state, and federal levels began recognizing the seriousness of the Love Canal neighborhood problem, and as they saw themselves on local and national television as well as state and national newspapers, legitimating the fears about local toxic chemicals that some had harbored for more than a decade, their sense of themselves changed.

Meetings of the newly created Love Canal Homeowners Committee in 1978, a group with over 600 members supported by dues, churches, the United Way, several volunteers, professional public health researchers, the United Auto Workers, and the New York State Teachers' Union, brought new meaning to the residents' predicament. Mothers at the meetings learned that a troubling number of other families in the area shared their personal sense of loss from a miscarriage or stillbirth. Everyone became painfully aware of an unusually large number of birth defects among local children. Residents collectively began to know they shared the role of blameless victims in a disaster that was far beyond their individual or even collective ability to remedy. Thus, the committee as a group identified public officials as the only recourse for an escape from their dire health and financial circumstances.

When Calspan Corporation, an engineering firm, advised Niagara Falls officials in 1977 about the magnitude of the toxic waste problem in the Love Canal neighborhood, however, the local officials voted not to take responsibility for the problem, even though the school board granted the land to the city. When the state health department announced a program to provide blood tests for local residents, 4,000 people arrived for the tests, only to find a mobile public health examination unit staffed by four medical workers. Those residents who managed to be tested received a twenty-five-page questionnaire about their medical histories with only 10 percent of the items asking anything about their children's health. When state health officials measured the air quality in residents' basements that same summer, in 1978, they sent back chemical reports that no one could interpret.

Led by Lois Gibbs, a mother whose children developed very serious health problems after enrolling in the 99th Street School, one with epilepsy and a second with a rare blood disease, members of the committee mobilized politically. The committee quickly learned how to wage a protracted protest through strategic, media attention-getting public events. When state highway officials announced their plan to excavate the chemicals in the canal, residents, mostly mothers, picketed the site for days, claiming that the officials' efforts would create more public health risks. A group of residents marched on the state capitol in Albany carrying empty, child-sized coffins. Gibbs and her coprotestors took every opportunity possible to announce official intransigency, indecision, and insensitivity through the local and national press and, more importantly for the development of public dread, through national television. By 1980 local residents' disillusionment with public officials they had unquestionably trusted for their entire lives reached a boiling point. They captured and held EPA officials hostage in the 99th Street School where the committee kept an office.

The hostage-taking event occurred in the spring of an election year, 1980. This desperate act on the part of the committee, of course, attracted substantial media coverage, both in the national press and major televison stations. The hostage taking also prompted official actions at the highest levels of government, although the political decision making about to unfold had been in process for at least one year. An EPA official called Gibbs by phone, telling her that President Carter would authorize federal funds for the relocation of 700 families away from the area. Senator Jacob Javits called to say that the federal government loaned New York state 15 million dollars to buy their homes. The state attorney general, soon thereafter, filed a 625 million dollar lawsuit against Hooker Occidental.

The public dread of toxic waste facilities, a part of American collective consciousness in the wake of media coverage of the Love Canal episode, surfaces again and again. A significant marking point in what environmental sociologist Robert Bullard calls the environmental justice movement occurred in 1982, only two years after the unprecedented resolution to the Love Canal episode.[32] Police arrested Washington, D.C., congressional delegate Walter Fauntroy, leaders of the Southern Christian Leadership Conference, and over 400 other demonstrators in 1982 as they protested North Carolina's decision to use the African-American community of Afton as "the final resting place" for 3,200 cubic yards of soil contaminated by polychlorinated biphynels (PCBs). Someone illegally dumped the contaminated soil along roads in fourteen North Carolina counties. For congressional delegate Fauntroy and his fellow protestors, the decision to put the PCBs in Afton was a racially discriminatory action that they suspected to be common, nationwide. Later empirical research by government and private research teams confirmed their suspicion.[33] The public protest in Afton, in conjunction with the empirical research, marked "the first major attempt to link the pursuit of an unpolluted environment to the mainstream civil rights agenda."[34] The Afton protest would not be the last for African Americans and others struggling to avoid locally undesirable land uses throughout the United States.[35]

More and more environmental sociologists are observing and writing about African-American mobilization for environmental justice, led by civil rights activists. Michael Bell, discussed earlier, envisions the field as the study of the causes and proposed solutions to conflicts between human groups and the biophysical environment. In the case of the environmental justice movement, both scholars such as Robert Bullard and civil rights activists insist that racism explains the origins of higher risks of exposure to toxic chemicals in the residential communities of racial minorities. Because state officials and commercial waste handlers envision minority communities as politically powerless, they target these places as sites for toxic waste facilities that few, if any, communities want. Civil rights activists, responding to official and corporate racism, strive to empower minorities in these communities to resist proposals targeting them with the toxics of profit-making industrial activities. Racial minority empowerment and public protest become civil rights activists' solutions to environmental injustice.

The victims of exposure to toxic chemicals at the Love Canal, like minority communities, faced official insensitivity and ineptitude. Learning about their collective predicament empowered the white working-class homeowners living near the Love Canal. The homes, representing their single largest asset, and their neighborhood literally presented the source of a dreaded exposure to toxics. Unless the homeowners successfully engaged the officials' help, a health-threatening environment of unmarketable homes entrapped them. Impelled by local women such as Gibbs, the homeowners' empowerment and repeated public protest led to an unprecedented resolution to their predicament through the federal home buyout. Once again we have exemplary environmental sociology with a grassroots movement leading to a resolution to a conflict between human groups and the biophysical environment.

In rural villages of the less-developed countries such as India, one also finds growing social empowerment and repeated protest, especially by rural women, over state-directed natural resource policies. These women mobilize to protect village commons and state forests so they can play their traditional roles as the family caretakers. Patriarchy and male-controlled economic development programs explain the origins of women's environmental protests in the less-developed countries.

These cases are exemplary environmental sociology because they portray the material embeddedness of people in forests and a built suburban environment. They also portray how changes in people's embeddedness within the natural world can affect groups differently according to their race, gender, and class. The field of environmental sociology, however—as we will show throughout this book—goes beyond social movements provoked by deforestation or class- and race-based exposure to toxic chemicals. Earlier in the chapter, for example, we examined important new work in the field on post–World War II globalization of environmental concern. The growing concern has impelled new international environmental agreements, new national ministries to regulate the environment in many countries, and more and more nongovernmental national and international environmental organizations working to see that these agreements are enforced around the world. People with environmental concerns organize, protest, and press powerful actors to act on their behalf through appropriate environmental decision making, creating potential precedent for future government action as well. Observing this globalization of environmental issues and the social dynamics that perpetuates the process is a crucial part of environmental sociology. The origins of the field will be discussed next.

THE ROOTS OF ENVIRONMENTAL SOCIOLOGY

We argue that there are five formative influences on the development of environmental sociology. Human ecology, developed in the 1920s at the University of Chicago and refined in the 1950s, came close to being part of environmental sociology. Human ecology's shortcomings, however, provided a driving force in the development of this new field. The rise and revitalization of the environmental movement is a second formative, but wavering influence. A third formative influence is the intellectual development of the New Environmental Paradigm (NEP) by pioneering environmental sociologists. Fourth, the important and long-standing interest in natural resources within rural sociology provided early momentum to form important, much needed professional organizations for environmental sociologists. Finally, urban sociology has been a formative influence on the field.

Human Ecology

Durkheim's first work, *The Division of Labor and Society* (1893), represents an especially important precursor to environmental sociology. How social organizational complexity evolves from human population growth and density represents the essential theoretical interest in *The Division*. According to Durkheim, the finite quality of the natural resources needed by a growing human population, sociologically, is crucial. Population growth can perpetuate intense interpersonal competition and conflict over scarce resources. Durkheim recognizes the problem of resource scarcity can be disastrous in societies where there is little, if any, opportunity for technological innovation or entrepreneurship, and, where social life involves routine, traditional activities and beliefs—a condition he calls "mechanical solidarity." Durkheim, however, optimistically recognizes the inherent ability of human societies to solve problems with natural resource scarcity through the process of dynamic density. Societies faced with the prospect of critical natural resource scarcity, according to Durkheim, can adapt, engaging the power of human communication and interaction ("intra-social relations") for innovation to escape life-threatening ecological conditions. The division of labor becomes more complex in this process because, through innovation, new roles and new economic opportunities develop. People adapt to natural resource scarcity, such as a fuel wood crisis, drawing upon existing social institutions: science, technology, and the fiscal resources of the state. As human populations encounter problems with natural resource scarcity in the process of industrial development, they thus draw upon cultural and institutional resources to produce more food, find new ways to move about in denser communities, and create new opportunities for earning a living. Although Durkheim's cause-and-effect relationship is unidimensional (the environment affects society through impinging scarcity, but not vice versa), *The Division* does represent an early attempt to consider the role of the environment in sociological analysis.

Sociologists Robert Park and Ernest Burgess, influenced by Durkheim and plant and animal ecologists, developed the field of human ecology at the University of Chicago in the 1920s.[36] Park and Burgess theorize that competition and cooperation are the key forms of human exchange for populations to organize, to adapt, and to survive within a dynamic, ever-changing environment. For Park and Burgess, how people compete and cooperate through

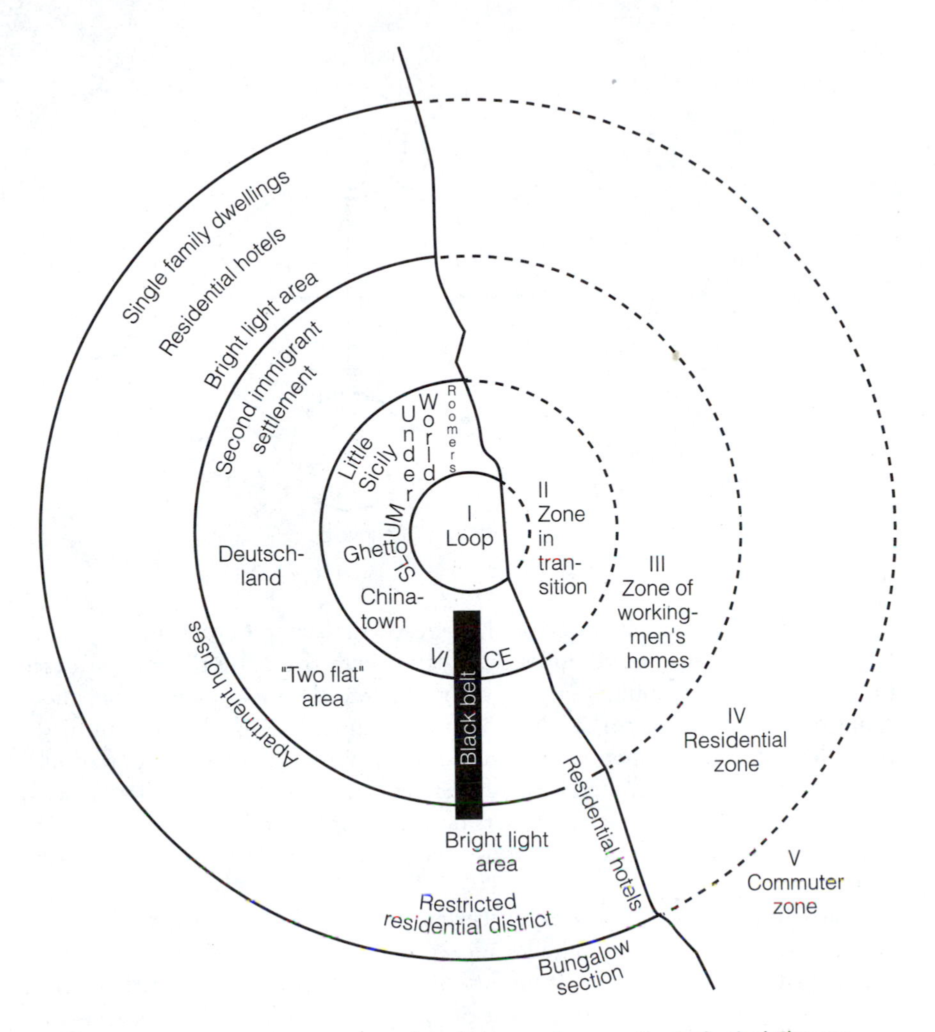

FIGURE 1.6 The Concentric Zone Model of Human Community Ecological Change

SOURCE: Burgess, 1925.

population distribution and redistribution under conditions of rapid industrialization, urbanization, and ethnic diversification are the imperative questions in the field.

Because Park and Burgess used Chicago as a natural laboratory for their ecological research, the field took a decidedly urban cast from inception, and the dynamism of metropolitan growth became human ecology's compelling research problem. Park and Burgess view the community as an interconnected system constantly changing with net migration and natural human population growth fueling a Darwinian struggle for the control of urban space. The expanding industrial organization of Chicago's population perpetuates the centrifugal drift of people away from the expanding commercial and industrial core, Zone I or the Loop in Figure 1.6.

FIGURE 1.7 The Human Ecological Complex

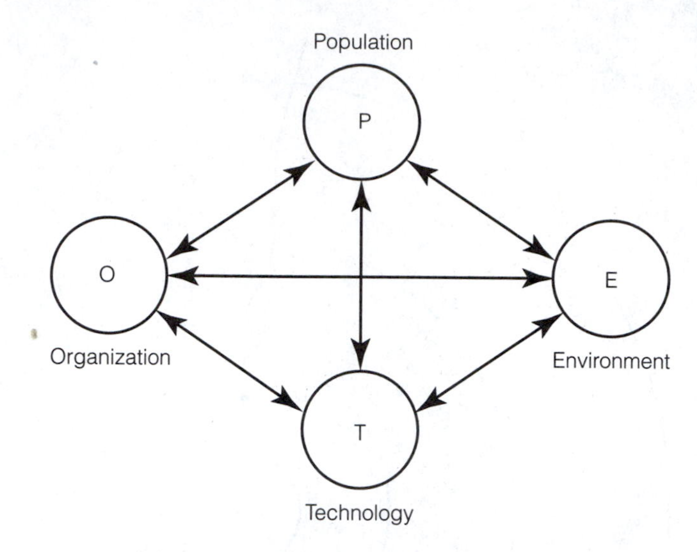

As the Concentric Zone Model suggests, the struggle to control urban land results in a segregated residential pattern with the "fittest" or wealthiest members of the community buying their way into the most sought after outlying residential neighborhoods. The poorest members of the community, those least able to compete, reside in the deteriorating neighborhoods about to be taken over by commercial land users in the expanding Zone in Transition (Zone II). Commercial land users dominate or control the choicest, most expensive central locations, the most geographically accessible part of the community.

Early human ecologists consider certain properties of the environment to be more salient than others. The location of communities, travel time, accessibility to and from particular community sites, and variation in population density are the environmental conditions of principal interest to these ecologists. Unfortunately, the influence of Durkheim's perspective—the argument that the adaptive power of human technology and social organization is sufficient to minimize the problem of resource scarcity—diverts early ecologists away from concern about pollution, resource scarcity, or other biophysical limits to the growth of urban industrial land uses in communities, growth unprecedented in human history.

Subsequent generations of human ecologists recognize the limitations of an exclusive focus on competitive cooperation in the spatial distribution of land uses in human communities.[37] They rework the conceptual basis of the field into the study of functional interrelationships among human population (P), social organization (O), environment (E), and technology (T)—the P-O-E-T complex, illustrated in Figure 1.7.[38]

Although this newer version of human ecology continues the study of the ways and means by which populations adapt for survival in the face of environmental constraints, these newer ecologists no longer restrict themselves to studying land uses in urban communities. The new ecology includes regions and entire societies. The crucial argument in this version of human ecology, again, is—the main parts of the "ecological complex" are interconnected and operate to sustain the population dependent on scarce resources. Changes in one part of the complex, therefore, cause other parts to change, as, for example, when farm automation causes the mass migration of black sharecroppers from the American South into urban communities.[39]

The human population, in this version of human ecology, is an aggregate of people varying in size, composition, rate of growth, and distribution in geographical space. Ecological organization refers to the life-supporting activities of the population—the routine, organized work that enables people to survive with finite natural resources, what the ecologists call sustenance activities. The concept of a sustenance organization includes occupational niches or functional roles, the distribution of a population in those roles, and the interdependencies among the niches as in the changing division of labor in U.S. agriculture over time.[40] By the technology of an ecological complex, ecologists mean the tools, techniques, and the built environment used by a human population.

Perhaps the most important and yet, ironically, least clearly specified concept of this revisionist human ecology is the environment. Some ecologists use the term, environment, as we do in this book, such as air quality[41] or natural resources.[42] Very recent work recognizes the environment as "physical elements contained in a population's territorial space" such as other plants and animals."[43] Most ecosystem scholars, however, more often use the concept of the environment in a Durkheimian sense, in reference to other competing human populations, a social definition of the environment.[44]

The neglect of the biophysical consequences of growth and change in human societies and the largely unquestioned, functionalist, Durkheimian assumption that social and technological changes are inherently adaptive are two shortcomings in human ecology that serve as an important basis for the development of environmental sociology.[45] Environmental sociologists do not necessarily accept the idea that the ecological complex is an adaptive or functional form of human social organization. Whether the activities of modern industrial societies are functional or sustainable from one generation to the next is a master problem in the field. As Andrew Weigert writes, "The mutual sustainability of life and culture is an empirical question, not an empirical given nor a theoretical certainty."[46]

The Rise and Revitalization of the Environmental Movement

Two forces, one political and the other intellectual, account for the ebbing and flooding of interest in environmental sociology. The contemporary politics of the environmental movement, in a real sense, energizes the field of environmental sociology. Between 1970 and the present, the level of scholarly activity in the field grows, declines in intensity, and grows once again in response to similar patterns in general public concern for the environment.[47] Since the dynamics of this ebbing and flowing of public interest and the revitalization of public environmental concerns since the mid-1980s in global terms will be discussed in Chapter 6, here we simply acknowledge the central political role of public opinion in shaping the field.

Important, ongoing scholarly efforts to correct a theoretical weakness shared by many sociological traditions, the omission of biophysical variables in traditional social theories, is a second new impetus for revitalizing the field. Important work in sociology, including classical social theory, is structurally oriented. This sociological work focuses on the relationship between social institutions such as advanced industrial capitalism or modern state bureaucratic policies and social inequality, a fundamental problem for justice in society. How these same institutions and institutional practices affect the environment, until environmental sociology developed, was an important, but unanswered question in sociology. Now sociologists such as Allan Schnaiberg are carefully examining relationships between advanced capitalism, social inequality, and a wide variety of environmental problems.[48] Other environmental sociologists such as Andrew Szasz write about challenges presented by post-modern society to governmental officials responsible for regulating the environmental actions of businesses and municipalities.[49] This scholarship, to be presented throughout this book, defines a large part of the contemporary intellectual architecture of environmental sociology.

Going beyond important work linking social structure, social inequality, and environmental issues, environmental sociologists are recognizing a real need to incorporate the subjective perceptions of various groups into the structural perspective. Accordingly, the field is beginning to reflect what Buttel calls the "cultural turn" in environmental sociology: social constructionism, ecofeminism, semiotics, and postmodernism.[50] Weigert's cultural analysis, "Lawns as weeds: status in opposition to life," illustrates this new trend. The concept of the lawn, as opposed to an unkept meadow or sward, historically stems from the landscaping practices of the European aristocracy. Scholars claim that Jefferson's landscape design for the University of Virginia, including a well-kept campus lawn, introduced Americans to the idea of a high status environmental tradition.[51] Emulating the high status landscaping of Jefferson and his European aristocratic predecessors, Americans work hard for a good, prestigious lawn. The proper lawn is homogeneous, consisting of nonnative, silky, dark green spiked leaves growing in a continuous, dense or lush pattern, kept short and well manicured. Weigert observes lawn ownership as "a system of appearances that announces the social and moral status of the lawn caretakers; enters their self-understanding as a part of their identity and a source of motivation; and partly defines the collective worth of the neighborhood."[52] The concept is deep enough in the American collective consciousness, according to Bormann, Balmori, and Geballe, to generate a 25 billion dollar lawn and turf industry.[53]

Another example of the cultural turn in environmental sociology would be Szasz's postmodern analysis of how media coverage of toxic waste accidents and other environmental problems leads to the creation of a series of cultural images in public consciousness: *icons*—the earth floating in space, nuclear power plant cooling towers, forty-gallon chemical waste drums (in the Love Canal), Canadian seal pups, and "the rain forest"—with emotive content that serve as rallying points, socially mobilizing the growing environmental movement, worldwide.[54]

Both new structural work on relationships between social institutions and the environment and cultural development such as the American love affair with lawns are important facets of contemporary environmental sociology. In the following chapters both aspects of the field will be discussed in contexts such as world population growth, food production, energy, and the environmental movement. There are, however, other forces driving the development of environmental sociology, including the New Environmental Paradigm.

Inventing the New Environmental Paradigm

Work by William Catton and Riley Dunlap serves as a third major intellectual force in the development of environmental sociology. Catton and Dunlap argue that alternative paradigms or perspectives in sociology are alike in their "shared anthropocentrism."[55] They insist that virtually all the various sociological paradigms contain a view of human society as the center of the natural world, with humans controlling and using the environment without regard to the natural resource-based limits to social growth. Conventional sociological paradigms, according to Catton and Dunlap, are united by a common worldview—the "Human Exemptionalism Paradigm" (HEP). While this work now appears with certain more recent extensions, we use the terminology commonly found in the now voluminous literature they stimulated.[56]

The HEP has four major assumptions:

1. Humans have a cultural heritage in addition to (and distinct from) their genetic inheritance and thus are quite unlike all other animal species.
2. Social and cultural factors (including technology) are the major determinants of human affairs.
3. Social and cultural environments are the crucial context for human affairs, and the biophysical environment is largely irrelevant.
4. Culture is cumulative; thus technological and social progress can continue indefinitely, making all social problems ultimately soluble.

Catton and Dunlap, in contrast to the HEP, advance a "New Ecological Paradigm" (NEP), which they see as a competing worldview and the basis for an environmental sociology critique of mainstream (HEP) sociological thought. The principal assumptions of the NEP include:

1. Even though humans have exceptional characteristics (culture, technology, etc.), they are but one among many species that are interdependently involved in the global ecosystem.
2. Human affairs are influenced not only by social and cultural factors, but also by intricate linkages of cause, effect, and feedback in a web of nature; thus purposive human actions have many unintended consequences.

3. Humans live in and are dependent upon a finite biophysical environment that imposes potent physical and biological restraints on human affairs.
4. However much the inventiveness of humans or the powers derived therefrom may seem for a while to transcend carrying-capacity, ecological laws cannot be repealed.

Catton and Dunlap use the HEP/NEP distinction to make two principal arguments. First, HEP/NEP is the principal paradigmatic divide in sociology. Second, the HEP is both obsolete and a contradiction with the realities of resource scarcity and natural environmental limits to social and economic expansion. Hence, the NEP is an important framework within which sociologists can understand the laws that shape the development of modern societies. The NEP can also assist policymakers in formulating policies that will enable societies to live within the limits of their survival bases.

The HEP/NEP schema of Catton and Dunlap has a curious influence on environmental sociology. Their work, to be sure, is widely read and often cited, especially in theoretically oriented papers. The HEP/NEP distinction is the point of departure for most important recent works in environmental sociology.[57] In particular, the Catton-Dunlap argument of the anthropocentric/unecological legacy of the classical sociological tradition is widely acknowledged by observers from a variety of theoretical stripes.[58] There has yet been relatively little fleshing out, however, of more precise theoretical statements and testable hypotheses. Perhaps the only major comparative macro-sociological study to work directly from the NEP is Catton's *Overshoot,* an elaboration of earlier work.[59]

The other major application of the NEP in empirical sociological research is with social survey research.[60] Empirical reseachers reconstruct the NEP as an empirically measurable set of cognitive beliefs and changes in beliefs. Dunlap and his colleagues indicate these beliefs can be measured in a reliable fashion.[61] Their correlates are reasonably stable across sample survey populations in Western societies throughout the world.

Some environmental sociologists question the centrality of the HEP/NEP distinction for the field.[62] The distinction is a valid one at an abstract level. In one instance they show how different traditional paradigms in sociology reflect the HEP.[63] What really needs to be done, and what is now occurring in environmental sociology, involves revising the major sociological paradigms to reflect the ecological interdependence between human societies and the natural world. Much of this book focuses on recent work building the NEP into major sociological paradigms and applying them to issues involving the environment, energy, and society.

The Influence of Rural Sociology

Another important influence on the development of environmental sociology is rural sociology.[64] Rural sociologists have a long-standing interest in questions of land use, outdoor recreation, and, in more recent years, the social and biological consequences of scientific advances in agriculture, notably biotechnology.[65] Rural sociologists are among the first responding to environmental problems with a sociological perspective. In fact, environmental sociology became institutionalized in the Rural Sociological Society, principally in its Natural Resources Research Group, long before other professional societies did so such as the Society for the Study of Social Problems (1973) and, last, the American Sociological Association (1976).

The rural sociological influence is important precisely because this field is the part of sociology closest for the longest period of time to the notion of society embedded in the materiality of nature.[66] Rural sociologists envision society, in other words, as vitally linked to finite natural resources. Their environmental focus appears in the context of the rural United States, especially places where persons of color live, as they become increasingly the preferred sites for locally undesirable land uses such as toxic waste depositories, prisons, and mental health facilities.[67] Rural sociologists' environmental interest is equally evident in their work on the less-developed countries. Nancy Peluso's work on conflicts between Indonesian villagers and state forest officials over human rights to control local forest resources is exemplary rural sociology directly contributing to environmental sociology.[68] Tom Rudel's work assessing the relative importance of human population growth and organized growth coalitions on Ecuador's tropical deforestation also exemplifies how rural and environmental sociology complement one another.[69]

The Influence of Urban Sociology

Urban sociologists, as with rural sociologists, have been developing work about land development and the social and environmental consequences of land development. Much of their work contributes to the environmental sociology of the built environment. William Michelson's *Man and His Urban Environment* is the pioneering work that integrated much of the published work in urban sociology that was relevant to the study of the built environment. Michelson shows how some urban sociologists study different degrees of congruence between peoples' lifestyles, shaped by social class, gender, and ethnicity, and the urban built environment.[70] Urban sociologists during the 1950s and 1960s began to investigate "how well the (built) environment actually accommodates the characteristics and behavior of people."[71]

An important problem with congruence, one that sociologists in the 1960s began to uncover, was the mismatch between the then newly created high-rise public housing projects in many U.S. cities and child-raising among urban poor families. Without adequate supervision by parents in the upper stories of the projects, rates of teen and preteen vandalism and other crimes began to rise. Building height was only one of many factors connected with the built environment of the projects: with apartments loading onto both sides of long corridors on a floor, residents had trouble discerning who belonged in the public space; stairwells allowed muggers to wait on a stair landing unknown to someone descending the stairs. The problem of social control in the housing projects, in combination with growing rates of unemployment and drug addiction, ultimately would lead the United States to begin demolishing high-rise public housing projects throughout the country by the 1990s.[72]

Urban sociologist Harvey Molotch's writing on the 1969 Santa Barbara oil spill also represents an important contribution to early environmental sociology.[73] Molotch's work on the oil spill is especially important because this highly publicized environmental crisis radicalized a professional class of residents in an affluent California community. People who historically perceived themselves to be immune from the environmental risks of modern industrial technology found themselves vulnerable and powerless as crude petroleum from a leaking Union Oil offshore drilling platform polluted their beaches and wildlife. The highly publicized spill, as with the Love Canal, provided an important stimulus for the American environmental movement. Molotch's work also set a precedent for sociologists to extend traditional work on social movements, such as the American Civil Rights movement, into work on the environmental movement.

Molotch's (1976) publication, "The city as a growth machine," placed in a leading sociology journal, is equally influential to the early growth of environmental sociology. Molotch's writing on growth machines includes a conflict theory over land development that can and often does affect the built environment in permanent ways. This class-based social conflict, in Molotch's theory, reflects deep divisions among community residents. The persistent goal for the rentier class, the group whose economic class standing, wealth, and power grows through their profits from land development, is to maximize the market or exchange value of their land holdings. Rentiers do so through the development of banks, convention centers, professional sports stadiums, highways, shopping malls, residential subdivisions, commercial and industrial office parks, apartment buildings, and so on. For other residents concerned about the quality of their built environment or the use value of the environment, the persistent land development efforts of the rentiers can mean rising taxes for new services such as schools, police, road maintenance, traffic congestion, more air pollution, noise, and growing problems with waste management. When and if a community conflict arises between the rentiers and their allies in banks, utility companies, and, of course, realtors and other community residents, as they inevitably do, the outcome of the dispute will decide the quality of the built environment and the wealth and power of the rentiers for years to come.

Environmental sociology, then, is very much a product of the intellectual context within which it emerged. The legacy of human ecology, growing public concern about the environment, the work of Catton and Dunlap on the HEP and NEP, and the involvement of rural and urban sociologists have all been formative influences on the study of natural, modified, and built environments as well as the social construction of environmental issues and problems. The kinds of questions raised by environmental sociologists will be discussed in the final part of this chapter.

KEY ISSUES IN THE FIELD

Environmental sociology began by raising issues about the social and environmental consequences of economic growth—issues such as how the public responds to the consequences of growth, such as pollution or a proposed landfill. These issues continue to be central to this academic enterprise. Closely related issues about social and environmental changes set in motion by shortages of natural resources, such as wood fuel in the less-developed countries, or the environmental movement, both nationally and internationally, also, are central to the field. Another long-standing interest for environmental sociologists is ascertaining the role of human population growth in creating environmental problems. In the material that follows we identify specific questions associated with these more generic issues central to environmental sociology. These questions will reoccur throughout the entire book.

Growth

1. On balance, do the social benefits of economic growth outweigh its social and environmental costs?
2. Must economic growth destroy the environment, or can alternative forms of growth—for example, decentralized growth, especially within the context of economic democracy, worker control, and self-management—significantly minimize these social and environmental costs?
3. Must economic growth continue for capitalist societies to be socially, economically, and politically stable?
4. Would the cessation of economic growth undermine capitalism? In other words, is capitalism an inherently expansionist system that must grow in order to survive?
5. Does capitalist economic organization inherently destroy the environment, as in the now dreaded problems with toxic wastes, or can capitalism be reformed to minimize environmental degradation?
6. Are environmental changes—pollution and resource scarcity—likely to result in qualitatively new forms of social organization due to the undermining of the institutions of economic growth?

Social and Environmental Change

1. What is the range of options for social structures in advanced societies that can be sustained by the environment and yet yield favorable living standards for all people? Is this range broad or narrow? How can this range of options best be conceptualized—in terms of renewable versus nonrenewable, centralization versus decentralization, capitalist versus socialist, or some other scheme?
2. Will these changes come about naturally and gradually or through massive political conflict or even revolution?

3. Can "technical fixes" (new technologies, chemical contraceptives, biotechnology, solar energy) replace the need for qualitative sociopolitical change?
4. To what extent are societies self-regulating, self-adjusting, or adaptive systems? What are the mechanisms of that adaptation?
5. Is the environmental movement a progressive agent of change or will the movement remain the victim of elitism and misplaced emphasis? Was the effort on the part of international environmental organizations to join forces with the Brazilian rubber tappers a harbinger of twenty-first century environmental politics?
6. Have the major environmental organizations, at least in the United States, come under the control of the very corporate interests who are responsible for many of the most serious environmental problems we have today?

International Questions

1. How do environmental phenomena shape international relations and structures?
2. Must the United States and other advanced societies heavily depend on less-developed countries for sources of raw materials to ensure the economic survival of the more-developed countries?
3. Do raw materials flows to the United States, Europe, and other more developed countries irreversibly deprive the less-developed countries of access to their raw materials, which are necessary for their own development?
4. Will resource scarcity foster worldwide struggles over access to limited resources as we saw in the case of Brazil and the Mendes tragedy?
5. What are the most desirable development paths for the less-developed countries, considering the finiteness of world resources?
6. Do the interrelationships between nations that export raw materials and those that import raw materials intensify social injustice in the Third World, such as the wood fuel crisis—perhaps with impending revolutionary consequences?

Population

1. What is the importance of population size and population growth for the emergence of environmental problems, including but not limited to food production?
2. Should zero population growth be a major policy focus or a minor concern in the United States and other more-developed countries?
3. Should zero population growth and family planning be a major focus or a minor concern, subordinate to economic development, in the less-developed countries?

The questions that environmental sociologists ask themselves are weighty. The questions do not lend themselves to simple theoretical or empirical solutions. These questions cut to the heart of the very nature of social change and the directions that societies will take in the decades to come. The chapters that follow will provide overviews of these issues. We must admit, at the outset, that we do not view our task as "solving" these issues or advancing hard and fast hypotheses. Although we will frequently interject our own opinions and evaluations, our purpose is to provide an overview and a reasoned sociological analysis of the work that has thus far been completed in the field.

SUMMARY

Environmental sociology, a field created a quarter century ago, necessarily casts a wide net. The field's scope is wide enough to include the political and economic forces at work in the tragic death of Chico Mendes; women struggling for the control of fuel wood in the foothills of India's Himalayan mountains through the Chipko Movement; and the environmental justice movement in the United States, organized by blue-collar workers of diverse racial and ethnic backgrounds, often women. It takes traditional sociological endeavors—the study of social stratification, race, ethnicity, and gender—and extends them into largely uncharted territory, the interrelationships between human society and the environment.

The question of how growth—demographic, energetic, and economic—influences the natural, modified, and built environments and how groups respond to changes in these environments caused by growth and change is the master problem in the field. While Durkheim may have had it right, arguing that human ingenuity and cultural traditions give society the ability to deal with the finite limits of the earth's resources in the process of growth, environmental sociology treats Durkheim's assumption as an empirical question. If society is inherently adaptive, why is there so much concern, worldwide, about population growth, hunger, deforestation, and climate change? What are the social class, ethnic, racial, and gender bases of those raising the concerns? What policies do these environmentalists advocate? How will these policies influence the opportunities for different groups to achieve what they want in their lifetimes? Environmental sociology searches for answers to these questions.

An essential, recently growing part of the field involves the development of social theories about the nature of society; how society changes; and how those changes are interconnected with the natural world. In the next chapter we will discuss how, historically, sociologists constructed theories without thinking about society-nature connections. The scholars who developed the classical theories in sociology did so in the nineteenth century, before the intricate and vital linkages between society and the environment became a scientific or public concern, before few could imagine the global environmental problems we now face. Leading environmental sociologists in the United

States, Canada, and Europe are now rethinking classical social theories, turning them "green" in the process. Emerging green social theories and their historical roots will be the subject of the next chapter.

CITATIONS AND NOTES

1. Revkin, 1990, p. 13.
2. Cohen, 1990, C 15.
3. Warren, 1990, A 9.
4. Dunlap and Catton, 1994, p. 23.
5. See, for example, Bell, 1998; Buttel, 1996.
6. Partridge, 1966, p. 763.
7. Bates, 1968.
8. Schnaiberg, 1972, 1980.
9. Worster, 1979.
10. Warming, 1909; Clements, 1905; Hawley, 1950.
11. Brown, 1954, p. 3.
12. Lovelock, 1979.
13. Botkin, 1990.
14. Worster, 1993.
15. Buttel, 1996.
16. Bell, 1998.
17. Bell, 1998, p. 2.
18. Agbo, Sokpon, Hough, and West, 1993.
19. Freudenburg and Grambling, 1994.
20. Brulle, 1995; Harper, 2001; Bell, 1998.
21. Bullard, 1994; Szasz and Meuser, 1997.
22. Dunlap, 1992; Dunlap, Gallup, and Gallup, 1993.
23. Hannigan, 1995, pp. 32–33.
24. Buttel and Taylor, 1992.
25. Frank, 1997; Frank, Hironaka, Meyer, Schofer, and Tuma, 1999; Frank, Hironaka, and Schofer, 2000.
26. Agarwal, 1986.
27. Levine, 1982. Also see work by Szasz, 1994; Bullard, 1994.
28. Merchant, 1992.
29. Agarwal, 1986, p. 30.
30. Agarwal, 1992, p. 151.
31. Levine, 1982.
32. Bullard, 1994.
33. U.S. General Accounting Office, 1983; Commission for Racial Justice, 1987.
34. Bullard and Wright, 1986, p. 175.
35. Bullard, 1994.
36. Park and Burgess, 1921.
37. Hawley, 1950; Duncan and Schnore, 1959; Kasarda and Bidwell, 1984.
38. Duncan, 1959.
39. Sly, 1972; Massey and Denton, 1993.
40. Murdock and Albrecht, 1998.
41. Duncan, 1961.
42. Gibbs and Martin, 1962.
43. Micklin and Sly, 1998, p. 57.
44. Sly and Tayman, 1977; Schnore, 1958.
45. Catton, 1980; Freudenburg, 1989.
46. Weigert, 1994, p. 82.
47. Dunlap, 1992; Dunlap and Catton, 1994.
48. Schnaiberg, 1994.
49. Szasz, 1994.
50. Buttel, 1996.
51. Bormann, Balmori, and Geballe, 1993.
52. Weigert, 1994, pp. 86–87.
53. Borman et al., 1993.
54. Szasz, 1994.
55. See Catton and Dunlap, 1978, 1980.
56. Olsen, Lodwick, and Dunlap, 1992.
57. Dickens, 1992; Murphy, 1994.
58. Benton, 1989; Martell, 1994.

59. Catton, 1976.

60. Dunlap and Van Lier, 1978, 1984; Dunlap, Gallup, and Gallup, 1993.

61. Dunlap et al., 1993.

62. Buttel, 1978, 1986; Humphrey and Buttel, 1982.

63. Catton and Dunlap, 1980.

64. Field and Burch, 1988.

65. Busch, Lacy, Burkhart, and Lacy, 1991; Kloppenburg, 1987; Kenney, 1986.

66. Buttel, 1996.

67. Bullard 1990,1994; Popper 1992, 1994.

68. Peluso, 1991.

69. Rudel with Horowitz, 1993.

70. Whyte, 1943; Young and Willmott, 1957; Gans, 1962.

71. Michelson, 1970, p. 31.

72. See Newman, 1973 for an elaboration of Michelson's (1970) work on congruence.

73. For work on the oil spill, see Molotch, 1970, and Molotch and Lester, 1975.

2

Social Theory and the Environment

Social theories are being recast to reflect the New Environmental Paradigm, as we will show in this chapter. Social theories are stories about how society is organized and how and why society changes over time. We equate the concept of a social theory with that of a paradigm. Sociologists work with paradigms that provide differing, unique conceptual maps for looking at human societies. Paradigms emphasize different principles or basic causal forces driving societal growth and change. The causal forces driving social change can be cultural values and beliefs shared by members of a society, social power, or social stratification with particular forms to the distribution of wealth and prestige in society. Social paradigms are synonymous with sociological perspectives or schools of social thought.[1]

Andrew Weigert's work on the lawn, discussed in Chapter 1, exemplifies how social theory works. Weigert begins with a distinction between the sward and the lawn.[2] Swards are living systems of plants, insects, and animals that naturally evolve in different ways depending on their particular environment: the climate, natural light, soil composition, rainfall, and so on. Lawns, however, are a socially constructed cultural form. They are what their owners learn to be socially prescribed, normative, desirable lawn characteristics: plant species composition, leaf color, density, and height. A traditional, high-status lawn is, according to Weigert, "composed of grass species only; free of weeds and pests; continually green; and (laboriously) kept at a low, even height."[3] The sward is a part of the natural world. The lawn is a social construction—a cultural and social product, part nature and part a reflection of what people learn the landscape of the lawn is culturally prescribed to be.

By the end of the twentieth century, the various manufacturers involved in the lawn care industry reported a total annual receipt of 2.8 billion dollars. Owners of high-status lawns require one-time investments in maintenance equipment: a tractor, mower, mulcher, thatcher, edger, "weed eater," aerator (rent that one), vacuum, blower, fertilizer applicator, roller, sprinkler, and sprayer. Add to these one-time equipment costs the yearly costs of fertilizers, insecticides, herbicides, water, gasoline, and oil, not to mention the value of personal labor, and one sees just how powerful a cultural form the lawn can be for American homeowners and the economy.

Weigert's cultural detective work indicates that to maintain the culturally prescribed height of the lawn, the American national consumption of gasoline is at least 580 million gallons annually in addition to human labor. The two-cycle gas engine commonly used to power the mower blade burns fuel inefficiently. The mower emits a considerable volume of hydrocarbons into the atmosphere (and all over you). Manicured grassland for lawns and golf courses, moreover, increases the demand for limited supplies of fresh water. These environmental effects do not even begin to consider the significance of the toxic lawn environment for human and animal health. The chemicals required to keep the lawn from becoming the antagonist, the sward, cause children, lawn care workers, gardeners, house pets, and wildlife to be exposed to toxic chemicals.

By juxtaposing the lawn and the sward, Weigert illustrates the clash between the natural laws governing ecosystems and a socially constructed cultural tradition prescribing the human normative standards for lawns. The biodiversity of the sward ensures natural regeneration through the combined power of solar energy, water stored in a highly variegated subterranean root system, and rich nutrient pools produced by natural decay. The various actors responsible for producing the lawn include lawn owners, hybrid seed companies, fertilizer producers, and lawn mowing and watering equipment manufacturers. Without the continuous capital investment for the fertilizers and biocide inputs necessary to compensate for natural processes present in the sward and absent in the lawn, the socially produced lawn as a cultural form naturally vanishes.

Weigert's work is a social theory, a story of how society is organized to meet a demand that is a product of American collective consciousness; a demand that affects the economy, our expectations about what constitutes a good neighbor, and the American landscape. Weigert's work meets sociologist Jonathan Turner's description of a social theory.[4] They are "stories about how humans behave, interact, and organize themselves." Social theories "seek to explain how and why social processes operate."[5] In this chapter we will indicate how environmental social theorists work, and, how various social processes affect the biophysical environment. These theorists also work to show how those effects, in turn, can change the organization and politics of human society.

Organized into five main parts, the chapter will identify the stories, paradigms, or theories about social structure and change of three classical social theorists: Emile Durkheim, Max Weber, and Karl Marx. Three compelling reasons guide our choice of these social theorists. First, Durkheim, Weber, and Marx are considered macro-level sociologists. Each theorist identifies and explains major, although different, structural trends in capitalist industrial societies. Since the imperative goal of environmental sociology is to explain the interrelationships between contemporary societies and their physical environments, the macro-level, structural focus of these theorists is apropos. Second, each of the theorists is both explicit and unique in explaining the major social forces that affect social structure and change. The three classical paradigms thus meet the criteria for paradigms discussed earlier. Third, the leading environmental social theorists in the United States, Canada, and Europe are respecifying work by Durkheim, Weber, and Marx to show the reciprocal causal links between social structures and the environment.[6]

The first part of this chapter will discuss Thomas Malthus's work on human population growth. While Malthus is not a sociologist, his work is important in the formulation of at least two of the classical theorists that we discuss. Malthus's work, now 200 years old, remains important as a driving force for the contemporary environmental movement, so we have more than enough reasons to include him.

The second, third, and fourth parts will both explain and reinterpret the classical sociological theories into conservative, managerial, and radical paradigms.[7] We will reinterpret the classical theories in this way for three reasons. First, many of the issues we discuss, such as the U.N. World Population Conferences or the movement for sustainable, intergenerationally replicable development, are unique to the modern world. The classical theorists rarely analyzed them. Second, we will, at times, incorporate insights that depart in substantial ways from the principal concerns of a particular classical theorist. Finally, the notions of conservative, managerial, and radical directly convey the political implications of each paradigm—a key quality in the way environmental sociology translates into public policy.

The fifth part will discuss how these paradigms provide insight about a globally crucial society-environment relationship: tropical deforestation. In doing so we examine work by the biologist, Norman Myers' *The Primary Source: Tropical Forests and Our Future.*[8] We also examine sociological work on tropical deforestation. We will discuss Rudel and Horowitz's research, *Tropical Deforestation: Small Farmers and Land Clearing in the Ecuadorian Amazon*[9] and Michael Redclift's *Sustainable Development,* a book containing work on Bolivian tropical deforestation.[10] By comparing these different theoretical approaches to deforestation, one can gain a better appreciation for Alford and Friedland's observation that no single sociological paradigm "is a complete and true one, in the sense that (the paradigm) alone can be an adequate source of concepts and hypotheses to explain any phenomenon."[11] Here and throughout this book we thus draw upon alternative, often overlapping environmental sociology paradigms.

THOMAS ROBERT MALTHUS (1766–1834)

Malthus's *Essay on the Principle of Population,* first published as a controversial book in 1798, played an important role in the development of Darwinian evolutionary thought, and the *Essay* continues playing an important role in environmental policymaking by powerful actors such as the United States government.[12] According to Malthus, "the power of population is indefinitely greater than the power in the earth to produce subsistence for man."[13] Because of this Malthusian principle, the problem of scarcity continually looms over society, perpetuating hunger, malnutrition, or the threat of such environmental problems.

A main proposition in Malthusian theory, based on his extensive study of human population censuses throughout the world, is that populations grow geometrically: 2, 4, 16, 32, 64, 128, 256, and so on. Having reviewed agricultural census reports worldwide, Malthus also argues that food supplies or "the means of subsistence" grow by arithmetic increase: 1, 2, 3, 4, 5, 6, 7, and so on. An acre of land might produce 100 bushels of wheat in one year, 110 bushels in the following year, 120 the next, in a linear progression that raises annual yields by a fixed quantity. Donella Meadows and her colleagues illustrate the differences in the two rates with a child who switches saving money from a piggy bank to a real bank in Figure 2.1. Malthus, in addition, argues that human labor and technology are subject to diminishing returns, with efforts to increase soil fertility producing less and less food per unit of effort through time.

The discrepancy between the rates of population growth and agricultural productivity is the heart of Malthusian theory. One dynamic leads to ever-increasing human numbers; the other leads to diminishing supplies of food. Although the visibility of these discrepancies varies over time and from one society to another, Malthus describes them as inevitable qualities. "The one is to the other as the hare is to the tortoise in the fable. To make the slow tortoise win the race, we must send the hare to sleep."[14] In the nineteenth century era of optimism, when the European economy was growing rapidly with colonial ties to the New World, Malthus's message was as welcome as the arrival of the bill collector.

Because the size of the human population cannot exceed the means of subsistence, Malthus reasons that two mechanisms continually operate to check the discrepancy between the rates of population growth and soil productivity. Figure 2.2 illustrates these Malthusian devices for "sending the hare to sleep."

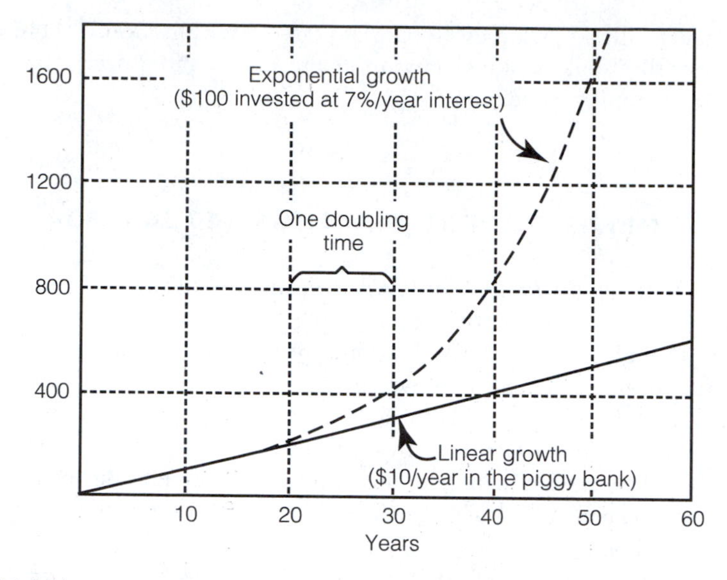

FIGURE 2.1 Linear versus Exponential Growth of Savings

NOTE: If a child puts $10 each year in a piggy bank, the savings will grow linearly, as shown by the lower curve. If, starting in year 10, the child invests $100 at 7% interest, that $100 will grow exponentially, with a doubling time of 10 years.

SOURCE: Meadows, Donella H., Dennis L. Meadows, and Jorgen Randers, *Beyond the Limits*, 1992. White River Jct, VT: Chelsea Green Publishing Co. *www.chelseagreen.com.*

Positive checks are miserable environmental conditions that increase human mortality: "unwholesome occupations, severe labor and exposure to the seasons, extreme poverty, bad nursing of children, great towns . . . the whole train of common diseases and epidemics, wars, plagues, famine."[15] Preventive checks, that Malthus considers to be social vices, are promiscuity, homosexuality, adultery, contraception, and abortion.[16]

Malthus, after five years of criticism, study, and travel throughout Europe, published another version of the *Essay* with a third solution to the imbalance between population and the means of subsistence, also evident in Figure 2.2. Moral restraint refers to continence, celibacy, or delayed marriage without premarital sexual relations. Moral restraint is a check without "vice" in Malthus's thinking for two reasons: (1) moral restraint precludes premarital sexual behavior that is normatively defined as illicit in early nineteenth century European society, and (2) moral restraint is an incentive for work. Marriage, for Malthus and others of his time, was a "natural state" that is legitimate when people have an income and the means of subsistence necessary to prevent positive population checks. Contraception was a "vice" to Malthus because this preventive check "permits sexual gratification free, as it were, and does not generate the same drive to work as would either a chaste postponement of marriage or children to care for."[17]

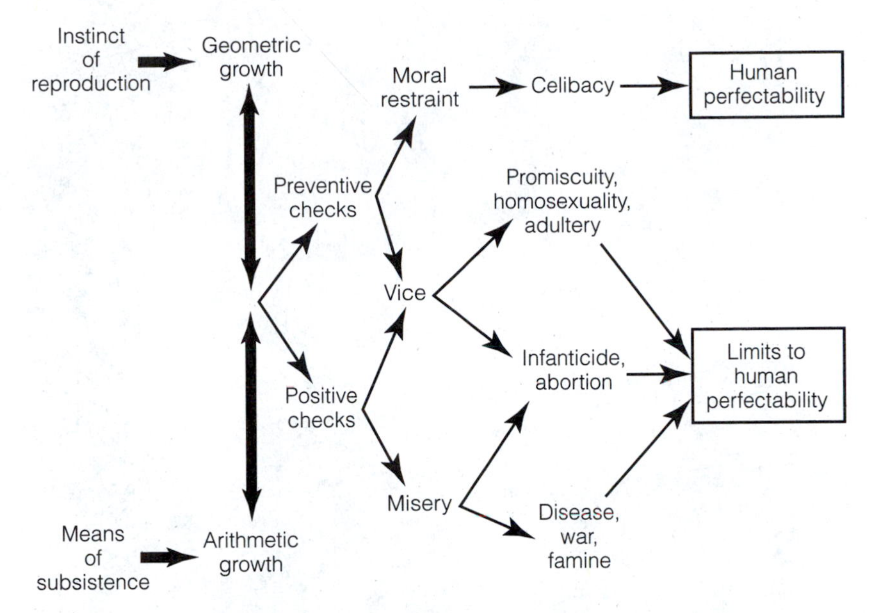

FIGURE 2.2 Malthus's Theory of the Sociological Consequences of Growth in Population and the Means of Subsistence

SOURCE: Humphrey and Buttel, 1982.

Malthus's work is controversial among nineteenth century social theorists and other intellectuals because the *Essay* challenges what William Catton calls the emerging "exuberance" of nineteenth century Western culture.[18] In an era when people believed the earth's natural resources to be limitless, Malthus presents a sobering counterpoint. The work of a British cartoonist in Figure 2.3 satirically captures the controversy. We will convey the controversy about Malthus through two of the three social theorists whose paradigmatic legacies are central to the remainder of this book.

As in earlier work, we follow the notion that we can best understand the differences in Durkheim, Weber, and Marx by using a common set of terms—culture, power, and social class—as a way of reducing their quite different theories to a common language.[19] We regard culture, power, and social class as the three master concepts in sociology. Table 2.1 indicates how the classical theorists define and use the ideas of culture, power, and social class.

EMILE DURKHEIM (1858–1917)

The distinctive element of Durkheim's sociology is his focus on culture as the basic causal force in society. Durkheim conceptualizes culture in terms of the

FIGURE 2.3 "Overpopulation," an English Cartoon Appearing in 1851

Table 2.1 Culture, Power, and Class as Paradigmatic Assumptions

Paradigm	Example	Basic Casual Force	First-Order Consequence	Second-Order Consequence
Conservative	Durkheim	**Culture:** Collective consciousness or shared values and goals	**Class:** The division of labor in society reflecting *dynamic density*	**Power:** Leadership in organizations; representation in government
Managerial	Weber	**Power:** Ability to achieve goals, even in the presence of opposition, through the acquisition of authority in large bureaucracies	**Culture:** Legitimating ideologies, beliefs, myths	**Class:** Accumulation/acquisition of "life chances," opportunities
Radical	Marx	**Capitalism:** Forces of production (profit motive); production technology	**Power:** The social relations of production; *alienation of labor*	**Culture:** Hegemonic beliefs ("all ships float on a rising tide"; economic growth is good)

SOURCE: Humphrey and Buttel, 1982.

shared values among the members of society. He argues that values such as the collective consensus on the desirability of democracy or free trade are the source of social cohesion and order in society. The first-order consequence of culture, for Durkheim, is class. Durkheim defines social class in terms of the differentiation of work roles in a complex society. He suggests that differential rewards for particular kinds of work are necessary incentives to maintain a smoothly functioning division of labor.

Durkheim views power in a more benign light than either Marx or Weber. Durkheim restricts his definition of power to leadership and the representation of interest groups such as trade unions, voluntary associations, and professional associations in politics. Power, for Durkheim, is a consequence of both culture and social class. Shared values define public goals; the role of the state is to implement these goals into law. As the social division of labor increases in complexity, driven in part by population growth, the differentiation of society into groups with an unequal distribution of economic resources becomes the basis for representation in the political sphere. The culture of a modernizing society, for Durkheim, thus produces a division of labor with differential rewards for people. The shared values of the population, along with the multiplicity of interest groups, lead to an exercise of power and to social stability, to gradual, evolutionary social change.

While Durkheim did not directly attack Malthus, his theory is decidedly anti-Malthusian. Durkheim's work truly reflects Catton's Human Exemptionalism Paradigm (HEP) discussed in Chapter 1. The evolving division of labor in society, with evermore complexity, for Durkheim, represents society's adaptation to population growth and the growing demand for natural resources.

Durkheim denies any parallel between Darwin's Malthusian view of the natural world and his own view of how human societies work. Writing about Darwin, Durkheim states, "As long as they (species) have more resources than they need, they can still live side by side, but if their number increases to such proportions that all appetites can no longer be sufficiently satisfied, war breaks out, and it is as violent as this insufficiency is more marked."[20]

Durkheim then argues humans are exempt from these natural limitations because they have an ever-evolving culture through an expanding division of labor. He writes, "It is quite otherwise if the co-existing individuals are of different species or varieties (occupations)." "The oculist does not struggle with the psychiatrist, nor the shoemaker with the hatter, nor the cabinet maker, nor the physicist with the chemist, etc. Since they perform different services, they can perform them parallelly (*sic*)."[21]

For Durkheim, the more people there are in a society, the more people can engage in dynamic density or what he terms "intra-social interaction." "The division of labor develops, therefore as there are more individuals sufficiently in contact to be able to act and react upon one another."[22] Durkheim argues that natural resource scarcity drives people to seek creative ways of adapting to the natural world in scientific laboratories, engineering institutes, government agencies, new scientific fields, and universities. Durkheim's thinking thus reflects Catton and Dunlap's HEP: Humans have a cultural heritage that makes them unlike all other animal species. Through this exemptionalist reasoning, Durkheim turns Malthus on his head.

The Durkheimian Conservative Legacy

The conservative paradigm, as with Durkheimian social thought, generally regards the basic composition of a society as having beneficial social institutions that evolve in time. The conservative paradigm is synonymous with the sociological notion of functionalism. Functionalists insist that recurrent forms of social behavior, such as the division of labor in society, persist because the behavior represents the best possible human adaptation to the social and physical environment. This regard for existing social arrangements as performing some function for society as a whole causes functionalists to be conservative about promoting major structural changes in society.

Conservatives, consistent with Durkheim, emphasize the importance of cultural values in leading to and providing for eventual solutions to problems with natural resource scarcity and environmental degradation. Conservative environmental sociologists argue that environmental problems are an unanticipated consequence of the pursuit of Western cultural values during the course of social modernization: industrialization, urbanization, an improved standard of living, and related social changes. These values include individualism, free market capitalism, democracy, and personal achievement. As individuals pursue these values, society changes through modernization.

Conservative environmental sociologists point out that culturally transmitted values associated with modernization developed during a historical period that William Catton calls "the era of exuberance."[23] In Catton's paradigm, these values developed when the New World was new and seemed to offer a "limitless" frontier of natural wealth. For Catton and Riley Dunlap, the modern collective conscience is contingent on a Dominant Western Worldview (DWW):[24]

1. People are fundamentally different from all other creatures on earth, over which they have dominion.
2. People are the masters of their own destiny; they can choose their goals and learn to do whatever is necessary to achieve them.
3. The world is vast, and thus provides unlimited opportunities for humans.
4. The history of humanity is one of progress; for every problem there is a solution, and thus progress need never cease.

If the DWW is the basic causal force in the modern environmental crisis, according to Catton, then the solution to that crisis is the development of a new form of collective consciousness. Existing social institutions such as education, religion, and science can serve as the principal agents impelling this change in collective consciousness to reflect the New Environmental Paradigm (NEP). For conservatives such as Catton, Western values, not capitalism or the decisions of powerful bureaucratic elites, are the basic causal force behind natural resource scarcity and environmental degradation. Catton's early work illustrates his perspective.[25] Writing about the wastefulness of the modern Western lifestyle, he states:

> . . .the steel industry has used trainload after trainload of iron ore and coal, gouged from great gashes on the face of the earth. So the steelmakers have been prodigal. But modern people have wanted steel in vast quantities, for buildings, machines, tools, playthings, weapons. The prodigality has been collective; it was not just a character defect in the corporate executive of United States Steel.

For Catton, the capital accumulation process is a response to culturally perpetuated collective needs and demands in society.

Deep ecology is another telling example of the conservative paradigm in environmental sociology. Arne Naess, a Norwegian philosopher and mountaineer, planted the cultural seed for this paradigm in an article, "The shallow and the deep, long-range ecology movement."[26] Shallow ecology, for Naess, is "the fight against pollution and resource depletion. The central objective of shallow ecology is to protect and/or improve the health and affluence of humans in the developed (modernized) countries." This ecology is shallow because the paradigm is anthropocentric. Shallow ecology aims to protect the environment for the purpose of reducing risks to human health and safety.

The deep, long-range ecology movement, according to Naess, is "a normative, eco-philosophical movement that is inspired by our experience as humans in nature and in part by ecological knowledge." As in Catton's advocacy of a shift in Western collective consciousness from the HEP to the NEP, Naess insists on a need to change the Western cultural concept of the self: from an anthropocentric self-conception to a biocentric self-conception. Naess insists that members of Western societies need not only to see themselves in relation to their fellow human beings, but also to see themselves in relation to other beings and entities such as rain forests and mountains. For Naess, ecocentric identification is the "ultimate norm of self-realization" and the only way Western societies can achieve deep, long-run ecological sustainability. While we will discuss the concept of sustainability in more detail in Chapter 7, here we use a common definition of the concept: recurrent forms of social behavior that can be passed on through generation after generation in any given society without permanently damaging the environment.

Environmental sociologist Bill Devall and, at times, his philosopher colleague George Sessions, make significant contributions to a more elaborate understanding of Naess's embryonic deep ecology paradigm. Devall published the principles of deep ecology in 1985. Deep ecologists such as Devall recognize the fundamental value of biodiversity, not because biodiverse environments offer scientists the opportunity to develop pharmaceutical products, but because the plants, animals, and other forms of life in biodiverse environments have intrinsic value. As Devall says, "human and non-human life have value in themselves."[27]

We argue that deep ecology is part of the conservative paradigm, in part, because of its explicit Malthusian stance. Devall's principles of deep ecology include "the flourishing of human life and cultures is compatible with a substantial decrease of the human population."[28] He also includes the principle, "present human interference with the nonhuman world is excessive, and the situation is worsening."[29]

Whether deep ecology is conservative or radical in content is debatable. A principle of deep ecology states, there is a need for change "in policies that affect basic economic, technological, and ideological structures."[30] In one sense, as we indicated earlier, there is eclecticism in the paradigms we are using. In most deep ecology writing, however, the absence of an explicit commitment to either a Marxian or a managerial paradigm, the explicit Malthusian stand, and the emphasis on ideological changes such as recognition of "a profound awareness of the difference between big and great" suggest that the deep ecology paradigm is an activist form of conservatism.[31] Several of the newer environmental organizations in the United States such as the Sea Shepherd Conservation Society, founded in 1977, and Earth First!, founded in 1980, politically act upon the principles of deep ecology.[32]

Devall emphasizes change in culture through new forms of collective consciousness, especially the shift from the anthropocentric to the biocentric self. While these changes certainly will lead to radical changes in the structure of institutions such as capitalism, the path to those changes, in Devall's work, starts with changes in culture, changes in Western collective consciousness, changes in the self, not changes in the structure of society such as a redistribution of wealth or power.

MAX WEBER (1864–1920)

The basic causal force in society, for Weber, is power. Weber defines power in terms of domination of society by a stratum of elites who are able to gain and maintain control over powerful bureaucracies, even against the will of others. Weber basically views large-scale formal organizations, such as state and corporate bureaucracies, as hierarchies of power, with groups at the top able to perpetuate their power and to advance their interests. In doing so, they use dominant cultural beliefs to convince their subordinates that their actions are in others' best interests; the actions of these elites are right and proper.

For Weber, culture—legitimating ideologies and myths—is the second-order consequence of power. Weber argues that the stability of a structure of power depends on the ability of dominant elites to convince their subordinates that their rule is in others' best interests, to convince their subordinates that their actions are right and proper. Legitimating ideologies used by elites might elaborate upon protecting free markets; elaborate upon protecting democracy; or, elaborate upon how a policy will protect jobs and a growing economy.

A twentieth century example of the Weberian power-culture interrelationship comes from the Eisenhower presidency in 1956. Eisenhower justified signing the U.S. Highway Act and building 40,000 miles of interstate highways as a national defense policy. With the interstates in place, Americans had a means to escape from major cities in the event of a military attack by the Soviet Union. Eisenhower legitimated his massive federal expenditure on highways, an act of power, using the culture of the Cold War and the fear among Americans at the time that a major East-West confrontation could occur, a cultural phenomenon.

Neo-Weberians such as Daniel Bell would not disagree with the idea that bureaucratic elites will use culture to legitimate their power.[33] For Bell, however, by the last quarter of the twentieth century and continuing now and in the future, the symbols and shared beliefs constituting culture become complex. In fact, an important part of cultural complexity involves expressive forms whereby individuals take on new identities and share them with others who oppose policies and cultural practices pursued by dominant elites: feminists opposed to patriarchy; peace activists opposed to U.S. military interventions; gay rights groups opposed to sexism; and environmentalists opposed to bureaucratic actions that present risks to the natural world and to human health and safety.

Weber defines social class as the accumulation of different life chances or opportunities. Weber insists that this accumulation of opportunities results from differential access to the machinery of state domination. Groups with access to state power have the ability to use the legitimate legal and financial power of the state to serve their own material ends, to enhance their class position in a society. Thus Weber's paradigm emphasizes that power, when stabilized by a legitimating culture, can lead to a class-based stratification of different opportunities for wealth and power for groups in society. Continuing with the highway example, Eisenhower's legislation literally paved the way to opportunities for the capitalist class—automotive manufacturers, tire makers, and oil companies—to accumulate vast wealth, to improve their life chances over a long period of time.

The Weberian Managerial Legacy

We view the managerial paradigm as one that emphasizes the need for substantial structural change in society to address environmental problems. The emphasis of decision-makers who use the managerial paradigm, however, is on assuming the continuation of capitalism, modern state bureaucracies, and other major social institutions, seeking reform from within. Managerial-oriented decision-makers work to change laws, grant state regulatory agencies more or less authority, or, work to improve the energy efficiency of major industrial producers.

Historian Richard Grove provides insight into why major environmental organizations, such as the ones in the United States today, commonly rely upon the managerial paradigm.[34] He does so historically, tracing the roots of Western environmentalism to managerial reform efforts by Caribbean tropical countries such as Tobago as early as 1764 and Mauritius in 1769. With the advice of plantation scientists, officials in these countries passed laws to protect native forests, thereby preventing soil erosion and retaining groundwater. By the nineteenth century, officials in India and parts of Europe initiated managerial efforts to protect what remained of their forests. This managerialism continued to expand in Western countries and now plays a central paradigmatic role in the thinking and efforts of major environmental organizations.

Because environmental groups, influenced by scientists, work to achieve new laws and regulatory agencies to protect the environment, environmental sociologists, among many others, study the process of state regulation. These scholars raise a crucial Weberian question: Can state officials in bureaucracies such as environmental regulatory agencies exercise power and regulate often powerful corporate polluters for the benefit of public health, safety, and the environment itself?

Neal Shover and his colleagues, for example, examine this issue in the context of the federal regulation of surface mining for coal.[35] Peter Yeager does so in the context of the regulation of water pollution in the United States.[36] Both studies, using multivariate statistical analysis, reach similar conclusions. They find direct relationships between the severity and repetitiveness of environmental violations by firms or municipalities and the exercise of sanctions by federal regulatory bureaucracies. Sanctions can be as mild as phone calls or warning letters, or, more serious official notices of violation, fines, and litigation. Smaller firms violate surface mining or water pollution laws more frequently than larger firms do. When larger firms violate environmental regulations, however, they pay lower fines. Larger firms pay less because, within their bureaucracies, they have environmental lawyers and scientists who can successfully manage to negotiate with public environmental regulators. Lacking these legal and scientific experts, smaller firms pay more for their violations. This pattern, in turn, accentuates inequalities in the profit margins of smaller and larger firms.

The question of whether the state can really rule and manage how individuals and groups affect the environment also appears in research on municipal bureaucracies. Environmental sociologist Ken Gould, for example, examines the ability of municipalities to enforce water pollution regulations in Canada and the United States.[37] The municipalities Gould studies differ in size, access to resources such as environmental organizations, and economic dependence on a single local employer, for instance, a paper mill or nuclear fuel processing facility. If these single industry communities also lack access to active environmental organizations, either within or outside the community, reasons Gould, they probably will have a more limited ability to enforce or manage pollution regulations.

Gould finds that firms dominating local municipal economies, both in Canada and the United States, through the control of 60 percent or more of local jobs and most of the municipal tax base, have recourse to numerous countermovements against environmental regulation. The employers can engage in what Gould calls environmental unconscious-making, for example. They deny emitting polychlorinated biphenyls (PCBs) into the local waterfront harbor. This unconscious-making strategy—the manipulation of culture by a powerful local employer—forces the municipality to wage a court battle that is cheaper for the firm than buying water pollution control equipment. The firm can also demand local tax rebates to offset their pollution control costs; threaten the municipality with the potential layoffs of workers at the plant; or, in yet another form of environmental blackmail, threaten to close the plant entirely.

Gould finds places with a more diverse employment base have more political autonomy and, thus, more managerial control capacity. This ability to manage more effective environmental regulation, according to Gould, is especially apparent in places that gain local employment from summer and fall tourism related to the Great Lakes beaches, fishing, and fall hunting. As the managerial paradigm in Table 2.1 suggests, greater control capacity means that the regulatory agency managers have both more political legitimacy in the community and more ability to exercise their legal and scientific authority through environmental management.

Canadian sociologist Raymond Murphy's *Rationality & Nature* is a contemporary, ecologically oriented extension of Weber's classic work.[38] Murphy argues that the contemporary environmental movement is an attempt by groups to re-rationalize the decision making of bureaucratic elites who control corporate and state organizations. These elite groups, historically, are trained (rationalized) in schools of law, business, and public administration. These schools rationalize potential bureaucratic elites so their decision making about production or state management is uncoupled from any thinking about the stewardship of the earth. Unprecedented numbers of people, consequently, are now exposed to environmental degradation posing unknown levels of risk to their health and safety. Environmental politics, for Murphy, is a field of social conflict where groups share different chances of benefitting from, contributing to, or being victimized by waste accumulation and environmental degradation.

In this account of environmental politics, divisions within society go beyond the traditional conflict between workers and capitalists identified in the classical Marxian/radical paradigm. The basic causal forces at work in environmental politics, moreover, are the principal contributors and beneficiaries of environmental waste generation. "Almost without exception," Murphy writes, "they involve bureaucracies, public and private, that decide to introduce products that degrade the environment."[39] Causal agents can be bureaucratic elites working as corporate producers, financial officers in banks that sponsor corporate loans, government developers of hydroelectric dams, highways, and airports, or, military installation officers, utility executives, and so on.

Managerial paradigm practitioners such as Murphy, like other scholars using the paradigm such as Gould, do not deny the centrality of the capital accumulation process to environmental degradation and the build-up of waste. Rather, the paradigm they use expands the theoretical domain of causality. Their work applies to other forms of bureaucratic domination and control, as well, such as capitalist and socialist state policies and programs, for instance, military weapons testing. These bureaucracies and the elites who control them gain power through the immense resources at their command. "The roots of environmental problems," Murphy writes, "are to be found in monopoly power itself—whether the private power in pursuit of profit in the capitalist market or Communist Party power in pursuit of bureaucratic domination and privilege—and anthropocentric, exploitive relationships with nature."[40]

An important consequence of this decidedly unnatural form of bureaucratic decision making is the growing risk of accidents or routine waste accumulation and environmental degradation. As the size and scale of corporate or state projects grow, environmental problems also escalate. "The manipulation of nature by means of science and applied science," according to Murphy, "gives capitalists and bureaucrats power, but it also, through the accumulation of waste, provokes the degradation of the natural environment and stimulates social conflict concerning the environment."[41]

As societies advance on the cumulative curve of unpaid waste costs and environmental degradation, classes of people exposed to environmental risks or potential risks energize social conflict through the environmental movement. Reflexivity and reflection are underlying processes in this growing social movement. "Modern, rational society is reflexive," according to Murphy, "in the sense that its social action turns back on the subject precisely because modern society ruins the equilibrium constructed by the dynamics of nature."[42] The environmental movement thus becomes a coalition of actors who reflect on the risks of victimization they encounter or could encounter in the future.

The environmental movement, in this process, becomes a struggle through green political parties to transform the environmentally degrading processes that threaten society or future society—the re-rationalization of society. Leaders of growing environmental organizations, increasingly, pressure bureaucratic elites in a wide variety of settings, including transnational corporations, to develop new accounting procedures that reflect the New Environmental Paradigm. The environmental movement, in doing so, must target state managerial agencies. Murphy insists, "only the state has the resources necessary to force information out of private companies and to set and enforce the rules needed to ensure that knowledge (about risk) is not monopolized."[43] Whether the state has the power to act independently, given the power of capital and state dependency on capital, is the imperative question in the managerial paradigm.

KARL MARX (1818–1883)

For Marx, class—defined as people's relationship to the means of production—is the basic causal force in society. Marx identifies capitalists' control over the means of production as the most significant element of social structure. Because capitalists own the means of production—factories and production technologies, they have the capacity to extract surplus value from the sale of goods. Labor, for Marx, plays a crucial role in creating the value of manufactured goods. Because capitalists own the means of production, however, they can set wages at a level lower than the full value of the goods labor produces. In a capitalist system, for Marx, the extraction of surplus economic value by capitalists leads to profit making and to the alienation of labor from the production process itself.

A first-order consequence of the class differences between capital and labor is social power. Government and law, in a capitalist society, are the agents of the dominant class; the state must enhance the interests of the capitalist class, even at the expense of the subordinate working class. Culture, for Marx, is a second-order consequence of class and power. Marx conceptualizes culture in terms of "hegemonic ideologies." Culture, for Marx, is the beliefs formulated by capitalists, state officials, and the church to rationalize and stabilize the political-economic system. Laborers, for example, endure exploitation by capitalists to work for their meager wages and, in the long run, to earn eternal salvation. Class relations, in the Marxian paradigm, thus, lead to power relations, that, in turn, necessitate a rationalizing and stabilizing culture.[44]

Marx and Friedrich Engels, his colleague, strongly criticize Malthus. By defining the poor as the victims of their own lack of moral restraint, the basic causal force of overpopulation and famine, Marx and Engels insist that Malthus acts as an apologist for the exploitive activities of capitalism with the complicity of state officials. In *Capital,* for instance, Marx writes, "The hatred of the English working class against Malthus is therefore entirely justified. The people were right here in sensing instinctively that they were confronted not with a man of science, but with a bought advocate, a pleader on behalf of their enemies, a shameless sycophant of the ruling classes."[45] Contemporary critics would put it more succinctly: Malthus was guilty of blaming the victim.

Engels is particularly critical of how capitalism affects the workplace environment.[46] In *The Condition of the Working Class in England in 1844,* he writes about the poor living in working-class London. The working class is "deprived of all means of cleanliness, of water itself, since pipes are laid only when paid for, and the rivers so polluted that they are useless for such purposes."[47] Capitalism as an economic system, for Engels, perpetuates a form of production fostering an uneven distribution of wealth. Social inequality becomes so severe that workers do not have even the basic necessities of life: decent housing, an adequate diet, clean air, or safe drinking water. The miserable conditions of nineteenth century industrial cities, for Engels, are not a result of overpopulated, overcrowded places; rather, the conditions of the capitalist system of production itself impel this human misery.

The Marxian Radical Legacy

In an analysis of the 1991 Persian Gulf War radical environmental sociologist James O'Connor states, "There was a political logic running more or less parallel with the economic logic, and (President) Bush's political motives peacefully coexisted with U.S. economic motives ."[48] O'Connor, in this statement, suggests bureaucracies such as states and capitalist markets have separate needs and inherent tendencies that cause the actors involved with them to perpetuate environmental problems and environmental social movements.

In this section we will examine what radical environmental sociologists think about the relationship between capitalism and the environment. Radical environmental sociologists, in contrast to those using the managerial paradigm, conceptually contract the theoretical domain of causal forces at work leading to the build-up of waste and environmental degradation. They do not take issue with the thinking of scholars such as Murphy, however. They recognize these paradigms as "parallel logics" that converge in the end with transformative agents—individuals acting collectively to enhance their own self-interests. The transformative agents could be labor unions, feminist organizations, or, in Murphy's thinking, the victims and potential victims energizing the environmental movement. Their collective action aims to restructure or redirect both state management bureaucracies and capitalist firms. These transformative agents, in doing so, aim to make bureaucratic elites, especially capitalist elites, environmentally responsible for their actions.

Radicals conceptualize environmental problems as inherent irrationalities in the capitalist mode of production.[49] Radicals insist that economic expansion is the primary lever by which capitalist societies resolve economic and social crises such as the Great Depression. The capitalist class and allies, such as state officials, deflect discontent with social and economic inequality by perpetuating economic growth necessary for the increased wages and rising material standards for the working class. Through this material, wage-based enfranchisement of labor, the capitalist class avoids overt repression of workers; protects their own privileged relationship to property; and, garners monetary profit in the process.

The radical paradigm ultimately implies that environmental destruction is necessary for the persistence of capitalism. At the level of the individual firm, for example, a company that does not develop new products or does not try to sell products for which there is no social need will find company profits decreasing or perhaps collapsing entirely. Concern for corporate survival perpetuates strategies that decrease the durability of products through planned obsolescence. More goods can be sold, thereby increasing the use of natural resources. A free-market economy also tends to promote individual, rather than social, consumption. A large component of the gross national product—the total amount of goods and services sold annually—depends on private production aimed at providing each household with a house, a garage filled with two cars, a moribund mass transit system, and so on. At the level of the economy as a whole, radical theorists argue that capitalism is an inherently expansionist system. When capitalist systems are not growing, they become unstable and in a critical state of depression or recession. Since economic expansion in advanced capitalist societies tends to be wasteful, the inevitable result is the excessive use of natural resources.[50]

Schnaiberg's influential book, *The Environment,*[51] and more recent work by Schnaiberg and his colleagues serve as important sources for elaborating the radical paradigm.[52] Two concepts are crucial: (1) the societal-environmental dialectic; and (2) the treadmill of production. The dialectic, involving contradictory social and environmental processes, explains the political-ecological dynamics of economic expansion in advanced industrial societies. Economic growth requires an increase in the extraction of resources, a process associated with additions and withdrawals from natural ecosystems. These additions and withdrawals, in turn, create incentives for increasingly transnational corporations to expand their sources of inexpensive labor and natural resources worldwide. Economic growth, in turn, provides the catalyst for the action of transformative agents such as the labor, women's, and environmental movements.

Schnaiberg presents resolutions to the societal-environmental dialectic in three separate syntheses. The most common historical resolution in the more-developed as well as the less-developed countries, according to Schnaiberg, is the economic synthesis.[53] Here bureaucratic elites simply disregard the antithetical relation between economic and ecological disruption. Damaging additions and withdrawals from ecosystems are not a part of the rational accounting of corporate decision making. As ecological problems mount, however, the economic synthesis gives way to a managed scarcity synthesis, as exemplified by the environmental policies of the past thirty years. The crucial feature of this synthesis is that only the most severe environmental problems that threaten to undermine production, public health, or both, receive attention by state environmental managers.

Another resolution to the contradiction between production and the environment is the ecological synthesis. A profoundly curtailed level of economic expansion is fundamental to the ecological synthesis. Natural resource supplies, the ecological conditions necessary to protect biodiversity, and human health considerations become paramount concerns in economic decision making in the ecological synthesis. Because Schnaiberg and his colleagues argue that the contradictory relations between economic expansion and environmental disruption have not yet reached a level of ecological seriousness to set off the political and economic forces necessary for the ecological synthesis, the dialectic in practice consists of swings between the economic and managed scarcity synthesis. They stress, in fact, that the historically inegalitarian nature of the capitalist political economy, where both capital and labor receive stable shares of benefits from additional production, perpetuates social and economic inequality. In times of economic downturns, workers and capitalists experience losses. The declining economic status and profit losses of capitalists and workers alike create a common interest among both classes to return to a purely economic synthesis.

The treadmill of production is a second major concept in the work of Schnaiberg and his colleagues. The treadmill represents the forces underlying economic growth in capitalist industrial society. The increasingly dominant role of monopoly sector firms both in terms of the consequences of their investments and their relations with government is the heart of the treadmill. The monopoly sector consists of several hundred large, capital-intensive, transnational firms and their increasingly nonunionized workers. The capital-intensive investments of firms in the monopoly sector tend to either displace labor or seek out cheaper labor in the less-developed countries.

The displacement of domestic labor places pressure on the state to pay welfare and unemployment benefits and to solve related social problems such as drug trading and domestic violence. The investments by monopoly sector firms also require significant public expenditures for research, public infrastructure, and education. Finally, these capital-intensive investments create ecological problems as well, requiring public expenditures for environmental restoration such as the 15 million dollar expenditure by the federal government to buy Love Canal homes, discussed in Chapter 1. The results are state spending, budget deficits, fiscal crises, welfare reform, the labor movement, environmental politics, and growing social inequality.

If investments in the monopoly sector lead to social and ecological problems, why would state managers not invoke policies to alter these investments? Schnaiberg's argument, the one that places him in the radical camp, is that state managers are constrained in implementing such policies because of the role of the state in capitalist society. The state faces two contradictory imperatives: to create the conditions for capital accumulation, and, to foster legitimation or public trust and social order. Thus, the political response to problems caused by capital-intensive economic growth are policies encouraging more expansion through road building, growing defense budgets, more research contracts, and space exploration that fuel the treadmill of production. As the treadmill revolves, environmental problems intensify, leading periodically to a more expensive managed scarcity synthesis, but with compelling political and economic pressures to return to the economic synthesis through deregulations, as during the 1980s with the Reagan and Bush administrations and, now, in the new Bush administration.

While Schnaiberg portrays the treadmill of production as a complex, self-reinforcing social institution, he does acknowledge that the treadmill has limitations or contradictions that may undermine the treadmill or make the treadmill subject to reforms. The trajectory of economic expansion after World War II, for instance, results in ecological problems, including a hole in the earth's protective ozone layer and the growing possibility of global climate change. Since the mid-1970s, this trajectory of economic expansion in the United States results, first, in seemingly insoluble crises of state fiscal stress and massive structural unemployment both in inner cities and rural natural resource-dependent communities. The trajectory of economic expansion, more recently, results in growing social inequality. Between 1974 and 1997 in the United States, for example, 60 percent of all wage earners, in constant dollars, annually increased their real income by no more than 0.78 percent, while the top fifth of all wage earners during the same period increased their earnings by 1.68 percent; the top 5 percent of all U.S. wage earners increased their earnings by over 2.5 percent.[54] The ideology of the treadmill of production, that "all ships float on a rising tide," apparently needs a qualification: Some ships are more buoyant than others.

Like managerialists such as Murphy, Schnaiberg insists that the state really is the only social institution capable of redirecting the course of economic growth in the more-developed countries and, increasingly, in the less-developed countries as well. Such an effort, however, requires both a collapse in public faith about the treadmill and the mobilization of political support for growth in directions other than capital-intensive technological change. Here the use of the radical paradigm, ironically, becomes conservative in the sense that Schnaiberg's hope boils down to a conservative cultural change in collective beliefs about the nature of appropriate production and the nature of economic growth.

In the early 1980s Schnaiberg identifies the appropriate technology movement, initiated by E.F. Schumacher's *Small Is Beautiful,* as having the greatest potential to provide a politically viable alternative to the treadmill of production, though his later comments became more pessimistic.[55] He now sees the early 1980s thrust of the appropriate technology movement, with its emphasis on cooperatively owned, small-scale, labor-intensive production technology powered by renewable (solar, for example) energy sources and cooperatively owned, as being too utopian to appeal to more than a small cross-section of the public at large. He argues that the movement must appeal to much larger fractions of labor and the middle class to have sufficient political resources to be a realistic alternative to the treadmill of production.

Schnaiberg and Gould, more recently, note that "sustainable development" is essentially an updated variant of appropriate technology movement, and that, like the appropriate technology movement, sustainable development strategies—economic development projects that promise to be lasting for generations to come—are not likely to challenge power relations or the logic of the treadmill.[56] Thus, in their view, if sustainable development allows people to provide for themselves and, in so doing, overlooks the problems of unequal, treadmill-based capital accumulation, this strategy will ironically buttress capital-intensive industrialization and perpetuate the societal-environmental dialectic.

SUMMARY OF THE PARADIGMS

Contemporary versions of the classical paradigms of Durkheim, Weber, and Marx will serve as the analytical frameworks to discuss the major aspects of environmental sociology throughout this book. We also include the work of Malthus because his work set off theoretical debates among social theorists about the causal forces that are linked with poverty and environmental degradation, and, as we will show, Malthusianism continues to play an important role in environmental policymaking.

More environmentally focused, contemporary social theories include Catton's revision of Durkheim's work and Devall's deep ecology. We designate these theories as conservative because of what they do and do not do. Catton and Devall advocate changes in Western collective consciousness to avoid ecological overshoot of global carrying capacity and the stormy, painful, positive checks of the Malthusian scenario. These changes will most likely lead to radical reorganization of the structure of society. Catton and Devall, however, do not focus on such radical structural changes. They concentrate instead on personal transformations from the anthropocentric to the biocentric self. For Catton and Devall, these transformations can occur through existing social structures such as public schools, colleges, universities, environmental organizations, and ecology institutes.

Managerial and radical environmental sociologists explicitly analyze the structures of bureaucratic states and corporations in advanced capitalist and socialist societies. For scholars using the managerial paradigm such as Murphy, bureaucratic elites in corporations and states share power through their command of immense and accumulating financial and legal resources. Elites use these resources to accumulate profits and advance bureaucratic careers. According to Murphy, in doing so, they engage in decision making that discounts the ecological costs of their activities, including the risks of accidents and the routine build-up of waste and environmental degradation.

This bureaucratic decision making, in turn, impels the environmental movement toward making state and corporate bureaucracies undergo a rerationalization process. Reflexive or negative ecological feedbacks from accumulating waste in the environment and environmental degradation cause various classes of victimized people to reflect on their exposure to risks and possible risks now and in the future. These reflexive and reflective processes provide the shocks that cause the environmental movement to try restructuring how bureaucratic elites exercise their power.

Radical or eco-Marxian scholars such as Schnaiberg do not disagree with managerial scholars such as Murphy. However, they question a fundamental assumption in the managerial paradigm. The imperative question is, can state bureaucratic elites, under growing pressure from environmental classes fearful of environmental victimization, free themselves from their dependence on the treadmill of production? Do state managers have the power necessary to redirect the course of state and capitalist decision making into active stewardship of the natural world? For radicals, such as Schnaiberg, the dependence of state bureaucratic elites on the fortunes of capitalist elites is so strong that only a new form of capitalism can make environmental stewardship possible. This radical change in the structure of capitalism would require a redistribution of wealth and significant power sharing in corporate boardrooms by workers, feminists, and environmentalists with experience and awareness of the risks that elites inflict on themselves, others, and nature. For Schnaiberg, such a radical transformation requires a vast change in Western collective consciousness. In this process Catton's argument that Western culture needs to shift away from human exemptionalism and toward the new environmental paradigm and Schnaiberg's ecological synthesis theoretically converge.

EXEMPLIFYING: STORIES OF TROPICAL DEFORESTATION

Because, as Alford and Friedland observe, no single paradigm adequately explains the complexity of the kinds of social and ecological changes included throughout this book, we elect to use three different sociological paradigms.

To illustrate how the use of alternative paradigms provides a far more comprehensive understanding of environmental problems, we focus on tropical deforestation. With an abundance of analytical work on tropical deforestation in the literature, we can identify conservative, managerial, and radical accounts of this ecological problem.

These explanations apply to one of the most important global ecological processes imaginable. Tropical forests cover only about twenty degrees of latitude north and south of the Equator, as one can see from Figure 2.4. Yet, these tropical ecosystems represent some of the oldest, most complex biological systems ever known. These forests, according to Myers, provide the habitat for two-fifths of all living plant and animal species on the earth.[57] Tropical forests serve as a primary source of knowledge about domestic plants and animals, about vitally important pharmaceutical and industrial chemicals, about biomass energy generation through methane or ethanol production, and about ecological processes themselves. Tropical forests provide an array of important environmental services. These forests regulate the discharge of rain water on the ground, slowing groundwater discharge into rivers and streams in rainy seasons and stretching the discharge flow during dry seasons. Rain forests also play an important role in regulating temperature in the tropics by reducing the warm solar reflection with their dense canopies.

The entire world took what may become a global effort with forest protection at the UN Conference on Environment and Development (UNCED), or the Earth Summit, in Rio de Janeiro, Brazil in June 1992. The 172 nations attending agreed to a set of fifteen Forest Principles, including Principle 8:

> Efforts should be undertaken towards the greening of the world. All countries, notably the developed countries, should take positive and transparent action towards reforestation, afforestation, and forest conservation, as appropriate.[58]

Because of the international agreement with the Principles, the world of forests, politics, and national development takes on a new context. Whether this step is an ecologically significant one, however, depends on the ability of national and international political bodies and other organizations to manage the challenges of caring for forest ecosystems. These challenges can best be understood through the conservative, managerial, and radical accounts about the basic causal forces at work in tropical forest deforestation and management.

In what follows we will exemplify how these sociological paradigms, implicitly or explicitly, appear in the literature on tropical deforestation. We will discuss work by Norman Myers, an internationally recognized biologist. We will also discuss two highly regarded sociological accounts of tropical deforestation. Rudel and Horowitz present research on deforestation in Ecuador's rainforest.[59] Michael Redclift, a British environmental sociologist, reports research on deforestation in the tropical rain forests of eastern Bolivia.[60]

A Conservative Account

Myers provides a major example of the conservative account of tropical deforestation derived from Malthusianism. Myers's work, as with many others throughout this book, is eclectic and, thus, not easily classified. The focal point of his work is slash-and-burn (swidden) agriculture in tropical rain forests. For centuries rain-forest dwellers cut and burned trees in the tropics to enrich soil for farming. Forest dwellers farm their cleared land until the soil is no longer productive. The forest farmers then shift to a new clearing, repeating their farming practices.

Swidden agriculture is sustainable; the practice can be repeated again and again, provided abandoned land remains fallow for a time, usually 15 to 20 years. Since the 1950s, however, the human populations of the less-developed countries, including those in the tropics, began to grow more rapidly. Farmers and forest farmers encountered growing competition for tropical forest land, reducing the fallow period and accelerating deforestation in the process.

Myers recognizes that the increased competition among farmers in tropical countries is an outcome of complex social processes. Strapped with foreign debt from national economic development efforts such as costly road building and the construction of large hydroelectric power projects, state officials in the less-developed countries, by the 1960s, began to encourage the commercial development of export crops such as soybeans, bananas, and coffee. Exporting these cash crops brings foreign currency to these less-developed countries that can be used for debt repayment to the more-developed countries such as the United States.

Large corporate commercial farmers often produce the export crops in the less-developed countries. The corporations, in doing so, buy out small subsistence farmers or displace those without legal title to their land. Once the small farmers are displaced, the farmers migrate into forests where they compete with traditional shifting cultivators for the best farmland.

State-based managerial policies, according to Myers, could be developed to reduce the rate of tropical deforestation. Public officials in tropical nations could create a more equitable farmland tenure system. Officials could grant legal title to land that has been farmed by tradition without legal land ownership titles for generations. Public officials also could improve small farmers' access to financial credit. Then small farmers would participate in the export economy now controlled by large corporate farmers. Myers describes the problem as "a maldistribution of existing farmlands, inequitable land-tenure systems, lack of property rights (for small farmers), inefficient agrotechnologies (farming methods), insufficient attention to the subsistence farm sector, lack of rural infrastructure, and faulty development policies overall."[61]

Even with these state management policies in place, nonetheless, rapid population growth will cause the problem of tropical deforestation to persist. The World Bank, for instance, projects population growth for the 2000–2010 period to be 18 percent in Brazil, 17 percent in Mexico, and 16 percent in the Philippines. "There is vast scope in population growth of the future," Myers

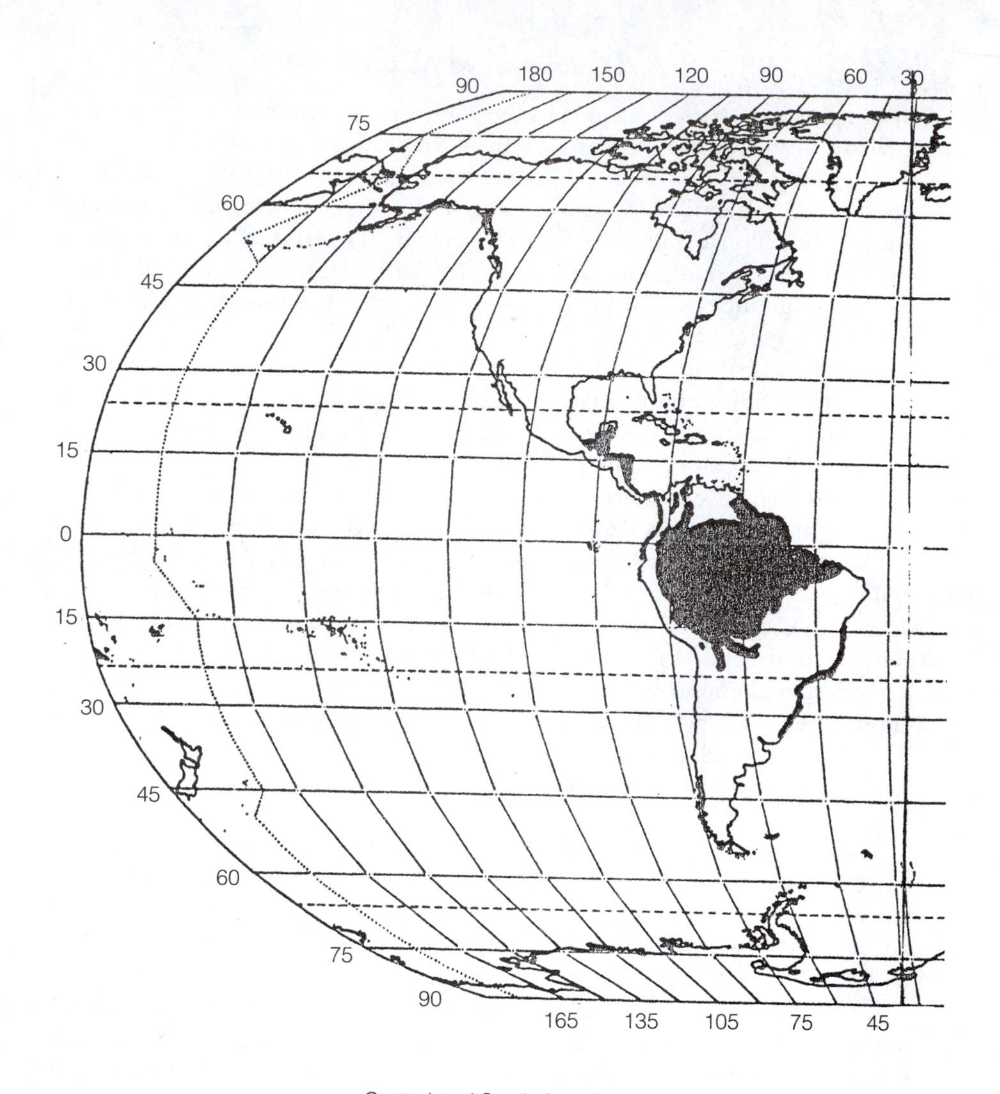

Central and South America

Belize
Bolivia
Brazil
Columbia
Costa Rica

Ecuador
El Salvador
French Guiana
Guatemala

Guiana
Honduras
Mexico
Nicaragua

Panama
Peru
Surinam
Venezuela

FIGURE 2.4 The Distribution of Tropical Forests among Countries of the World

SOURCE: Reprinted from Norman Myers, *The Primary Source: Tropical Forests and Our Future*, Copyright 1992, pp. xxxiii–xxxiv, with permission from Elsevier Science.

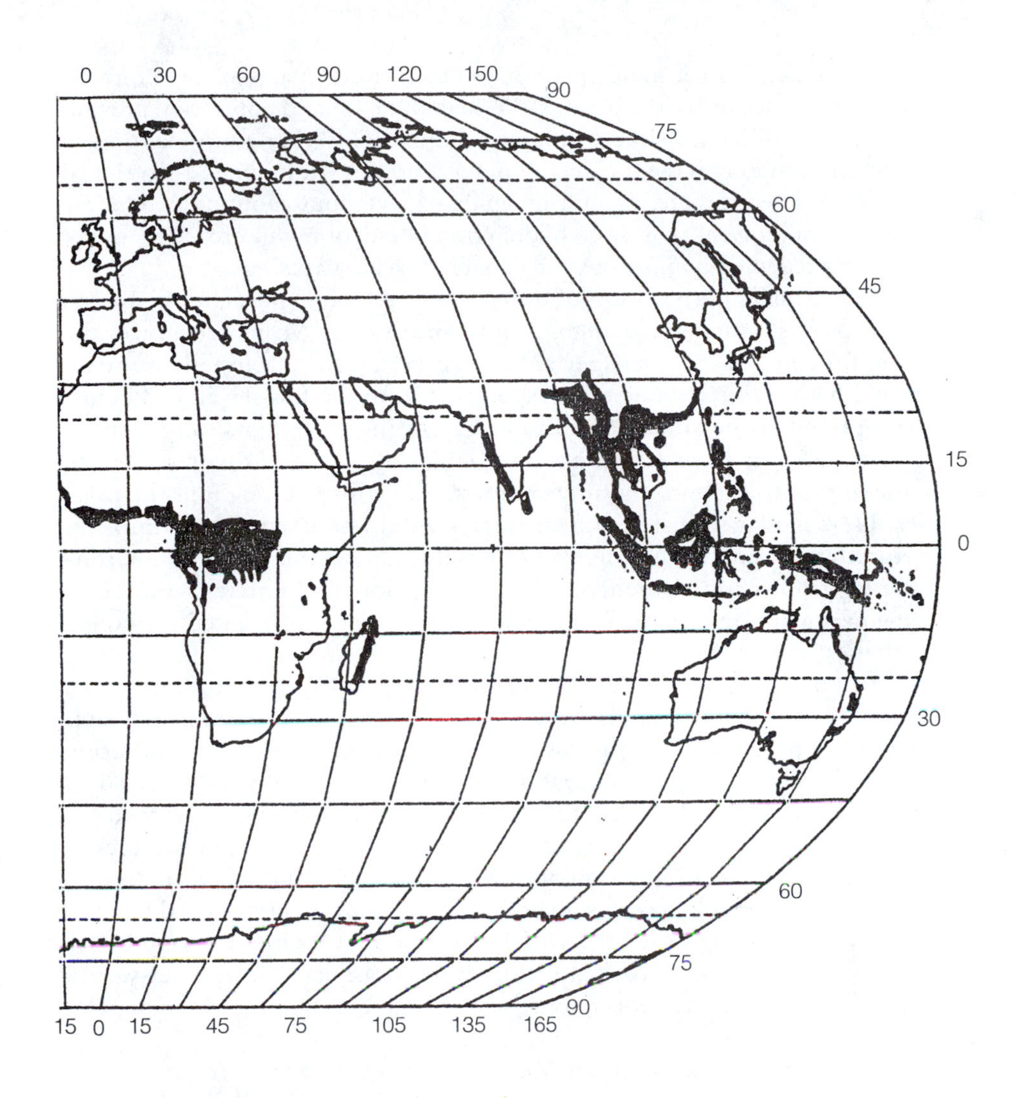

Africa

Burundi
Cameroon
Central African Republic
Congo
Ethiopia
Gabon
Ghana
Guinea
Ivory Coast
Kenya
Liberia
Madagascar
Nigeria
Rwanda
Sierra Leone
Tanzania
Uganda
Zaire

Asia

Bangladesh
Brunei
Burma
India
Kampuchia
Laos
Malaysia
Melanesia
Nepal
Pakistan
Papua New Guinea
Philippines
Sri Lanka
Thailand
Vietnam

writes, "for still larger numbers of shifted cultivators to accelerate deforestation in many sectors of the biome."[62] Where there is rapid population growth, a country will have a large young population. This young population age structure compounds the problem of population growth in the less-developed countries. Even if there is a shift in family expectations about desired family size so couples want only two children, the population will grow because of the unprecedented number of young couples having two.

Why is this interpretation of tropical deforestation conservative? In one sense, Myers's implicit paradigm is not conservative. Myers recognizes the need for state-based, managerial actions: (1) more-developed countries would have to buy conservation easements from the less-developed countries priced at the annual cost of not harvesting their remaining tropical forests; (2) both state and nongovernmental organizations would need to raise money to support grassroots reforestation efforts throughout the tropics. New managerial state bureaucracies would be needed to administer reforestation, and both nongovernmental organizations such as environmental groups and residents in the less-developed countries would find new opportunities for work and new locations for forest-based economic activities.

An important line of conservative reasoning in Myers's work, however, links his analysis to Malthusian reasoning two centuries earlier. In a later edition of his *Essay,* Malthus saw an escape from the misery and suffering caused by rapid population growth. If countries could expand their commercial and industrial sectors, there would be more avenues for upward social mobility for the poor and working classes. With more opportunities to consume the material goods of an economically growing urban, industrial society, people would forego some of their childbearing either for their own material well-being or for the well-being of their children through better education. Accelerating the cultural tradition of capitalism, in other words, with an expanding market economy would lead to a solution to the population problem.

Myers's work contains similar Malthusian reasoning. He writes:

> Unless these large-population countries (for example, the Philippines, Indonesia, and India) proceed faster than hitherto with economic development that provides opportunities other than forest farming for landless people to make a living (i.e., through urban manufacturing and the like), the numbers of forest farmers will continue to grow at a rate far faster than that of national populations.[63]

So Myers argues for an extension of the Western cultural development tradition with capitalism as a key way to deal with the basic causal force of tropical deforestation, population growth and competition between forest farmers and the shifted cultivators. Myers's argument relies on the cultural tradition of Western capitalism, an extension of the Durkheimian conservative paradigm.

A Managerial Account

Environmental sociologists Rudel and Horowitz argue that Myers's account of the causal mechanisms leading to deforestation is one of several different paths to deforestation,[64] as suggested by Figure 2.5. Through immiserization rural cultivators without a secure legal title to their farmland can be transformed into poor, landless proletarians working for very low wages as tropical forests are cleared by wealthy rural elite planters. Combined with rapid population growth, immiserization forces displace rural subsistence farmers into deforestation, so they regain arable land in a process identical to Myers's account.

The dynamics of tropical deforestation, according to Rudel and Horowitz, change after 1960. Even earlier, in the 1950s, state-based multilateral lending bureaucracies provided loans for the development of public infrastructure in the less-developed countries. The loans, continuing today, support road building, the construction of massive hydroelectric generating stations—reservoirs that spill water to spin turbines and generate electricity, and seaport facilities. The less developed countries use this public infrastructure as an incentive for private foreign capital investment.[65] The World Bank and regional multilaterals such as the Inter-American Development Bank, the Asian Development Bank, and the African Development Bank are the key lenders for building this public infrastructure. While private investors fund these bank development projects, the investors take the financial risk because the banks have collateral: the national treasuries of the more-developed countries supporting agencies such as the World Bank.

Once these major development projects came to fruition, beginning in the 1960s, according to Rudel and Horowitz, tropical deforestation accelerated. As Figure 2.5 suggests, lead institutions, growth coalitions, and free riders caused tropical deforestation after 1960. The lead institutions or bureaucracies such as the governments of the less-developed countries or private energy and logging companies use their substantial financial resources to build roads into forests, gaining access to tropical timber, oil, or other valuable natural resources.

"In one sequence of events, documented on all three continents," according to Rudel and Horowitz, "loggers construct a network of logging roads and extract the most valuable timber from the area."[66] Because the resource extraction companies do not control access to their roads, colonists "ride free," settling at the edge of vast stands of otherwise closed tropical forests. Some colonists work independently. Contrary to Myers's forest farmer argument, Rudel and Horowitz observe that most of these free riders will fail to make a living from logging. They must raise food as well. The colonists, with financial backing from wealthy relatives, urban investors, or government colonizing agencies—what Rudel and Horowitz call growth coalitions—are more likely to persist with deforestation. They use wages from their financiers to buy food and other necessities as they log and blaze trails in the tropical forest.

There is a causal sequence in Rudel and Horowitz's theoretical modeling of deforestation that involves private capital investment, lead institution road

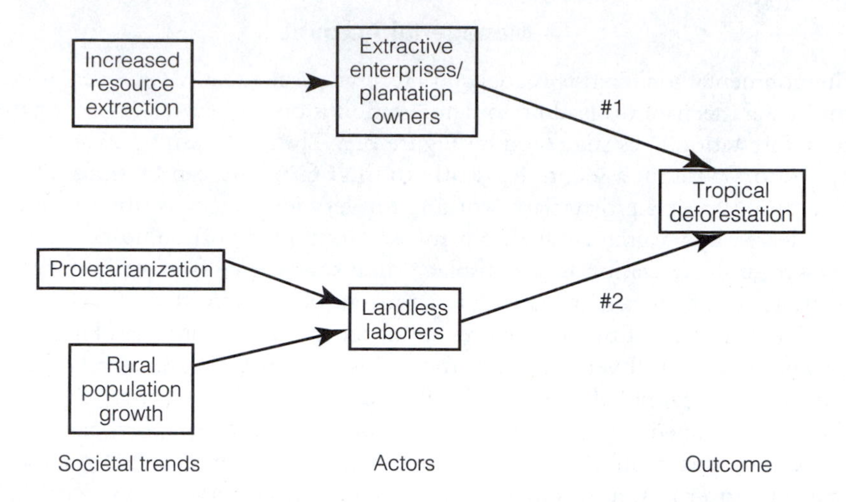

The Immiserization Model

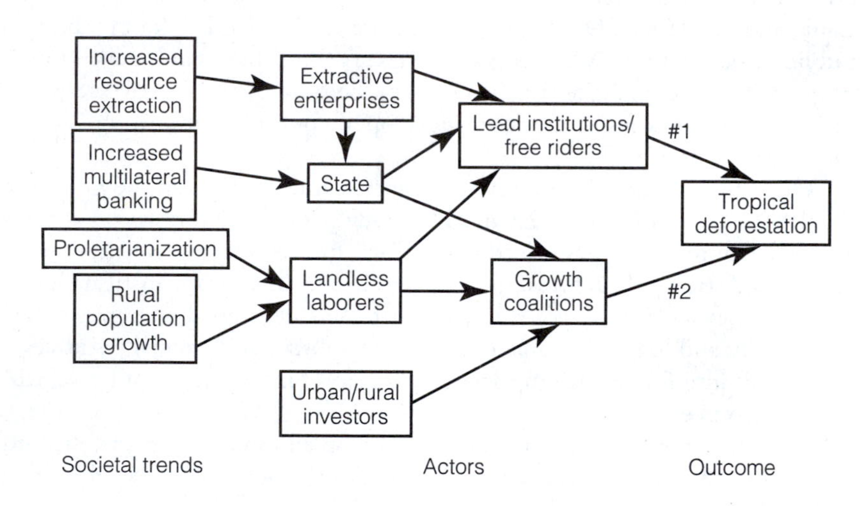

The Growth Coalition-Lead Institution Model

FIGURE 2.5 Rudel and Horowitz's Theoretical Models of Tropical Deforestation

SOURCE: Rudel with Horowitz, 1993, p. 38.

building, free riding, deforestation, and profit-making that parallels radical analysis. This sequence of events often begins with post–World War II multilateral bureaucratic lending backed by Western nations. The royalties paid to government bureaucracies in the less-developed countries by energy and logging companies for natural resource extraction rights, moreover, become the source of revenue for nation-states to play the role of lead institution. The Transamazon Highway in Brazil or Indonesia's Transmigration Colonization Program are exemplary state-based managerial programs that cause tropical deforestation. These governmental bureaucracies manage colonization in their forested frontiers, not always for profit, but often "to secure sovereignty over those regions."[67] Thus, statesmanship and state-managed development programs, according to Rudel and Horowitz, precede private capital investment as the basic causal force in tropical deforestation. We will save additional comparison between this managerial account and the other accounts of tropical deforestation for the closing discussion of this section.

A Radical Account

Redclift develops work on deforestation in Santa Cruz, Bolivia, located in the Eastern Lowlands of the Amazon Basin.[68] His work exemplifies the application of the radical paradigm. Bolivian tropical deforestation, in Redclift's account, is an artifact of "the inexorable pressure to open up more land to the market."[69] Capitalist international trade is the root of that pressure. Less-developed countries such as Bolivia promote commercial logging, mining, ranching, and agriculture for export commodities. With the financial returns from the exports, less-developed countries' government and private investors seek to modernize their economies through urbanization and industrial growth. Less-developed countries such as Bolivia export commodities in their tropical forests to escape from international debt incurred to finance their economic development efforts. Here Redclift's radical account overlaps with Rudel and Horowitz's account and part of Myers's work as well.

The process of tropical deforestation is conflict-ridden not only in Bolivia, but in Brazil as exemplified by the death of Chico Mendes, and in Indonesia where peasants killed state forest officials with shovels and machetes.[70] In Bolivia the social conflict involves armed confrontations between landless *colonistas,* indigenous tribal people, and powerful national and international economic elites, often backed by the military, over land tenure rights. These conflicts are common, especially on the Bolivian frontiers adjacent to Brazil, Peru, Chile, Argentina, and Paraguay.

In Redclift's radical analysis, efforts on the part of Bolivian state officials to gain political legitimacy involve social welfare efforts for the poor. These official efforts, however, are often subverted by contradictory alliances between powerful national or international capitalists and public officials. As early as 1962 Bolivian officials embarked on a colonization program financed by the Inter-American Development Bank. The goal of the Bolivian program was to settle almost one-half million landless *campesinos* into the lowland tropical frontier. By the early 1980s, nonetheless, 3 percent of the lowland owners controlled over 50 percent of the tropical forest land in parcels ranging in size from 500 to over 10,000 hectares. The maldistribution of private capital and the maldistribution of private ownership of tropical forest land, unfortunately, go hand in hand.

Since the productivity of the land in the long run is less important to these forest land owners than the short-term value of the tropical forest land as a commodity to be converted into capital, the conflict-ridden deforestation process involves extensive ecological degradation. Among the *colonistas* that means, with a noteworthy exception, they inevitably encounter what Redclift identifies as the *barbecho* crisis: (1) the loss of soil fertility within five years after initiating swidden agriculture; (2) rapid invasion of weeds among the stumpage; and (3) an acute shortage of labor or capital to deal with these possibly life-threatening ecological problems.

Redclift identifies four possible resolutions to the *barbecho* crisis in frontier Bolivia. Modernization of the *colonistas'* farms is one possible resolution to the crisis, although a costly one. The frontier farmers, in the modernization process, use tractors to remove tree stumps and clear their land. They then raise a monocrop such as state-subsidized rice. The expensive tractors, herbicides, and pesticides necessary for modern monocrop farming, however, mean financial debt for the already poor *colonistas.* The plowed fields, without ample vegetation, in addition, causes serious problems with soil erosion because of the heavy rainfall endemic to rain forests. Redclift also questions the future of state subsidies to cash crop farmers with the growing foreign debt crisis in Bolivia.

A second option for the *colonistas* is either the sale or abandonment of their frontier farms. This resolution, unfortunately, is a common one in Bolivia. The option is unfortunate for ecological as well as social reasons. Land sales or abandonment are unfortunate ecologically because this option simply shifts responsibility for environmental problems associated with modern agriculture in the tropics from one class of farmers to another. This second option also puts landless *colonistas* in a dilemma. Either the *colonistas* resume the arduous task of clearing more of the rain forest, or they become low-wage farm workers. Becoming low-wage workers is more likely, since, as Redclift says, "few men over the age of 40 even attempt it (recolonization), and fewer still return to try a third time."[71]

Some *colonistas* successfully solve or avoid the *barbecho* crisis through small-scale, ecologically sustainable agriculture. Successful Bolivian *colonistas* tend to be married couples with four to six children living in homogeneous ethnic *nucleos* or small farm villages. The *nucleos* consist of thirty-five to forty households sharing a common school, general store, and a water supply. Mixed, labor-intensive farming is the key to the success of these *nucleos.* People in the villages raise annuals such as rice and corn interspersed with rows of nitrogen-fixing plants: clover or chick peas. *Nucleo* farmers also raise perennial cash crops such as coffee, cocoa, or plantains, and chickens, sheep, and pigs fed with household food scraps.

While Bolivian frontier *nucleos* suggest that sustainable agricultural systems offering equal opportunity to poor *colonistas* are possible, the future of these *nucleos* and their ecologically sustainable practices is uncertain. Many Bolivian *colonistas* or would-be *colonistas* will become part of the growing ranks of the frontier semi-proletariat. They engage in multiple survival strategies: clearing the rain forest, engaging in part-time farming, selling depleted land to elite planters and ranchers, and engaging in the highly lucrative, illegal preparation of coca leaves for processing into cocaine. A night's work preparing the coca leaves into a paste for cocaine pays a low-wage farm worker the equivalent of a week's pay in agriculture.

Redclift's work with the radical paradigm points toward a process of tropical deforestation driven by large-scale commercial farmers and ranchers. National and international capital and a growing supply of low-wage labor along the Bolivian tropical frontier support the farmers and ranchers. Given the ecological problems associated with modern tropical agriculture, especially soil erosion, the landed elite increasingly controlling the frontier share a common interest. The landed elite persist in state lobbying for public subsidies in the form of agricultural price supports, tax incentives for the mechanization of farming, road building, and other forms of public infrastructure development. Successful lobbying by the Bolivian elite perpetuates the treadmill of production through land clearing deeper and deeper into the Bolivian frontier rain forests.

SUMMARY

How do these alternative paradigms, as they are applied to tropical deforestation, compare with or go beyond traditional human ecology discussed in Chapter 1? What can we learn about the nature of these paradigms from comparing them in the context of tropical deforestation? Both of these questions will be addressed briefly here because the discussion continues in subsequent chapters as well.

How does the application of our three paradigms to the problem of tropical deforestation compare with traditional human ecology? The applications of the conservative, managerial, and radical paradigms in analyzing tropical deforestation both compare favorably with and go beyond traditional human ecology. Our paradigmatic analyses are comparable because they ask a question shared by traditional human ecologists in sociology: What kinds of social organization do human populations form in adapting to a particular environment? They also go beyond sociological human ecology in a common way. The paradigms or their practitioners recognize the vital importance of the biophysical properties of tropical rain forests in shaping human organization. This shared approach is perhaps the single most important characteristic distinguishing environmental sociology from sociological human ecology, a field typically defining the environment in terms of competing human groups within a given spatial boundary. Environmental sociologists and other tropical deforestation analysts such as Myers, a biologist, certainly do not assume that the forms of social organization developing in the tropics are adaptive. In fact, they are threatening to species indigenous to the rain forests as well as the human communities, indigenous and otherwise, who, along with the agencies they control, are the key actors in the deforestation process.

The tropical deforestation scholars we discuss also differ in the aspects of tropical deforestation they highlight. The issue, for Myers, is what could happen to life on the earth, given the linkages between tropical deforestation and species extinction. For Rudel and Horowitz as well as Redclift, all sociologists, the significance of tropical deforestation is interpreted differently. For Rudel and Horowitz, the nature of tropical forests with their massive height, density, and wild rivers limits the paths of deforestation to areas made accessible by organized, often state-financed road building. For Redclift, the question is whether ecologically sustainable forms of human social organization will survive in less-developed countries such as Bolivia. Competing alternative forms of social organization such as short-term profit-making with modern agriculture, semi-proletarianization of unsuccessful *colonistas,* and the cocaine trade, among others, offer known alternatives for human survival, often with life-threatening costs to tropical forest lands.

What can we learn about the three paradigms from their application to the study of tropical deforestation? Which has the most credence? In the case of tropical deforestation, we would argue, like Rudel and Horowitz, that a combination of conservative Malthusianism and state management best explain the historical origins of tropical deforestation. State-controlled land management aimed at economic development precedes private domestic and international investment in logging, mining, ranching, and agriculture. State managers play what Rudel and Horowitz call the role of a lead institution. The managers seek multilateral and bilateral government aid for the development of public infrastructure, especially public road building. State managers, in doing so, pave the way for deforestation by capitalist entrepreneurs on the frontier.

In addition, tropical forests, by definition, are the disputed domain of states that have colonized noncapitalist, indigenous tribal cultures. How state managers initially choose to distribute forest land tenure rights is the basic causal force at work in shaping who gains access to the forest and who benefits economically, at least in the short-term, from deforestation and subsequent land uses. Moreover, tropical deforestation occurs, in part, to settle national frontiers, thus providing concrete evidence for the political domain of a tropical developing nation. Power, not profits, are at stake in early tropical deforestation.

Once the infrastructure necessary for private capital investment is in place in a tropical country, something like Redclift's analysis is theoretically more applicable. Whether capitalism and tropical deforestation go on unabated, even in the presence of new, state-based managerial efforts such as the signing of Forest Principles by 172 nations at the Rio Conference is a key issue here. How the growth of environmental movements in the less-developed countries will affect tropical deforestation also is a key issue, not only for environmental sociology, but also for the future direction of the environmental movement. We will return to these issues in Chapter 6 on the global environmental movement.

In the next chapter we will examine another basic causal force at work in tropical deforestation—population growth. In doing so, we will raise a question that spans the time between the writing of Malthus (1798) and Myers (1992): Can the adoption of Western cultural traditions of capitalism and the growth of urban industrial social and spatial organization provide an escape from poverty in the less-developed countries in ways that will encourage smaller families? What will be the costs of these social changes for the environment?

CITATIONS AND NOTES

1. Olsen, Lodwick, and Dunlap, 1992.
2. Weigert, 1994.
3. Weigert, 1994, p. 86.
4. Turner, 1998.
5. Turner, 1998, p. 1.
6. For contemporary work on Durkheim, see Catton, 1980; for Weber, see Murphy, 1994; and for Marx, see O'Connor, 1998.
7. Here we are following work by Alford and Friedland (1985) and our earlier work (Humphrey and Buttel, 1982).
8. Myers, 1992.
9. Rudel with Horowitz, 1993.
10. Redclift, 1987.
11. Alford and Friendland, 1985, p. 390.
12. Humphrey and Buttel, 1982.
13. Himmelfarb, 1960, p. 9.
14. Bonar, 1966, p. 407.
15. Himmelfarb, 1960, p. 160.
16. Petersen, 1975.
17. Petersen, 1975, pp. 154–155.
18. Catton, 1980.
19. Humphrey and Buttel, 1982.
20. Durkheim, 1893, p. 266.
21. Durkheim, 1893, p. 262.
22. Durkheim, 1893, p. 257.
23. Catton, 1980.
24. Catton and Dunlap, 1980.
25. Catton, 1976.
26. Naess, 1973.
27. Devall and Sessions, 1985, p. 70.
28. Devall and Sessions, 1985, p. 70.
29. Devall and Sessions, 1985, p. 70.
30. Devall and Sessions, 1985, p. 70.
31. Devall and Sessions, 1985, p. 70.
32. Nash, 1989.
33. Bell, 1976a, b.
34. Grove, 1992.
35. Shover, Clelland, and Lynxwiler, 1986.
36. Yeager, 1991.
37. Gould, 1991.
38. Murphy, 1994.
39. Murphy, 1994, p. 180.
40. Murphy, 1994, p. 154.
41. Murphy, 1994, p. 215.
42. Murphy, 1994, p. 24.
43. Murphy, 1994, p. 147.
44. Turner, Beghley, and Powers, 1998, p. 107.
45. Marx, 1890, p. 147.
46. Engels, 1892; excerpt in Parsons, 1977, pp. 200–202; see work by Foster, 1997, 1999 for an excellent elaboration on early Marxian ecology.
47. Parsons, 1977.
48. O'Connor, 1998, p. 219.
49. O'Connor, 1994; Schnaiberg and Gould, 1994; Gould, Schnaiberg, and Weinberg, 1996.
50. O'Connor, 1994.
51. Schnaiberg, 1980.

52. Schnaiberg, 1994; Schnaiberg and Gould, 1994; Gould, Schnaiberg, and Weinberg, 1996.
53. Schnaiberg, 1994.
54. Wilson, 1999.
55. Schumacher, 1973; Schnaiberg, 1983.
56. Schnaiberg and Gould, 1994.
57. Myers, 1992.
58. UN, 1992, p. 292.
59. Rudel with Horowitz, 1993.
60. Redclift, 1987.
61. Myers, 1992, p. 37.
62. World Bank, 1992, p. 37.
63. Myers, 1992, p. 157.
64. Rudel with Horowitz, 1993.
65. Sachs, 1992.
66. Rudel with Horowitz, 1993, p. 30.
67. Rudel with Horowitz, 1993, p. 34.
68. Redclift, 1987.
69. Redclift, 1987, p. 12.
70. Peluso, 1991.
71. Redclift, 1987, p. 126.

3

Population and the Environment

More than one million years passed before the size of the world's human population reached 1 billion, sometime during the first quarter of the nineteenth century. World population then tripled in size to approximately 3 billion by 1960, and with great speed another billion people were added in the fifteen years between 1960 and 1975.[1] Between 1975 and 1990 the world experienced the addition of yet another billion individuals. Today, world population exceeds 6 billion.

While UN projections estimate that the world will add another 1 to 2 billion more individuals by 2015, we are beginning to see that the "population explosion" is slowing, as world fertility rates continue to decline. The question now becomes: At what pace will the world population growth rate decline to replacement level; what will be the ultimate size of world population; and how will this population be distributed among and within nation-states? The decline in world population growth, the causes of this growth, and the consequences of population growth for the environment, energy, and society will be the topics of this chapter.

Research on world population growth and the consequences of population growth is important to environmental sociology because the subject invites different paradigms with competing policy implications. Controversies about Malthusianism, the argument that poverty and human misery are rooted in population growth and natural resource scarcity, go back more than a century. Marx and Engels bitterly attacked Malthus, arguing that he blamed the victims of an exploitive capitalist Industrial Revolution, thus protecting the interests of an elite, privileged class in Western societies. Contemporary work that empirically demonstrates a positive relation between population growth and subsequent increases in agricultural production in less-developed countries since World War II adds an important dimension to this controversy.[2] The ecological argument that these recent increases in agricultural productivity will not be sustainable in the long run because of environmental problems such as soil erosion and water pollution will be developed in this chapter and in Chapter 4 as well.[3]

The chapter is organized in four main parts. An overview of world population will be presented in the first part. This part also includes a well-known, contemporary debate about whether population growth is the key cause of environmental problems. Then we will discuss demographic transition theory, a conservative paradigm about how populations grow in relation to social and economic change over time. In the third part we will discuss the historical development and contemporary efforts to revise neo-Malthusian population policy—thinking that dominates state-based, managerial efforts to control population growth in most countries throughout the world. The fourth part will discuss critiques of neo-Malthusianism. We will use feminist versions of the conservative and radical paradigms to explain how critics in recent years have made significant efforts to revise international population policy.

WORLD POPULATION: AN OVERVIEW

Of the approximately 6 billion persons living in the world today, 4.5 billion, or nearly 80 percent, live in the less-developed countries of Latin America, Africa, and Asia. The average annual rate of population growth in the more-developed countries, including Japan, North America, Western Europe, and Oceania, is about 0.4 percent. Populations in these countries are doubling in size every 175 years. Annual population growth in the less-developed countries averages 1.9 percent. Less-developed countries are doubling in size every thirty-seven years. Since the less-developed countries contain most of the world's people, and since their people continue to have large families, the numerical dominance of the less-developed countries will continue. Figure 3.1 illustrates this very important point.

A statistical picture of the world's population is presented in Table 3.1, where the population sizes and rates of growth in the more-developed and the less-developed countries are estimated by major subdivisions and continents. The table documents the wide disparities in population growth rates between the less-developed and the more-developed countries. Latin America, Africa, and South Asia continue to exhibit rapid, but recently slowing, population growth rates, while Europe and North America generally have the lowest rates of growth.

Table 3.2 reports the projected population sizes for aggregate regions to the year 2025 as compiled by the United Nations. The UN report uses high,

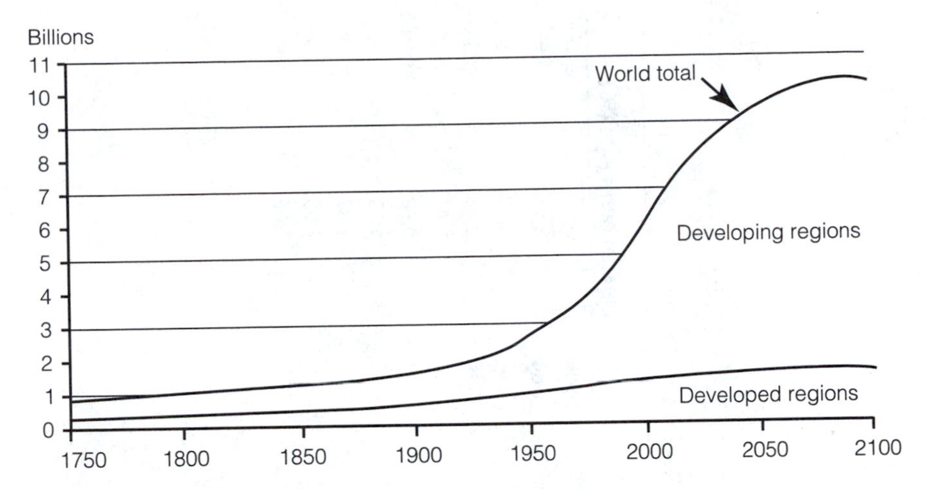

FIGURE 3.1 World Population Growth for the More- and Less-Developed Countries, 1750–2100

SOURCE: Merrick et al., 1986. Used by permission of the Population Reference Bureau.

medium, and low variants for these projections, representing relatively pessimistic to relatively optimistic assumptions. The high variant for total world population in 2025 is 8.979 billion, nearly one and one-half times the present size. The medium and optimistic variants are 8.294 and 7.603 billion, respectively, with the low, or optimistic, variant indicating a growth rate of just over 30 percent over the next quarter century.

Although the low or optimistic variant seems to have gloomy implications for the issues of environment and energy discussed throughout this book, recent trends in world population growth rates show declines sufficiently large to offer hope that some progress is being made in curbing the population explosion. As evident in Table 3.1, annual world population growth rates declined from 1.84 percent in the 1970s to 1.57 percent in the first half of the 1990s.

While the peak in absolute world growth will occur in the next two and a half decades, most scholars foresee a continuing decline in world growth rates.[4] Most of the decline, obviously, is expected to come in the less-developed countries. East Asia is farthest along in this transition, for controversial reasons to be discussed later in the chapter. Recent years have seen a more rapid fertility decline in Latin America than was projected, while fertility in South Asia remains higher than projected.[5] Africa, especially Sub-Saharan Africa, is the slowest region to make the transition toward low fertility rates.[6] These numbers mask great disparities between countries; Turkey, for instance, has about one-half the fertility of most Arab states.

With near consensus among the scientific community that world population growth rates will continue to decline, attention has turned to how fast

Table 3.1 Populations 1970, 1980, 1990, 1995, and Rates of Growth in Major Areas and Regions of the World

	POPULATION (MILLIONS)				ANNUAL GROWTH RATE (%)		
Region	**1970**	**1980**	**1990**	**1995**	**70–80**	**80–90**	**90–95**
World Total	3697	4444	5285	5716	1.84	1.73	1.57
More-developed regions	1002	1080	1143	1166	0.74	0.57	0.40
Less-developed regions	2695	3364	4142	4550	2.22	2.08	1.88
Europe	656	693	722	727	0.55	0.41	0.20
U.S. and Canada	226	252	278	293	1.09	0.98	1.10
Oceania	19	23	26	29	1.91	1.23	1.50
Asia	2147	2642	3186	3458	2.07	1.87	1.70
Africa	364	476	633	728	2.68	2.85	2.80
Latin America and the Caribbean	283	358	440	482	2.35	2.06	1.80

SOURCE: *World Population Prospects: The 1994 Revision.* 1995. New York: The United Nations. Used by permission of the United Nations.

Table 3.2 Population of the World and Major Areas in 1990 and 2025, According to High, Medium, and Low UN Projection Variants (in millions)

		2025, ACCORDING TO			
Area	1990	High Variant	Medium Variant	Low Variant	Range between High and Low Variants
World total	5285	8979	8294	7603	1376
More-developed regions	1143	1332	1238	1155	177
U.S. and Canada	278	407	369	333	74
Europe	722	763	718	679	84
Japan	124	130	122	116	14
Australia–New Zealand	20	32	29	26	6
Less-developed regions	4141	7647	7056	6448	1199
Africa	633	1602	1496	1382	220
Asia (excluding Japan)	3186	5248	4838	4412	836
China	1155	1667	1526	1363	304
India	851	1502	1392	1286	216
Latin America and Caribbean	440	784	710	642	142
Oceania (excluding Australia–New Zealand)	6	13	12	11	2

SOURCE: *World Population Prospects: The 1994 Revision.* 1995. New York: The United Nations. Used by permission of the United Nations.

this decline in population will occur as well as how the growth will be distributed within nation-states. One area of particular interest is the increased, rapid urbanization of the world's population. About three-fourths of the more-developed countries, presently, and one-third of the less-developed countries are urbanized. By 2005 one-half of the world's population is projected to be urban, and by 2025 about two-thirds, with most of this shift occurring in the less-developed countries.[7] Urbanization in the less-developed countries holds tremendous implications as to the nature of anthropogenic demands placed on ecosystems, as well as the distribution of the costs and benefits of these demands. The spatial distribution of population, age structure, and pace of growth are all key areas on which more attention will be focused in the future.

How Important Is Population Growth for the Environment?

The question of the linkages between human population growth, the consumption of natural resources, and the environmental impact of human population growth is centuries old. Environment sociologists Tom Dietz and Gene Rosa claim the question really started to receive systematic attention by Thomas Malthus in his *Essay* (1798). "Malthus posed a pivotal question in his first essay," Dietz and Rosa write, "What effect does population growth have on the availability of resources needed for human welfare?"[8]

This section discusses a debate about the linkage between population, natural resources, and the environment. We will do so, first, by looking at work by three scientists who popularized the debate over the past three decades: Paul Ehrlich, Julian Simon, and Barry Commoner. We will also do so by discussing empirical work by two environmental sociologists: Tom Rudel and Alan Mazur. Their work is significant because they use empirical evidence to give some closure to this important debate.

The Ehrlich/Simon/Commoner Debate Biologist Paul Ehrlich played a catalytic role in the rise of the contemporary environmental movement. His early book, *The Population Bomb*—first published in 1968—became a national best-seller in the United States. Ehrlich, writing in the Malthusian tradition, suggests that world population growth is the basic causal force for pollution and world hunger. Two-thirds of the earth's inhabitants in the less-developed countries subsist in traditional agricultural economies. Residents in the more-developed countries have industrial economies where agricultural productivity is at a maximum and the population continues to grow, but ever more slowly. The national and international consequences of increasing human numbers in terms of health and nutrition, war, and environmental quality, according to Ehrlich, are of such basic importance to the continuation of life on earth that all nations should take immediate decisive action.

The international policies recommended in *The Population Bomb* involve the formation of a political alliance among the world's less-developed countries to promote family planning, agricultural development, and industrialization. Ehrlich insists that this alliance formation is especially important in countries that still have demonstrable potential for avoiding famine and death and reaching zero population growth. Ehrlich, moreover, acknowledges the need for an even greater diffusion of conservationist values toward the natural world.

Some social scientists, however, do not subscribe to the neo-Malthusian argument that managerial efforts on the part of governments are warranted to slow population growth and stimulate economic development. The classic example is Durkheim's work, discussed in Chapter 2. Durkheim argues that population growth itself is a catalyst for invention, or dynamic density. Population growth, for Durkheim, fosters new means of production; the growth increases social differentiation through occupational specialization; a growing population brings forth more efficient means of communication and transportation.

Among the contemporary social scientists endorsing this anti-Malthusianism is the economist, the late Julian Simon. In *The Ultimate Resource 2,* published in 1996, Simon insists that problems with population growth such as hunger, crowding, and energy shortages are short-term problems that present challenges to human ingenuity. As these Malthusian problems arise, they create incentives—through rising prices—for reduced consumption and for entrepreneurs to find new ways to expand the supply of a natural resource. Rising prices motivate entrepreneurs to explore for more reserves or to invent substitutes for the resource in short supply. "The bigger the population of a country," Simon writes, "the greater the number of scientists and the larger the amount of scientific knowledge produced."[9] The ultimate resource, according to Simon, is the human population: "skilled, spirited, and hopeful people who will exert their wills and imaginations for their own benefit, and, so inevitably, for the benefit of us all."[10]

A third well-known participant in this debate is biologist Barry Commoner, director of the Center for the Biology of Natural Systems at Queens College. Commoner is the author of numerous books such as *Science and Survival* (1966), *The Closing Circle* (1971), *The Politics of Energy* (1979), and the frequently revised *Making Peace With The Planet* (1992). Critical of Ehrlich's Malthusianism, Commoner is also critical of Simon's euphoric view of free-market economies and their environmental impacts that threaten nature and human life.

Commoner insists that pollution is caused in many ways. Air conditioners, large automobiles, and aluminum containers have been widely used since World War II, and Commoner links these products to pollution and energy shortages. The manufacturing of aluminum or the increased use of air conditioning, for example, increase the need for electricity from coal or oil-fired electrical generators. These convenient, but energy-intensive, products not only contribute to energy shortages but also cause pollution when sulfates from a coal-fired electrical power plant pollute the air or a supertanker such as the Exxon Valdez spills petroleum into the ocean.

A frequently used example by Commoner is the post–World War II growth in the use of synthetic pesticides in U.S. agriculture. Between 1950 and 1967, U.S. farmers increased their consumption of synthetic pesticides by 266 percent or 168 percent per unit of crop production. In the same time period, U.S. population increased 30 percent, and crop production per capita increased by 5 percent. Soil runoff from farms, in turn, enrich plant life in lakes and streams to a point where the demand for oxygen becomes so great that plants and the marine life depending on the plants die at an abnormally high rate. The problem, for Commoner, is not that U.S. farmers wanted to consume the synthetic pesticides. The problem is that chemical companies produced the synthetics for profit without regard to their ecological impact and the health risks they pose to farm workers. Commoner, therefore, lays the principal blame for ecological problems not so much on population growth, but rather on ecologically risky technologies and the power of corporations to profit by promoting these technologies.

The Ehrlich/Simon/Commoner debate, as we will show throughout the book, continues with the larger environmental movement. The debate raises the questions: ecology for what end and in whose interest? Commoner believes that environmental reform must include social and environmental justice. He thus embraces a radical strategy for change—the redistribution of wealth and power—as necessary for the United States to address environmental problems. Simon, to the contrary, advocates the pursuit of conservative, human exemptionalist economic growth policies. The ecological problems

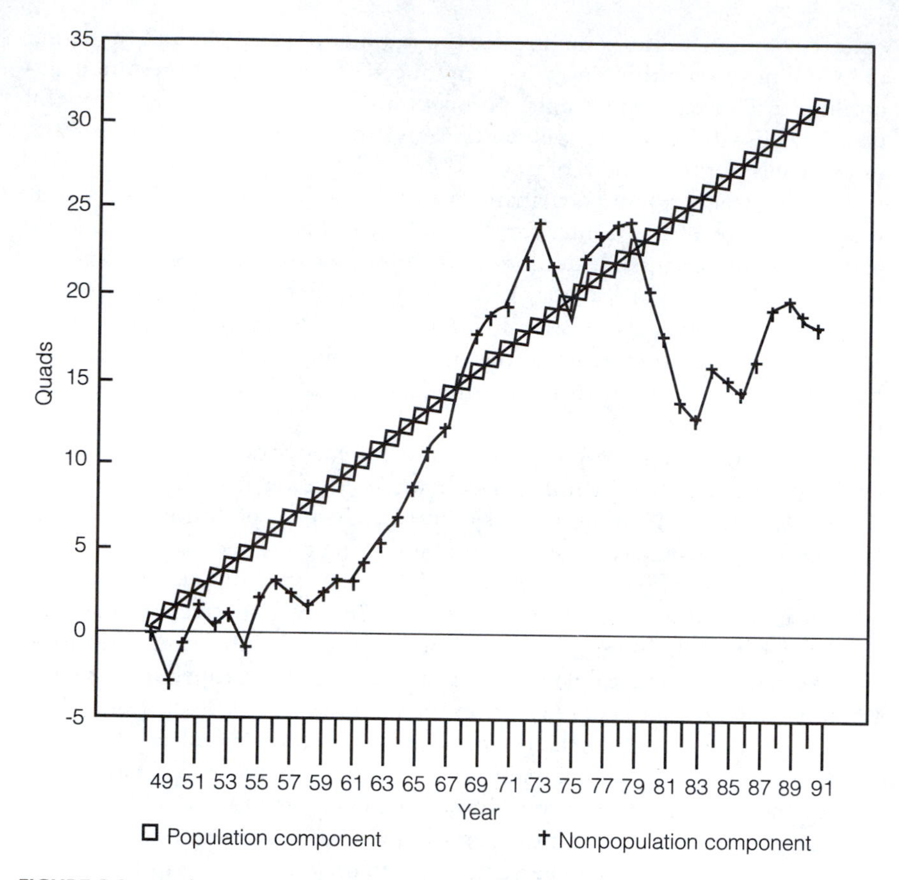

FIGURE 3.2 Population and Nonpopulation Source of Total Energy Consumption for the United States, 1948–1991

SOURCE: Mazur, 1994. Used by permission of Human Sciences Press, Inc.

this strategy may present will decline in time as new technologies and technologists learn to expand the economies in more enlightened ways. For Ehrlich, state-based management through subsidized family planning, worldwide, is essential to the survival of humans and the natural world. Is there evidence to suggest who has the most valid argument about the role of population growth in causing environmental problems?

Empirical Insights from Mazur and Rudel Alan Mazur recently examined the role of total population growth and other sources of change in U.S. energy consumption, 1947 to 1991.[11] Figure 3.2 presents Mazur's basic results. Total U.S. energy—mostly oil and electricity—are plotted by year in total quads (quadrillion units of consumed energy). Nonpopulation sources of change in energy consumption include factors such as the price of oil, average automobile gas mileage, and the efficiency of industrial energy consumption.

Mazur's work uncovers an important insight into the Ehrlich/Simon/Commoner debate. Nonpopulation sources of change in U.S. energy consumption, over this forty-five year period, are more important than total U.S. population growth. Following the two oil shocks in the 1970s, for example, total energy consumption quickly fell. This observation bolsters Simon's claim that markets make a difference in the conservation of natural resources.

Figure 3.2 contains a second, and equally important, insight into the Ehrlich/Simon/Commoner debate as well. Since the two oil shocks, especially after 1978, Americans became more conservative energy users. As they did so, however, total population growth remained as a driving force in total energy consumption. As Americans became more conservative energy users, the number of those users became a decisive factor in how much energy the country consumed. Ehrlich's Malthusianism may prove to be right, after all.

Tom Rudel's analysis of deforestation in thirty-six tropical countries also provides insights into the Ehrlich/Simon/Commoner debate. Rudel, using multivariate statistical analysis, uncovers two basic patterns for tropical deforestation. In countries with relatively small, scattered rain forests—smaller countries such as Burundi or Haiti—population growth is the major causal force in deforestation. "Population growth," Rudel writes, "contributes to deforestation directly (by increasing the population which clears the land) and indirectly (by increasing the demand for wood products in a country)."[12]

The dynamics of deforestation differ in countries with vast expanses of rain forests. This dynamic is especially the case in countries with relatively high per capita GNP or total goods and services sold per year. State managers in countries such as Brazil, working with more taxable wealth, can build extensive road systems and grant concessional loans to logging companies. Echoing the reasoning of Commoner, Rudel finds circumstances where powerful actors impel deforestation. Rudel also finds, however, that even in less-developed countries with some taxable wealth and large forests, population continues to have a positive, statistically significant impact on deforestation.

Thus, between the work of Mazur and Rudel, we can find evidence, under certain conditions, where each of the arguments of Ehrlich, Simon, and Commoner hold true. The sharp decline in American energy consumption after the oil shocks, obviously, illustrates Simon's free-market orientation. Ehrlich's Malthusianism is evident both in the analysis of American energy consumption and tropical deforestation. The work of powerful state and corporate actors is found to be more important than population growth in explaining tropical deforestation in wealthier less-developed countries with large rain forests. Here state managers have the resources to perpetuate large-scale deforestation. Population growth, as in Mazur's work, remains as a significant causal force in deforestation as well. How world population growth can be reduced and some of the important debates about population growth policies are the subject of the remainder of this chapter.

DEMOGRAPHIC TRANSITION THEORY, OLD AND NEW

Demographers frequently use the notion of a demographic transition to explain trends in population growth. The theory is fundamental enough in the study of population to prompt one contemporary population scholar to write, "The conventional wisdom of the theory has had a deep impact and guides the work of programs of international organizations, technical assistance decisions by government, and popular analyses in the media."[13] The statement suggests there are important policy implications both of the theory and its contemporary variants. We will discuss them after analyzing the theory and renewed scholarly work on this fundamental part of the relationship between population and the environment.

The Theory of the Demographic Transition: A Conservative Paradigm

The conceptual scheme of the transition is straightforward and, much to the chagrin of its critics, relatively simple. Societies can be considered as being in one of three stages in the process of modernization: (1) primitive or pre-transition social organization (Stage I) where mortality fluctuates widely around a relatively high average value, and fertility is correspondingly high; (2) transitional social organization (Stage II), characteristic of many less-developed countries today, where mortality is declining, fertility remains high, and the population exhibits high rates of natural increase; and (3) modern society (Stage III), chiefly characterizing the more-developed countries, where mortality stabilizes at a relatively low average value, fertility is approaching the level of mortality or even going below the mortality rate, and a stationary or declining population is possible now or in the near future. Because the paradigm predicts that people in all societies eventually will adopt Western cultural values such as capitalism, higher education, and a preference for nonagricultural work—parts of the modernization process—the theory is a conservative one involving the worldwide convergence of human cultures.

The earliest work on the theory defined modernization as the growth "of huge and mobile city populations" or "urban industrial society."[14] The technological changes accompanying the industrial revolution, through a process of Durkheimian dynamic density, put people into contact with strangers, improved their education, and took away the "props" responsible for high fertility—religious doctrines, moral codes, laws, and marriage customs favoring large families. As a consequence of this socioeconomic change, people experienced a fundamental change in their worldview. Caldwell begins his extensive work by calling this process Westernization, or the emergence of what Notestein describes as a change in human consciousness with the emergence of the rational, secular, scientific worldview.[15]

Modernization thus involves industrial development in society with changes in the nature of work and a shift in the spatial organization of popula-

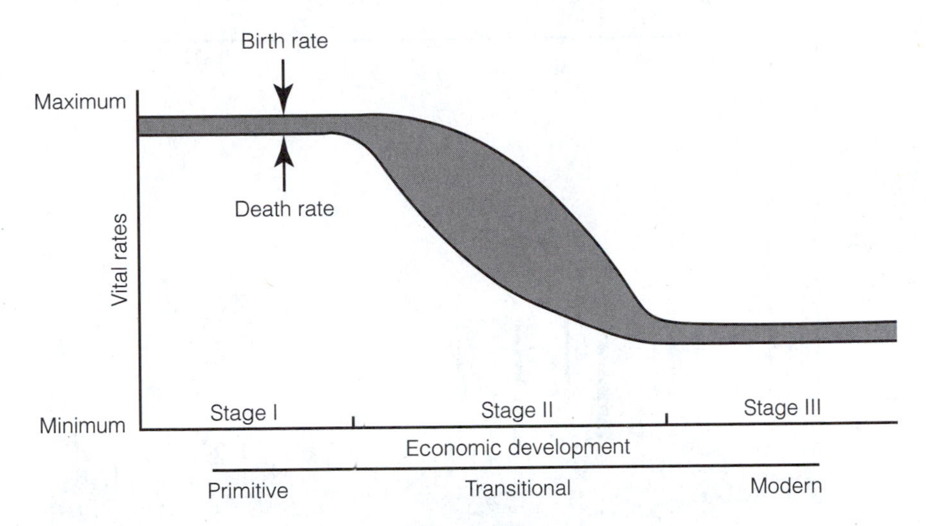

FIGURE 3.3 The Theory of the Demographic Transition

tion from decentralized, sparsely populated rural communities to large, concentrated, densely settled urban communities. The modern personality is less traditional, with openness to new experience, increased independence from parental authority, optimism about the efficacy of science for health and for freedom from natural resource scarcity, and greater ambition for oneself and one's children.[16]

Figure 3.3 depicts the nature of the demographic transition in terms of its stages, typical birth and death rates, and the resulting population growth. Obviously, the longer a society remains in Stage II of the transition, the larger the population of a society will be.

History of the Demographic Transition

The history of the demographic transition is double edged. The theory is a history of a mortality transition from high to low death rates, and a history of a comparable fertility transition. Next we will discuss both transitions, including recent thinking about the causes of both.

The Mortality Transition Pre-industrial populations, historically, fluctuate dramatically in total numbers of persons, largely due to oscillations in mortality resulting from epidemics of contagious diseases, inclement summer or winter weather, and changes in the cost and supply of food.[17] Figure 3.4 shows these oscillations based on the historical reconstruction of population growth in 404 English church parishes from 1514 to 1851. The bars on the graph represent the percentage of parishes where mortality in a given period was significantly higher than expected. The decades between 1550 and 1560 and 1720 and 1730 are especially dramatic in the devastating effects they had on

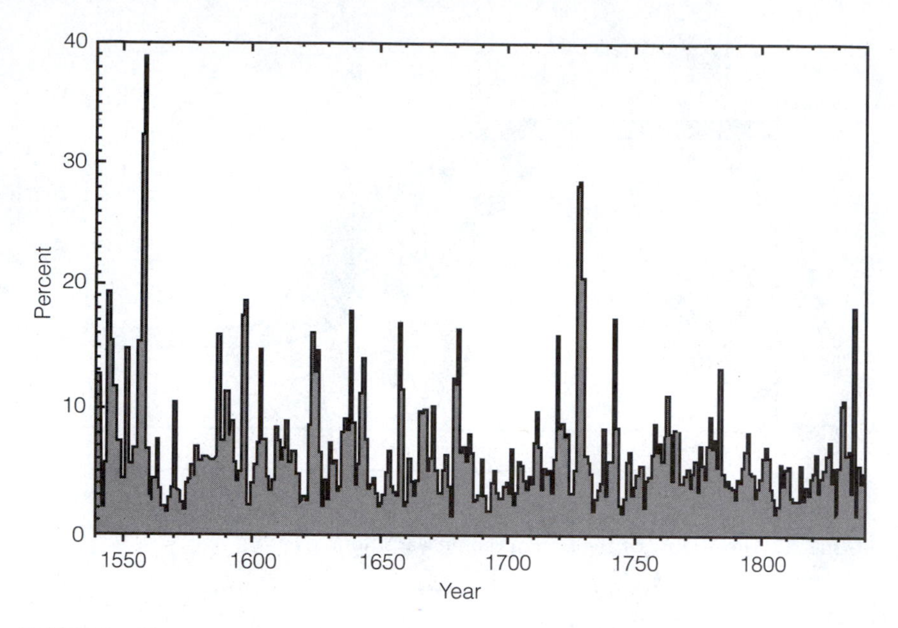

FIGURE 3.4 The Percentage of English Parishes Experiencing a Mortality Crisis, 1541–1851

SOURCE: Wrigley and Schofield, 1981. Used by permission of Edward Arnold Ltd.

England. The peak in the mid-sixteenth century is a plague in London; the eighteenth century peak is a widespread smallpox epidemic.[18]

McKeown's work traces death rates in selected European countries since they adopted central statistical offices for reporting vital events such as births, marriages, and deaths.[19] By the onset of the nineteenth century, death rates declined in Sweden and France, as evident in Figure 3.5. Death rates declined in England and Wales (E and W) by 1821, and by 1851 in Ireland.

From these data and trends in fertility, McKeown argues that the onset of European population growth is a result of declining mortality alone. Wrigley and Schofield's reconstruction of trends in fertility and mortality in England suggests that both declining mortality and rising levels of childbearing played a role in the onset of European population growth.

An important question, in any case, centers on the origins of improved health in Europe decades before Pasteur in 1860 opened a new chapter in medicine through the discovery of the microbial basis for human sickness and death: the germ theory of disease. For McKeown the initial decline, at least a half century before Pasteur's work, was linked to improved diets through increased trade and the diversification in the supply of nutritious foods. In addition, the nascent Industrial Revolution was tied to the mass production of cotton textiles. People gained access to affordable, washable clothing, rather than traditional woolens, at the same time they were enriching their household diets.

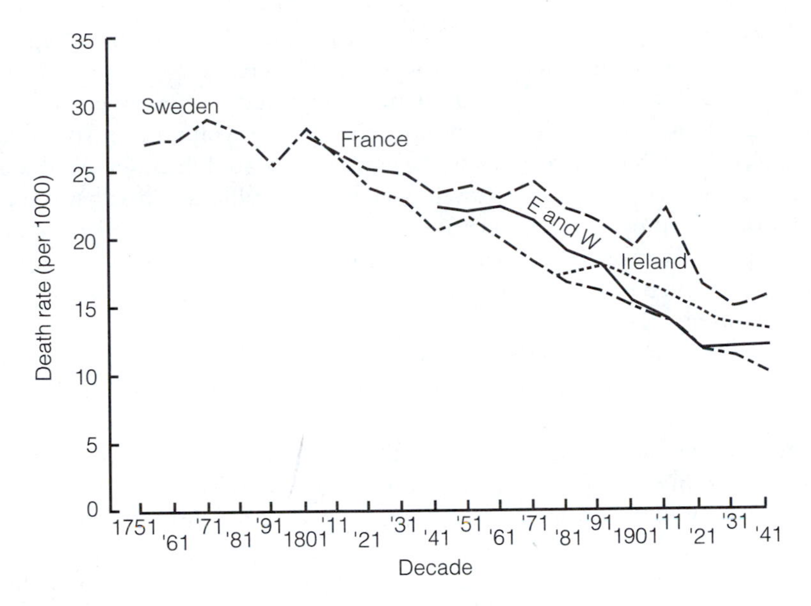

FIGURE 3.5 European Death Rates from the Time of the First National Vital Registration System

SOURCE: McKeown, 1976, p. 28. Used by permission of Academic Press.

While modern vaccines against contagious diseases began to play an important role in the Western mortality transition by the beginning of the twentieth century, a revolution in public health accelerated the decline in mortality after 1870. Simon Szertzer's research shows a pronounced rural-urban difference in the life expectancy for Britain as the Industrial Revolution drew migrants into new cities such as Liverpool, Manchester, and Leeds.[20] By 1830 residents of British industrial cities could expect to live, on average, twenty-nine years, while the national average life expectancy was forty-one. One paid a truly grave price for the higher wages of these early industrial cities.

By the 1850s the twelve-year age differential in life expectancy between industrial cities and Britain as a whole started to decline. Residents of industrial cities could expect to live thirty-four years; the national life expectancy remained forty-one. In 1890 British residents could expect to live forty-six years; those in cities had an average life expectancy of forty-two.[21] The improvements in urban life expectancies and public health, shared by industrial cities growing in Western Europe and the United States, stemmed from a set of breakthroughs in epidemiology, even before Pasteur's germ theory was widely accepted by the medical community. These findings carried over into an early urban environmental reform movement.

As industrial cities grew in the nineteenth century, water-born diseases such as typhoid fever and cholera, highly contagious smallpox, and tuberculosis in evermore densely settled working-class slums reversed their recent downward trend occurring elsewhere in England. The epidemiologist William Farr of the British Registrar-General's Office, charged with keeping track of births and deaths in Britain, discovered the growing discrepancy between the national and urban industrial life expectancies. Farr and his coworkers, in addition, traced unusually high incidences of cholera and typhoid to parts of cities such as Manchester where a private water company sold drinking water contaminated by human and animal waste. They also established links between air pollution caused by coal-fired factories and coal-heated homes and respiratory diseases such as bronchitis. Farr publicized his findings nationally about the greater incidence of highly contagious tuberculosis in the deteriorating housing of working-class slums.

By the 1860s, as Pasteur worked on the microbial basis of diseases, urban reformers worked hard trying to improve the environment of industrial cities. National and municipal officials, in what today would be called the environmental justice movement, garnered political support from the working class that suffered most from the growing contagion. The officials promised working-class households cleaner streets, centrally controlled public water, sewers, and safe food. The mayor of Birmingham successfully retained his office, for example, through "gas and water municipal socialism." Municipalities throughout Britain formed public health departments. After the Adulteration of Food Act (1870), trained public health officers began regular inspection of meat and milk. Loans from the British Treasury to cities for constructing costly public water and sewerage increased from a total of 11 million pounds between 1858 and 1871 to over 87 million pounds by the turn of the century.[22] After the Notification of Birth Act (1907), Lady Health Visitors from local health departments contacted families with newborn infants to teach them about hygiene for babies, including sterilization of bottles for their milk.

Szertzer says that the early Industrial Revolution and the mortality transition was a dialectical one.[23] Before residents of the newly emerging industrial cities lived long enough to appreciate their rising incomes, they collectively endured severe health risks associated with crowded living arrangements, unsafe factory work, and highly contagious diseases. Epidemiological research and the political aftermath of the research were important steps in urban environmental reform and the mortality transition.

The Fertility Transition Demographers have learned more about the early fertility transition through the European Fertility Project. Researchers in that project, using church records, reconstructed the childbearing histories of married couples in 579 European provinces. The observation at the heart of the findings was that between 1870 and 1910 in the vast majority of European provinces, married couples began to limit the size of their families. Prior to 1870—prior to the fertility transition—couples practiced repeated childbearing in marriage until the woman's physiological capacity to reproduce was exhausted—the natural fertility regime. Couples, increasingly after 1870, began to control their childbearing after reaching their desired, and ever smaller, family size. As one project participant observes, "In 1870 few provinces showed signs of the onset of the fertility transition; by 1930 few did not."[24]

There is evidence to suggest that modernization, as defined by the early demographic transition theorists, played a role in the early fertility transition. The modernization occurred at a time when mortality levels for people fifteen years old or younger declined, largely as a result of urban public health improvements;[25] the fertility decline often started in larger cities of European provinces;[26] and started among wealthier households.[27] The transition, nonetheless, quickly spread through linguistically homogeneous regions to places varying in size and wealth, among other characteristics. Such findings prompted scholars to write, "Although the European experience confirms a loose relationship between socioeconomic modernization and fertility decline, it also suggests that there was an important innovation-diffusion dimension to the reproductive revolution that swept the continent."[28]

Research on the history of population growth in Belgium is especially telling about this innovation-diffusion dimension, a cultural dimension, to the fertility transition. Ron Lesthaege compares seventy pairs of neighboring communities, never more than six miles apart, on either side of an east-west line in Belgium separating two linguistic cultural traditions: the French-speaking Walloons and the Dutch-speaking Flemish.[29] The two regions, on average, were comparable in social and economic characteristics such as the demographic size of villages, the percentage of residents in agriculture, and the average size of the manufacturing workforce. In addition, there were no statistically significant differences in these variables for each contiguous pair of communities on either side of the language line.

In the 1864 to 1868 period, married Belgian couples within specific age groups living in each of the seventy neighboring pairs of communities had comparable numbers of children. By the 1898–1902 period, however, couples living in sixty-two of the communities in the French-speaking Walloon region had statistically significant, lower levels of childbearing, irrespective of the social and economic characteristics of these communities. As Lesthaege observes, "In this narrow strip, 100 miles long but only 6 miles wide, we find a real demarcation line between two obviously noninteracting demographic regimes."[30] Something other than change in the socioeconomic modernization of the villages brought about the fertility transition.

For Lesthaege and others such as Susan Watkins these changes in childbearing in French-speaking Walloonia manifested a cultural or ideational transformation diffusing through Europe, led by France.[31] The changes included how individuals thought about themselves, their marital partners, and their children—a cultural change. People began to realize that their material well-being and that of their families was linked to the number and spacing of children, and couples had the right to remain childless, if they so desired.

New Contributions to Transition Theory

The research on the European fertility transition, like some recent work in environmental sociology discussed in Chapter 1, encouraged writers in population studies to take a cultural turn. They began to focus on the dynamics of ideational change—the changing ideas and beliefs shared by Europeans, and how those ideas influenced population growth. Industrialization and urbanization played a role in the fertility transition, but changes in European marital patterns and childbearing were not limited to couples in cities. During the late nineteenth century, in England and across the Continent, peoples' expectations about what they could achieve in their lifetimes, or what they might do for their children, so they could benefit from the Industrial Revolution, began to change.

Using this historical work and other recent research in population studies, scholars are making revisions to transition theory with important implications for contemporary population policy. This cultural turn, ironically, brought Notestein's work back into the discourse about the dynamics of population growth: the emergence of the rational, scientific worldview in Western culture.[32] A key scholar in this revision of transition theory is anthropologist John Caldwell. His work and the work's implications for contemporary population policy will be discussed next.

Caldwell's Work on the Fertility Transition The demographic transition, for Caldwell, involves a fundamental social change from a fertility regime where couples are rational to have unrestricted marital fertility to one where couples are rational to have restricted fertility. Using extensive interviews of married women in rural and urban Nigeria, Caldwell finds that large familes persist in modern cities such as Ibadan and Lagos. Even in these modern industrial, urban environments, most married couples continue believing that "each new pair of hands helps to feed the extra mouth."[33]

"The fundamental issue in demographic transition," Caldwell writes, "is the direction and magnitude of intergenerational wealth flows."[34] Caldwell defines wealth broadly to include food, precedence in eating, clothing, the control of home space and facilities, and access to transportation. Wealth means power and access to services, the right to be pampered, to make unchallenged decisions, to decide about what one will work on, and the right to control one's leisure activities.[35]

Wealth flows from the young to the old in pre-transition societies, and from females to older males. Shannon Stokes effectively illustrates the principles in Sub-Saharan African marriage and family customs.[36] Marriage is not a contract between two individuals, in these patriarchal African cultures, but rather an exchange of a woman's productive and reproductive rights to the husband's extended family. Men's parental rights such as claims on everything their children produce in farming, also a reflection of traditional patriarchy, are separate from the responsibility for children's economic support. Women have the duty to provide economic support for children, although the support may not come from a child's biological mother. Extended kin groups within this pre-transitional culture practice bride price. The groom's family exchanges money or some other form of wealth for their son's rights to her productive and reproductive rights. The exchange of the bride price between the groom's and the bride's families, however, is often not finalized until the birth of the first of many children.

Caldwell does find a small percentage of Nigerian women (for example, 0.7 percent of Ibadan's 750,000 residents) who crossed the "divide"—women over forty in monogamous marriages using contraceptives to limit their family size. Caldwell finds these women have the most formal education of any Nigerians, and they have husbands, in many cases, who work in nonmanual occupations. Reporting other findings about these women, Caldwell writes, "They are less concerned with ancestors and extended family relatives than they are with their children, their children's future, and even the future of the children's children."[37]

Two forces are at work, according to Caldwell, to cause these women, as well as their husbands, to reverse the flow of wealth in their families: formal education and the mass media. Caldwell argues that the effect of mass education on the family is a primary determinant of the timing of the fertility transition. Education alters familial structures, increasing the costs of children. Caldwell argues that the fertility-reducing effect of mass schooling is unlikely if education is restricted only to males. The education of only half the community does not have the same effect as the education of most of the community. In his own words, "When only a fraction of the population has been to school, there remain strong forces to maintain family morality as the basic morality of the society."[38]

For Caldwell, there are at least four ways education changes familial structure. First, school "reduces the child's potential for work inside and outside the house."[39] Education changes the family, most obviously, because school reduces the number of hours the child has available to provide labor. Also, the child may be alienated from traditional chores because of her/ his new learning and status. Further, parents and other relatives may feel that the child needs to conserve energy for succeeding in school; that traditional chores may not prepare educated children for nontraditional employment and status; and the parents and relatives may be apprehensive about alienating a child who demonstrates the ability to succeed in the nontraditional world of the West.[40]

The costs of children increase both directly because of fees, uniforms, and stationery; indirectly, children's costs increase because attending school places greater demands on families to better clothe and feed their school children. Schooling also lends additional respect among parents for their school children as they are socialized into the literate world.

That literate world introduces the child to values of the Western middle class. The educational system is largely imported intact from the West. Most public school texts in Western Nigeria come from England or are modeled after these texts. The family they convey to be ideal is the Western nuclear family "with strong conjugal ties and the concept of concern and expenditure on one's children."[41] Caldwell argues that the Westernization of cultural beliefs and attitudes (not just more secular economic rationality) is the basic causal force in the fertility transition.

Contemporary Westernization of less-developed countries such as Nigeria, according to Caldwell, also results from imported media content from newspapers, radio, television, and movies. The process is very much like the changes taking place in French-speaking Belgium during the late nineteenth century. The role models the media projects are wholly imported from the West, along with the message of the ideal nuclear family. This media places a large emphasis on romance, sexuality, and sexual fidelity in monogamous marriages. This greater emphasis on sex between husbands and wives, in a society where people traditionally considered conjugal relations as less important than extended family relations, tended to increase conjugal emotional bonds and further nucleate the family.

While Westernization has not taken hold as deeply, as yet, in much of Africa, where it has, in Southeast Asia, Japan, and urban Mexico as well as urban Brazil, the consequences are revolutionary in scope and ultimately lead to a controlled fertility regime. A growing sense of legitimacy for intimacy within marriage, the nuclear family, the transport of that family away from the extended families in rural areas to cities, and rising standards of mass education for both boys and girls are among the revolutionary cultural forces in the Westernization process. Westernized families are smaller in size, reflecting a couple's deliberate control of childbearing. Smaller families allow more wealth to flow toward children, especially for the children's Westernized education.

Policy Implications of the New Transition Theory

In an analysis of theories about human fertility, George Alter makes what might seem to be a puzzling observation: Some theories about human fertility imply that the modernization of society "is the best birth control."[42] Alter's statement is part of what now appears to be an unnecessarily polarized debate beginning in the 1960s and 1970s over the appropriate form of population policy. There were fertility theorists who thought modernization would lead to the demographic transition and other theorists who thought that national family planning programs are sufficient to lower birth rates.

For those who believed development is the best birth control, the massive expenditures of public funds for family planning through organizations such as the U.S. Agency for International Development or the UN Fund for Population Action would not substantially reduce world population growth. These managerial or state-based programs, initiated by the Johnson administration in 1965, focused on research for safe, effective contraceptives and the operation of family planning clinics to help married couples achieve their desired family size.

The UN Fund for Population Action depiction in Figure 3.6 captures the family planning perspective. The emphasis is on "target populations"—a medium projection with a world total population of just over 10 billion or a lower projection of 7.5 billion. World population in this figure is a direct function of average family size; projection 1 occurs if the average number of children per couple reaches 1.9 by 2020–2025. How this target is to be achieved

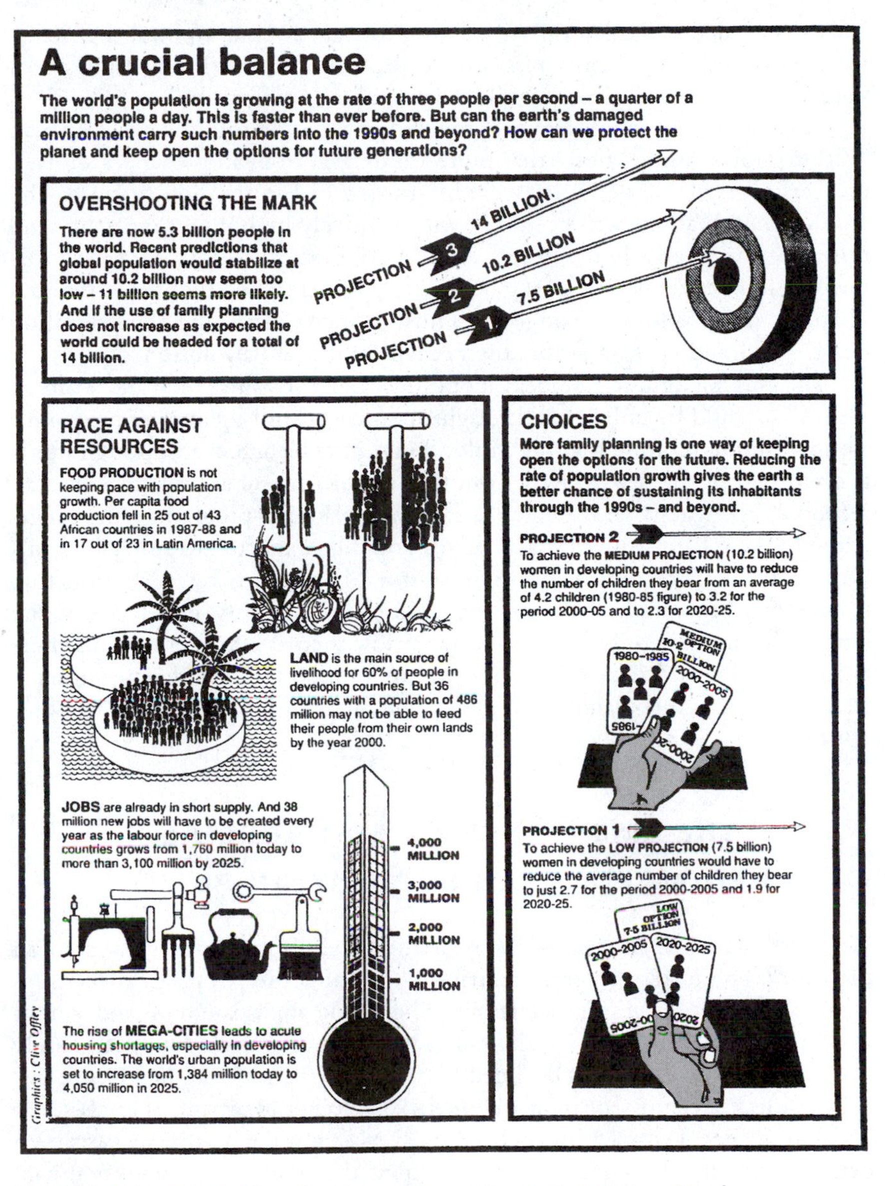

FIGURE 3.6 The Family Planning Perspective of the United Nations Fund for Population Action

SOURCE: Sadik, 1990, p. 1. Used by permission of the United Nations Fund for Population Action.

and what structural conditions in society motivate couples to limit their family size, according to family planning logic, is a result of the distribution of state-funded contraceptives and information about using them.

The European Fertility Project provided new insight into the debate. Susan Watkins, for example, notes that more modern European provinces showed the earliest signs that couples were beginning to limit childbearing by the 1870s. Then Watkins writes, "those with relatively high infant mortality and illiteracy and relatively low levels of industrialization and urbanization followed within a few decades, long before they . . . reached the same threshold of social and economic change."[43] Thus, France's birth rate started falling nearly one hundred years before the French Industrial Revolution.

The cultural diffusion of ideas about new roles for young men and women, ideas that could be enhanced through organized family planning programs, along with the promise that such roles could now be achieved through modernization, tips the debate away from the modernization argument. In fact, a recent study suggests, in the absence of family planning programs, "the population of Asia, Latin America, and Africa could be expected to reach 14.6 billion by the end of the twenty-first century instead of the 10 billion that is currently projected."[44] At the beginning of the twenty-first century, the long-standing debate between the transition theorists and family planners faded into the past. The historical roots and contemporary changes in neo-Malthusian or family planning programs will be discussed in the next section.

THE EMERGENCE OF MANAGERIAL, NEO-MALTHUSIAN POPULATION POLICY

We know that both changes in Western culture and social structure, based on the European Fertility Project, contribute to changes in people's expectations about gender, marriage, and the family. The young adult daughters and sons of the European upper classes led this cultural revolution. While European family size among the middle class declined between 1870 and 1900, family size declined at least a century or more before 1870 among the urban upper classes of London, Venice, Lyon, Milan, and Genoa.[45] Massimo Levi-Bacci suggests even before young members of the upper class knew the economic benefits of urban, industrial growth, these European elites began limiting their family size. Upper class couples did so as a way to concentrate the flow of wealth to the next generation, an idea consistent with Caldwell's theory of the fertility transition.

One important sociological consequence of the Victorian middle-class lifestyle, with the accompanying desire for smaller families, was the increasing popularity and demand for contraceptive information. To scholars such as McLaren[46] and Soloway,[47] this phenomenon is more than an artifact of an expanding middle class; the demand was the beginning of the women's movement. As one scholar observes, the growing interest and use of birth control techniques, especially interest in female contraceptive methods such as cervical caps and diaphragms, reflected "the gradual emergence of greater female independence, self-awareness and expanded expectations."[48]

Middle-class women, transformed by this social movement, not only gained a new sense of themselves, but they gained the desire to have working-class women join in this new form of independence. This growing interest became the basis for the International Family Planning Movement, beginning in the late nineteenth century in England. The International Family Planning Movement, in turn, shaped the way leaders in many countries thought about the relationship between population and the environment to this very day. The roots of the international family planning movement and the influence of the movement's central paradigm, neo-Malthusianism, will be the subject of this section. We will also discuss how women's organizations throughout the world began changing key policies in neo-Malthusian thinking through international population conferences beginning in the 1990s.

Contraception and Family Planning

The late nineteenth century collapse of the natural fertility regime in Europe exploded into public view in England with the nationally sensational Bradlaugh-Besant trial of 1877. Charles Bradlaugh, a member of the British Parliament, social reformer, and reputed "free thinker," staked his freedom on behalf of Mrs. Annie Besant. An early feminist, divorcee, and proprietor of a bookstore in Bristol, Besant was arrested and jailed for selling Knowlton's *Fruits of Philosophy,* a pamphlet describing condoms, diaphragms, contraceptive sponges, and douches. Bradlaugh objected to the court's censorship of Mrs. Besant's entrepreneurial and free speech rights. To challenge the court's decision under the British Obscene Publications Act, Bradlaugh republished Knowlton's pamphlet with Mrs. Besant and notified the police of their activity.

Bradlaugh and Besant were charged, committed, tried, convicted, and subsequently acquitted for their conduct, which received widespread news coverage. The sales of *Fruits of Philosophy* rose from approximately 1,000 copies annually before the trial to 125,000 copies in 1877, and another 60,000 copies sold shortly after the trial. Mrs. Besant subsequently published her own book on contraception and family planning, *The Law of Population,* which had extensive circulation in England, and 110 editions.[49] The trial was to be "a breach in the walls of Victorian prudery and sexual repression."[50]

The Bradlaugh-Besant trial, obviously, contributed to the diffusion of knowledge about contraception. The trial also played an important role in the evolving feminist social movement for world family planning through the establishment of the Malthusian League. The league was originally established to provide legal aid for persons who promoted the distribution of contraceptive information such as Bradlaugh and Besant. After the trial, however, the goals of the Malthusian League expanded to take advantage of existing desires in English society for women's and men's rights to family planning and contraceptive information and technology. By the next decade the movement mobilized by the organization of more personnel to persuade others about the benefits of family planning for women and their families.

Progressive medical doctors and scholars in the Malthusian League attempted to persuade the British Medical Association (Association) to sanction the prescription of contraceptives as legitimate preventive medicine. Their efforts met with strong opposition from the Association, including the professional censure of Henry A. Allbut, M.D. for publishing *The Wife's Handbook*. Dr. Allbut's book was endorsed by the Malthusian League as a most readable guide with illustrations and the addresses of suppliers for condoms, quinine solutions, and pessaries or cervical caps. *The Wife's Handbook* sold over 500,000 copies before going out of print in 1927.[51] For conservatives in the Association, prescribing contraceptives perpetuated cancer and nymphomania in women and heart palpitations and memory loss among men.[52]

Because the Malthusian League met such strong opposition in their efforts until the beginning of the twentieth century, league leaders operated family planning clinics on their own. Twelve such clinics opened in England and Scotland by 1927. Thus, the controversial ideas of Malthus came to life as a social movement with concrete service agencies and professional family planners working to help couples achieve their desired family size. Unlike Malthus, however, these family planners advocated the deliberate use of contraceptives as preventive checks to control population growth—neo-Malthusianism.

The growth of birth control clinics had scientific as well as practical significance for the issue of population. The Birth Control Investigation Committee (Committee), organized in 1927 by the twelve British family planning clinics, promoted scientific research on the validity of Malthus's and the Association's moral opposition to the use of contraceptives. The formal objective of the Committee was "to investigate sociological and medical principles of contraception: the possible effects of the practice on physical and mental health; and the merits and demerits of contraceptive methods."[53] Although the practice of family planning was a foregone conclusion among upper- and middle-class American and European families (including medical doctors and their wives) by the time scientists began contraceptive research, the work conducted by the Committee moved prohibitions against preventive checks outside the realm of science. The family planning movement, as part of the women's movement, institutionalized Malthus's concern for population growth and the many consequences of population growth. The movement did so without Malthus's conservative value judgments about the effects of contraceptive behavior on men and women—through managerial efforts: formal organizations such as family planning clinics and the scientific community.

Contraceptive information became popular in the United States early in the twentieth century, but once again with much public controversy. Central to this controversy was the revolutionary work of the early feminist reformer, Margaret Sanger, shown in Figure 3.7. Having lived in Europe, Sanger brought the work of the Malthusian League to the United States, opening the first birth control clinic in Brooklyn in 1916. Even though Sanger served thirty days in prison for her work, she continued undaunted.

Sophia Smith Collection, Smith College.

FIGURE 3.7 Margaret Sanger's Portrait in the 1930s

By 1921 she and her neo-Malthusian sympathizers organized the American Birth Control League, the predecessor of what is now the American Planned Parenthood Federation, to establish clinics nationwide. In 1927, her organizational efforts brought about the first World Population Conference in Geneva, long before the birth of the United Nations. In the 1950s, in her seventies, Margaret Sanger raised the money needed for pharmaceutical research that would lead to the development of the birth control pill.[54] When looking at the origins of American population policy, few are more important than Margaret Sanger.

The legacy of Sanger's work on birth control spread worldwide. In 1980 the International Planned Parenthood Federation (IPPF) consisted of fifty-nine national associations. Today, clinics operate in over 150 nations with regional headquarters in Nairobi, Kenya; Tunis, Tunisia; New York; and London. The London office, IPPF's world headquarters, also serves as the regional headquarters for East Asia, South Asia, Southeast Asia, and Oceania. Other agencies contributing funds and personnel to family planning programs throughout the world include philanthropic organizations such as the Ford and Rockefeller Foundations, The Population Council, and political bodies such as the UN's Fund for Population Action, The World Bank, and governments in the United States, Sweden, Great Britain, Norway, Japan, and the People's Republic of China. According to Donaldson and Tsui, the total annual expenditure on family planning programs is now about 500 million dollars.[55]

The Role of the United States in the Neo-Malthusian Managerial Paradigm

The neo-Malthusian paradigm, of course, reflects many of the ideas contained in Malthus's *Essay.* In the neo-Malthusian managerial perspective, like the NEP, the proponents recognize that people live in a world of finite natural resources; that population growth is the basic causal force in the production of natural resource scarcity and environmental pollution; that economies cannot grow indefinitely, in part, because technology is subject to the law of diminishing returns; and, that natural resource scarcities can stimulate Malthusian positive checks, including national or international conflicts.[56] As a result of world population growth, correspondingly, neo-Malthusians promote managerial public and philanthropic financing of family planning programs throughout the world to avoid the potentially destructive consequences of Malthusian positive checks.

The United States is the world leader in financial support for neo-Malthusian policies. U.S. leadership in the family planning endeavor is now at least forty years old. Of the 500 million dollars annually spent since 1985 on family planning programs, the United States contributes at least forty percent. The other two leading financial supporters are Japan, averaging 5 to 10 percent annually, and Norway with an average contribution of about 5 percent. Even during the conservative Reagan and Bush administrations, the United States, by far, led in financing these family planning programs.[57]

U.S. leadership in international family planning reflects a long-term commitment to the economic development of the less-developed countries, especially since World War II. The assumed link between family planning and economic development efforts has many sources. Betsy Hartmann and Peter Donaldson, however, both identify the preeminent, formative role of William H. Draper, Jr. in making such a link.[58] Appointed by President Dwight Eisenhower to advise Congress about U.S. military assistance throughout the world, Draper gained additional recognition as Undersecretary of War and Undersecretary of the Army. In 1959 Draper reported to the U.S. Senate Committee on Foreign Relations: "The population problem, I am afraid, is the greatest bar to our whole economic aid program and to the progress of the world."[59] This statement gained bipartisan support from presidents, secretaries of state, and UN population policymakers for decades to come.

Draper's statement reflects a neo-Malthusian managerial perspective on economic development in the less-developed countries within the context of the East-West Cold War after 1945. Rapid population growth, reasoned Draper, prevents economic development in the less-developed countries because any foreign aid simply lets a developing country tread water with existing levels of agriculture, industrial production, as well as health and educational services. In Draper's thinking, moreover, high and growing rates of poverty in the less-developed countries are associated with political instability, creating in-roads for communism. For the Population Crisis Committee, a powerful lobby group in Washington started by Draper, the

Table 3.3 Adult Female Literacy Rates and Total Fertility Rates for Selected Countries, 1990

Country	Adult Female Literacy Rate	Total Fertility Rate
Yemen Arab Republic	3%	7.0%
Burkina Faso	6	6.5
Afghanistan	8	6.9
Oman	12	7.2
Sudan	14	6.4
Honduras	58	5.6
Singapore	79	1.7
Thailand	88	2.6
Canada	93	1.7
Chile	96	2.7
Hungary	98	1.8

SOURCE: World Resources Institute, 1990–1991, p. 256. Used by permission of the World Resources Institute.

less-developed countries are sources of oil, minerals, and other natural resources essential to the U.S. economy and defense; or, they occupy strategic locations in gaining access to those natural resources. Thus, American state-based support for family planning reflects a commitment to economic development of the less-developed countries for improved international trade and a secure future for U.S. natural resources.

The Women's Movement and Neo-Malthusianism Managerialism

Neo-Malthusian policy analysts, in addition to emphasizing national economic development, now recognize the importance of education, especially the education of women, in reducing family size. Table 3.3 presents striking evidence for the role of women's education in women's total childbearing for less-developed countries varying in levels of female literacy. Women in countries such as Afghanistan or Burkina Faso (with literacy rates less than 15 percent) average six to seven births, while highly literate Thai women average fewer than three children. More educated women marry later, raise healthier children likely to reach adulthood, and choose to work outside the home rather than playing the role of the traditional homemaker. These trends, inferring from the work by Mazur and Rudel—discussed earlier—suggest that improved educational attainment for women could also lead to environmental improvements: reduced pressure on rain forests, for example.

Improved education in general and greater educational opportunities for women are likely to have an even more central role in neo-Malthusian policy in the future. In the Third International Population Conference on Population and Development in Cairo, Egypt, in 1994, women's organizations from around the world pressed to make gender equality in education the cornerstone of international population policy. Groups such as the Women's Caucus, including more than 400 women's organizations from sixty-two countries, worked to achieve their goal. The caucus lobbied for "a new definition of population policy . . . giving prominence to reproductive health and the empowerment of women while downplaying the demographic rationale for population policy."[60] While economic development and family planning remained important goals at the closing of the Cairo Conference, the context of those goals changed and reflected the growing power of the international women's movement.

Ten years prior to the Cairo Conference, at the UN Conference on Population in Mexico City, the U.S. delegation's official position included a number of issues related to the status of women. Antagonism to this position from the most conservative of President Reagan's appointees in the delegation, however, kept these issues largely off the agenda supported by the United States. But even so, partly as a result of women's lobbying and partly out of fear over the "gender gap" in the upcoming presidential election, feminists from the United States joined forces with feminists in other national delegations to create a separate section in the final Mexico City conference action plan on the role and status of women in population planning.

The success of the feminist agenda at Cairo resulted from a number of factors, including a different conference structure, better organization, and a more sympathetic Clinton administration. At the Cairo Conference, for the first time, nongovernmental organizations participated in the Conference proceedings and all the preparatory meetings for the Conference. This new opportunity for the nongovernmental organizations provided members of the feminist movement with an opportunity to press their agenda; and, they were well prepared to exploit that opportunity. In all, the Women's Caucus worked to place intense pressure on Conference participants to advance issues important to women that would reshape neo-Malthusianism.[61] The UN Decade for the Advancement of Women (1975–1985) spurred the feminist movement to act globally.[62] Many women's organizations came into being in the less-developed countries by the Decade for the Advancement of Women, often with considerable help from Western feminist groups, and these groups joined together in a process of coalition building.[63]

At the UN Conference on Environment and Development or the "Earth Summit" two years earlier, in 1992, officials avoided discussion of population issues. In the accompanying, yet separate, NGO Forum, however, the debate over population, the environment, and development occurred and a degree of consensus formed.[64] This NGO Forum represented a breakthrough event for NGOs, setting the stage for the 1994 Cairo Conference.[65]

Apart from this landmark NGO Forum, 1,500 women met in 1991 in Miami to draw up a document, "Women's Agenda 21," that they presented to the Earth Summit.[66] Women attending the Conference also convened their own priming event at the 1992 Earth Summit forum entitled "Female Planet." This flurry of meetings put the feminist movement a step ahead in building consensus and planning for Cairo, so their influence on the proceedings was substantive.

Before the second major preparatory meeting for the Cairo Conference, the International Women's Health Coalition circulated a document, "The Women's Declaration on Population Policies," that basically outlined the feminist stance on neo-Malthusianism. Prepared by women on all five continents, numerous organizations and individuals around the world signed off and worked together to push for the initiatives called for in this declaration. While there was considerable opposition to this declaration, the end of this Conference saw the formation of a fragile consensus and the inclusion of two new chapters in the final plan to accommodate the feminist agenda on neo-Malthusianism.

Important to the success of the feminist agenda was the election of President Clinton. This election brought in an administration concerned about population, environmental, and women's rights issues. Clinton, upon taking office, immediately reinstated funding for the UN Fund for Population Action and the International Planned Parenthood Federation, deleted from the budgets of the Reagan and Bush administrations. The official government stance stood largely in line with the feminist, pro-choice policies concerning family planning and abortion rights. Also, many women sympathetic with feminist ideals could be found in high-level policy and management positions in population-related institutions and agencies by the time Clinton took office.[67]

The final Program of Action adopted at Cairo states "programs that are demographically driven, and are intended to act directly on fertility, are inherently coercive and abusive of women's right to choose the number and timing of their children. Such programs should be replaced by others that 'empower' women by increasing their educational levels, providing them with satisfying jobs . . . and otherwise raising their status in the family and community."[68] In her newspaper, *The Woman Rebel, No Gods, No Masters,* Margaret Sanger (1914) writes: "The aim of this paper will be to stimulate working women to think for themselves and to build up a conscious fighting character." In Cairo, they did.

CRITIQUES OF THE NEO-MALTHUSIANISM MANAGERIAL PARADIGM

Because women's organizations are becoming active in shaping national and international managerial efforts to reduce population growth, we examine feminist conservative and radical critiques of neo-Malthusianism next. We will mostly omit the feminist managerial critique, since the efforts by groups such as the Women's International Health Network and the Women's Caucus at the Cairo Conference represent that critique exactly. We say mostly omit because some of the policies pursued by conservative ecofeminists are the very ones endorsed by women's groups at Cairo: achieving equal educational attainment for young women and men, for example—a cultural phenomenon. Women's groups at Cairo used a managerial, state-based strategy to redirect national and international population policies along conservative lines. Thus, we see the overlapping nature of our paradigms, once again. With population growth playing a direct role in environmental problems such as energy consumption and deforestation, these critiques by feminists also represent a search for more and more effective environmental policies.

The Feminist Critique of Neo-Malthusianism

Since the resurgence of the women's movement in the late 1960s, some feminist writers in sociology developed a critique of what they designate as the population control movement. The critique is premised on two general arguments. They insist, first, that neo-Malthusianism narrowly focuses on the need to reduce birth rates among the poor in the less-developed countries as a means of reducing poverty and malnutrition. Feminist critics prefer to place population policies in a more encompassing effort to strive for national economic development that also achieves gender equality and environmental sustainability. Feminist writers on population policy such as Betsy Hartmann do not argue that large families have little to do with poverty or natural resource scarcity.[69] They do argue that poverty and resource scarcity have other causes as well: the loss of access to commonly owned land for farming and wood gathering in the less-developed countries, gender inequality with women controlling very little of whatever marketable commodities a family has, and natural resource exploitation in less-developed countries with profits flowing to corporations based in the more-developed countries.

Critics such as Hartmann insist, second, that the "population controllers" tend to be male medical personnel, funded by more-developed counties' managerial agencies such as the U.S. Agency for International Development or pharmaceutical companies. These agencies' main goal is distributing contraceptive techniques and information. Feminist critics of the population control movement insist there should be more focus placed on cultural changes in societies—cultural changes that ensure women's reproductive health, basic health care for everyone, nutritional programs, and gender equality in educational attainment and employment opportunities. Since the first of these arguments will be covered in the section on the radical perspective on population and the environment, we will focus on the second of the main arguments here. Because the second critique is contingent on basic feminist theory, some of those assumptions and arguments will be included first in the next section.

Cultural Feminism and the Critique of Neo-Malthusian Managerialism

Cultural feminism rests at the core of most feminist perspectives. Cultural feminists analyze the worldview shared by members of a given society.[70] Worldviews explain, justify, and maintain relationships of domination and subordination. For the feminist scholar Carol Warren there are three significant features of oppressive cultural worldviews: (1) value-hierarchy or up-down thinking that places higher value on what is perceived to be of superior value; (2) value dualisms or unconnected pairs, where one member of the pair has higher value or status than the other; and (3) a logic of domination, where subordination is justified because one party has the more highly valued characteristic and the other does not.[71] Because animals do not have the ability to reason, for instance, they are regarded as inferior to humans; or, since men are said to think more scientifically than women, men are regarded as superior to women.

Cultural feminists believe that shared traditional beliefs and expectations that define male characteristics as superior to female characteristics, the culture of patriarchy, is a basic, fundamental causal force in most societies. Patriarchy oppresses women in sexual ways by defining them as beings whose primary function is to bear and raise children or to satisfy male sexual desires. A patriarchal worldview is one that takes male-dominated beliefs, values, and attitudes and assumptions as the standard; patriarchy gives higher status to what has been traditionally identified as male than to what has been traditionally identified as female.

Important to the feminist critique of neo-Malthusianism, cultural feminists argue that in order to liberate women, there must be an end to the male control of women's bodies by dismantling the cultural tradition of patriarchy. At the Cairo Conference, for example, women made significant steps toward dismantling the culture of patriarchy. They brought all the delegates to agreement that countries throughout the world should strive to educate women for gaining control of their sexual and reproductive rights. The women delegates convinced the participating countries and nongovernmental organizations to work for ensuring public access to an array of health services for women, including safe, legal abortion services in their home countries. They also brought the delegates to agreement that countries need programs to build women's leadership capacities, beginning with equal education for girls and boys.

These policy recommendations in the final World Population Plan of Action, among others, aim to dismantle the cultural tradition of patriarchy. Patriarchal thinking leads to normative dualisms and a logic of domination and ultimately the oppression of women.[72] Therefore, feminism seeks to eliminate patriarchal thinking that leads to other forms of oppression such as racism, classism, and heterosexism—resulting in other forms of domination.

Betsy Hartmann's work follows the logic of cultural feminism.[73] Hartmann insists that the population control programs developed around the world in the past twenty-five years are coercive. Such programs involve circumstances where poor families with inadequate sources of food have to participate in family planning programs involving sterilization, intrauterine devices, or birth control pills in exchange for incentives such as food and money. The programs, for Hartmann, divert foreign aid from more general maternal and child health programs; they also inadequately fund health care for participants who experience health complications from the use of the contraceptives.

Bangladesh's early efforts with family planning, from 1975 to 1985, exemplifies these patriarchal biases in the design of programs. Officials in the Bangladesh program assigned quotas for local birth control clinics to enroll participants for sterilization. Incentives for participation included a new sarong, food, and the equivalent of two weeks pay. All married couples participated after the birth of their third child. The goal ("target") of this effort was to reduce births from 43 to 32 per thousand people annually by 1985.

The Swedish International Development Authority, the UN, and the U.S.-based International Women's Health Network, in 1985, lobbied to terminate financial support for the sterilization program in Bangladesh. The Bangladesh program put poor women and children in a position where they had little choice but to participate in exchange for the incentives. The women and children would have to do so—whether they understood the consequences of sterilization or not. The lobby groups also insisted that the funding diverted foreign aid away from more generic programs to improve child health, education, and nutrition.

Indonesia's initial family planning effort between 1970 and 1990 reduced the birth rate from 45 to 22 per thousand persons annually, with contraceptive usage increasing from 2 percent to 58 percent among all married couples. Indonesia, as a result, is said to be the "family planning showcase" of the less-developed countries. Local women such as the wives of village political leaders, in this effort, act as a link in a social network of regional family planning clinics by providing contraceptives and contraceptive information. The women village leaders explained the benefits of longer spacing between births to village women whom they recruited. Village women who joined the program received access to financial credit and basic maternal and child health services. While this program was not committed to an aggressive sterilization program, as in Bangladesh, the program is not without critics. According to Hartmann, Indonesia places more emphasis on family planning than the government does on improving general health care through more Indonesian doctors, nurses, and hospitals.[74] With Indonesia having the highest maternal mortality rate in Southeast Asia, 450 maternal deaths per 100,000 live births each year, Hartmann's criticism is a fair one.

As an alternative to the population control movement, cultural feminists, like their contemporaries at the Cairo Conference, advocate national development efforts that create a balance for women to gain control over their own sexual and reproductive capacities. Cultural ecofeminists also want more distributive justice in the control of wealth among women and men.[75] Cultural feminists encourage women's advocacy groups to provide grassroots cultural education programs about how women can gain control over their own sexual and reproductive capacities. One such program provides an alternative to population control through offering an appropriate array of health services, including abortion. The program aims to combat hierarchical relations encouraged by discriminatory attitudes and practices of all kinds. The program also aims to build women's leadership capacity and to form linkages with government agencies, nongovernmental organizations, and international funding agencies. Such efforts go beyond population control. The programs endorsed by cultural ecofeminists aim to change culture and help to create societies where there is no patriarchal domination of women, minorities, and children and to create conditions where population growth and environmental degradation will decline.

The Radical Feminist Critique of Neo-Malthusian Managerialism
While radical feminist scholars subscribe to the tenets of cultural feminism, they go beyond these tenets. Radical feminists recognize the importance of capitalism in creating international class-based inequality in wealth and political power. Capitalists and their transnational corporations have the ability to alienate the world's workers from the full return on their labor. Radical feminists, however, extend the radical paradigm. Because men as a group control and derive value from the labor of women, radical feminists view men as a class that exploits women as a class.[76] These feminists, essentially, enlarge the theoretical domain of the radical paradigm. These feminist scholars focus on how men exploit women in sexual, procreative, emotional, and domestic labor relationships. Thus, they extend the concept of class to include gender relations throughout the global economy.

Radical feminists, like cultural feminists, are critical of neo-Malthusian managerialism. Neo-Malthusian population policies and programs often share patriarchal biases. Placing the burden for family planning on poor women in Bangladesh, discussed earlier, exemplifies this bias. Radical feminists' critique of neo-Malthusian policies, however, goes farther than the cultural feminists' critique. Radical feminists show how poverty and population growth are inextricably rooted in capitalism. A second radical feminist argument deals with the ways in which capitalism shapes the population control movement. Both aspects of this radical critique are covered next.

In the wake of Ehrlich's *Population Bomb* (1968), *The Limits to Growth* report in 1972 to the Club of Rome, and numerous world conferences on population and the environment from the Stockholm Conference in 1972 to the Rio Conference on Environment and Development and the Cairo Conference on Population, the world's leaders in both the more-developed and the less-developed countries defined poverty and the environmental crisis as the result of a single, unitary cause—overpopulation. Cultural feminists are critical of the conventional solution to the population problem because of the patriarchal bias in many family planning programs. Radical feminists agree, but they insist that there is more to the limits of conventional neo-Malthusianism. Blaming patriarchy fails to recognize the role of male-controlled modern technology, industrialization, or the affluence of the more-developed countries. To blame patriarchy is to ignore that 70 percent of all energy produced in the world is consumed by 30 percent of the population living in the United States, Europe, and Japan. In the words of one radical feminist, "There is no mention of the exploitive and colonial system we are living in, no mention of the prevailing development paradigm (more GNP/capita is better) or of the wasteful production and consumption patterns of the industrial societies which are producing most of the environmental destruction."[77] Box 3.1 elaborates on consumption patterns.

Radical feminists agree with cultural feminists who argue that neo-Malthusianism leads to patriarchal population control policies. Maria Mies, for example, emphasizes the patriarchal bias on the part of public and private organizations in the more-developed countries that fund experiments

BOX 3.1 Focus on the United States: Population and Consumption

To understand the impact of population on the environment, consumption is a key issue. The environmental impact of one person living in the United States, where rates of consumption are among the highest in the world, differs from the impact of one person living in a less-developed nation. On average, a person living in the United States has an environmental impact that is thirty to fifty times greater than an average person living in a developing nation.

Huge disparities in the consumption of energy, food, and natural resources, such as metals, exist between industrialized and developing nations. For example, meat consumption in the United States in 1995 averaged 119 kilograms per person. In Bangladesh, the average was only 3 kilograms per person. Countries with the highest consumption rates also generate the most environmental costs—pollution and waste. The United States is the top producer of municipal waste per capita in the world and the leading producer of carbon dioxide emissions, producing 24 percent of the world total carbon dioxide emissions.

The United Nations 1998 Human Development Report highlights the inequalities in consumption. "[Twenty percent] of the world's people in the highest-income countries account for 86% of total private consumption expenditures—the poorest 20% a miniscule 1.3%" (2). The report breaks consumption into some specific categories:

	RICHEST FIFTH OF THE WORLD	POOREST FIFTH OF THE WORLD
Meat and fish consumption	45%	5%
Energy consumption	58	4
Telephone lines	74	1.5
Paper consumption	84	1.1
Vehicle ownership	87	<1

Consumption doesn't just differ cross-nationally. Over time, consumption rates have increased dramatically. For example, the world produced 249 million cars in 1975 compared to 456 in 1993; in 1980, 6,286 billions of kilowatt-hours were used, in 1995 this increased to 12,875. "Americans own twice as many cars, drive two-and-a-half times as far, and use 21 times more plastic than their 1950 counterparts" (Motavalli 32).

Examining the consumption of resources illustrates that where population growth occurs matters for the environment. It also shows that blaming population growth for environmental problems leaves out a critical factor: differences in levels of consumption.

SOURCES: *Human Development Report.* 1998. New York: Oxford University Press. Motavalli, Jim. 1996. "Enough!" *E The Environmental Magazine* VII (2):28–35. United Nations Development Programme. 1998.

with chemical contraceptive implants such as Norplant among women in less-developed countries. Because of concern about the possible carcinogenic side-effects of Norplant, the more-developed countries withheld the drug from commercial consumption in their own countries.[78] This practice by the pharmaceuticals, for radical feminists such as Mies, is more than developing a contraceptive that places the burden for family planning on women. The pharmaceuticals exploited women in less-developed countries to see if their drug was safe for profit-making. Mies also writes that early experimentation with amniocentesis and ultra-sound scanning took place with women in the less-developed countries. This medical technology, designed to observe prenatal health, also determines the sex of a fetus. In countries such as India, where women pay the family of their husband-to-be a dowry before marriage, at a great cost to their own families, there is an incentive to "breed male" through the abortion of females, something possible with the development of this procedure and inexpensive abortion services in parts of India.

Significant patriarchal consequences of the population control movement are evident in post-Maoist China since 1979. While China worked to reduce population growth even before this time through a " later-longer-fewer" policy, encouraging later marriages with more years between births, the nation became even more aggressive in 1979 with the one-child population policy. The program, under the direction of the male-dominated Communist Party, delegates administrative responsibility for implementing the one-child policy to local family planning committees in cities and in rural villages. Committee members recruit married couples to sign a one-child certificate. Those couples who meet their contractual obligation receive monthly cash supplements for their child until age fourteen. Rural couples receive additional land for farming; urban couples receive more space in public housing. Only children receive priority in medical care and admission to day care facilities as well as public schools. Children with brothers or sisters receive no such privileges and their parents lose 5 to 10 percent of their income annually until the youngest reaches age sixteen.

Given that the Chinese live to pass on their lineage through the male descent line, scholars, not surprisingly, began finding an increase in the ratio of male to female births by 1985. International criticism of China's human rights violations, including the coercion of women, unequal rights for male and female fetuses, and difficulties with the implementation of this effort on a grand-scale led China to a de facto two-child policy for families if the first child is a girl.[79]

Radical feminists also critique what has become the second major effort in Western modernization policy, namely efforts to help women in the less-developed countries become a part of their newly emerging agricultural, manufacturing, and service economies. These efforts are consistent with demographic transition theory. Modernizing the economies of the less-developed countries provides employment opportunities for women. To take advantage of these opportunities, women also become involved with family planning. Barbara Herz's statement at a meeting of the International Monetary Fund in Berlin is exemplary: "It is expected . . . that raising the status of women through education and income generating activities—which the World Bank since 1988 calls 'investment in women'—will not only lead to a reduction of the birth rate but will at the same time solve the problem of poverty."[80]

Being critical of programs to invest in women would seem anomalous, especially for radical feminists. Yet, radical as well as cultural feminists support efforts at family planning that are part of more comprehensive efforts to create equity in gender, race, and class relations; they oppose patriarchal efforts and population control that jeopardize the health of women and children. Feminists are critical of investment programs that bring women into the emerging low-wage production systems in the less-developed countries. Such investment programs transform women from independent subsistence farmers to low-wage producers of export crops or manufactured goods to relieve the debt burden in their countries. These programs do not create gender equality—the end goal of both radical and cultural feminists. Referring explicitly to efforts to improve the education of girls in the less-developed countries, Betsy Hartmann writes, "In my more cynical moments, I also worry that educating girls is seen as a strategic component of incorporating more women into the low-wage, insecure service sector, where most new jobs are being created."[81] Poor women, in the more general argument, are being integrated into the global capitalist economy with most of the benefits flowing largely to the male-controlled corporations in the more-developed countries of Europe, North America, and Japan.

Feminists, in the truly radical feminist critique of modernization, insist that the less-developed countries should resist programs such as population control and investments in women sponsored by the World Bank and other development agencies. Radical feminists, instead, search for ways to end dualistic, hierarchically structured social organization such as Western patriarchy of which global capitalism is the latest manifestation. They support, for example, the Chipko Movement in the foothills of India's Himalayan mountains. Chipko is a multi-class environmental movement led by women who surround mixed hardwood forests to protect them from state-supported efforts to convert the land into commercial tree plantations. The programs would privatize an environment that Indian women have shared in common as a source of food, fuel, and fodder. Chipko leaders encourage women and peasant farmers to pull out the seedlings of state-supported social forestry programs in the Himalayan foothills, and to replace the seedlings with native tamarin and mango that bear fruit and more foliage for their fertilizer-producing domestic animals. The protest protects the ecology of the region and the ability of women to raise their children and care for their families as they have done for centuries.[82]

SUMMARY

The residents of the less-developed countries are numerically dominant in the world today and their rate of population growth, although declining, is substantially higher than that of the more-developed countries. Because of this population growth, neo-Malthusians argue that the prevalence of landless, unemployed, poorly fed people will become increasingly apparent in the less-developed countries in the future. Even if young persons entering their childbearing years decide on this very day to limit their family size to two children, world population would grow by several billion because of demographic momentum. More young people now survive to their childbearing years than ever before, and their childbearing will produce a quantum leap in the size of the world's population.

All but the most conservative human exemptionalists recognize the importance of population growth for intensifying natural resource scarcity, especially in a mid-twenty-first century world of 7 to 10 billion people. Alleviating world poverty and social inequality, rather than reducing fertility per se, is the key to solving natural resource scarcity problems, according the proponents of most paradigms discussed throughout this chapter. While scholars differ in the means of dealing with these social problems, neo-Malthusians, cultural feminists, radical feminists, and even some conservative human exemptionalists insist that enhanced opportunities for the poor will provide positive incentives for reducing family size.

Conventional neo-Malthusians in the past advocated private and state-managed programs in family planning to reduce population growth for the sake of modernization. Because of the growing international environmental movement and the women's movement, increasingly, this mainline view is being attacked from the right, center, and left. Conservatives, including cultural feminists and some ecological Malthusians, argue that the concept of modernization, developed since the 1950s, is shallow. The modernization concept fails to recognize the significance of economic development efforts for the natural world, focusing solely on development benefits for humankind. Cultural feminists go farther, arguing that development programs involving population control efforts narrowly focus on birth rates rather than on comprehensive programs to improve hospitals, to expand health care services, and to improve education—especially women's education—in the less-developed countries. These critical views of mainline neo-Malthusianism are beginning to gain attention not only in intellectual circles, but also in the international deliberations of nations about world population and environmental policy. Criticism of narrowly focused neo-Malthusian policies by women's organizations was an important part of the deliberations at the 1994 UN World Population Conference in Cairo.

The proponents facing the greatest challenge in making their voices heard in national and international deliberations on population are the Marxists such as the radical or socialist feminists. There are at least two principal reasons why their views are problematic. First, they work in a political atmosphere profoundly influenced by the collapse of the former Soviet Union and the socialist economies of Eastern Europe. To talk about Marxian population and environmental policies since the late 1980s inevitably requires a start with apologetic qualifications such as endorsements of socialist policies that do not follow the Soviet-style socialist production system. Their radical policies involve a search for new and more socially based forms of organization.

Second, they advocate that the less-developed countries de-link or resist integration into the global capitalist economy. Many of the governments targeted by radical policy analysts, however, are deeply in financial debt to the more-developed countries. As a means of debt relief for more development efforts, multilateral lending agencies such as the World Bank and the International Monetary Fund encourage the less-developed countries to make structural adjustments before they can be eligible for more foreign aid. The less-developed countries must lower protective trade barriers, set limits on wages, increase the production of exports (including foods and oil), devalue their currencies, and so on. While some of the major environmental groups manage to have some impact on the kinds of projects the lending agencies will support in the less-developed countries, neither these groups nor the lending agencies themselves have anything close to a sympathetic ear for socialist policies.

Yet, new social movements, exemplified by Chipko in northern India, that lead women to protect their traditional subsistence economies, resisting efforts at modernizing the rural economies of the less-developed countries, could serve as a role model for the growing millions of poor peasants throughout the world. Women who benefit even less than men in conventional modernization efforts through national road building, port facility construction, and land extensive hydroelectric power projects are prime candidates for such social movements in the less-developed countries. The loss of commonly owned wood and pasture land through these conventional economic development projects directly affects poor women. Now rural women, especially those who are poor, must travel longer distances to raise food and collect the fodder and fuel they need to carry out their traditional roles as mothers and family caregivers. Whether a widespread sense of class consciousness will occur under circumstances where women are spending more and more time literally struggling to stay alive remains to be seen. What consequences this consciousness might have for world population growth remains as a crucial question in the study of population and the environment.

An equally important issue is whether the conventional modernization programs supported by neo-Malthusian groups such as the U.S. Agency for International Development or the World Bank are ecologically sustainable, a problem that will reoccur throughout the remainder of this book. The Worldwatch Institute, the World Resources Institute, and other environmental organizations, increasingly, express scientific concern about modern grain production technology, soil erosion, and the development of vast hydroelectric production facilities in temperate and tropical forests and their impact on wildlife habitats and the homes of aboriginal people in the less-developed countries. For at least some of the major environmental organizations (to be discussed in Chapter 6), the population control approach is appealing. Neo-Malthusian policies emphasize ideational changes in culture with respect to marital sexual practices and family size. For radical feminists and environmentalists, such conventional population policy avoids an attack on the root cause of the world's population growth—widespread social inequality that leads to poverty and the need for families to have more hands to survive. For people in the less-developed countries and a growing number of women's organizations worldwide, the conventional neo-Malthusian approach to the population problem, one involving population control with little development, is neither just nor promising to the unprecedented numbers of youth in the next generation.

CITATIONS AND NOTES

1. Keyfitz, 1991; Levi-Bacci, 1992.
2. Hayami and Ruttan, 1987.
3. Meadows, Meadows, and Randers, 1992.
4. Bongaarts, 1994; World Resources Institute, 1994; Keyfitz, 1991; Levi-Bacci, 1992.
5. World Resources Institute, 1994.
6. Bongaarts, 1994.
7. World Resources Institute, 1994.
8. Dietz and Rosa, 1994, p. 278.
9. Simon, 1981, p. 202.
10. Simon, 1981, p. 348.
11. Mazur, 1994.
12. Rudel, 1989, p. 336.
13. Caldwell, 1976, p. 321.
14. Notestein, 1945, 1953.
15. Caldwell, 1976.
16. Easterlin, 1983, p. 563.
17. Wrigley and Schofield, 1981.
18. Wrigley and Schofield, 1981
19. McKeown, 1976.
20. Szertzer, 1988, 1997.
21. Szertzer, 1997, p. 77.
22. Szertzer, 1988.
23. Szertzer, 1988.
24. Watkins, 1991, p. 171.
25. Matthiesen and McCann, 1978.
26. Sharlin, 1986.
27. Levi-Bacci, 1986.
28. Knodel and van de Walle, 1986, p. 417.
29. Lesthaege, 1977.
30. Lesthaege, 1977, p. 112.
31. Lesthaege, 1983; Watkins, 1991.
32. Notestein, 1953.
33. Caldwell, 1976, p. 349.
34. Caldwell, 1976, p. 344.
35. Caldwell, 1976, p. 560.
36. Stokes, 1995.
37. Caldwell, 1976, p. 352.
38. Caldwell, 1980, p. 249.
39. Caldwell, 1980, p. 249.
40. Caldwell, 1980, 1985.
41. Caldwell, 1976, p. 356.
42. Alter, 1992.
43. Watkins, 1986, p. 440.
44. Bongaarts, Mauldin, and Phillips, 1990, p. 307.
45. Banks and Banks, 1964; Levi-Bacci, 1986.
46. McLaren, 1992.
47. Soloway, 1982.

48. Soloway, 1982, p. 136.
49. Peel and Potts, 1969; Seccombe, 1983.
50. Soloway, 1982, p. 53.
51. Soloway, 1982.
52. Peel and Potts, 1969.
53. Peel and Potts, 1969, p. 11.
54. Chesler, 1992.
55. Donaldson and Tsui, 1990.
56. Homer-Dixon, Boutwell, and Rathjens, 1994.
57. Donaldson and Tsui, 1990.
58. Hartmann, 1995; Donaldson, 1990.
59. Piotrow, 1973, p. 39.
60. McIntosh and Finkle, 1995, p. 1.
61. Ashford, 1995.
62. Mies, 1986.
63. Crane, 1994; Finkle and McIntosh, 1994; McIntosh and Finkle, 1995.
64. McIntosh and Finkle, 1995.
65. Ashford, 1995.
66. Oliveira, 1994.
67. McIntosh and Finkle, 1995; Ashford, 1995.
68. McIntosh and Finkle, 1995, p. 227.
69. Hartmann, 1995; also, Mies 1994.
70. Merchant, 1992; Warren, 1987, 1993.
71. Warren, 1987.
72. Warren, 1993.
73. Hartmann, 1995.
74. Hartmann, 1995.
75. Hartmann, 1995; Mueller-Dixon, 1993.
76. Jaggar, 1983.
77. Mies, 1994, p. 45.
78. Mies, 1986, 1994.
79. Hull, 1990; Kaufman, Zhirong, XinJian, and Yang, 1989.
80. Cited in Mies, 1994, p. 55.
81. Hartmann,1995, p. 19.
82. Shiva, 1988; Agarwal, 1994.

4

The Struggle over Hunger

Social and Ecological Dimensions

Human survival requires the energy supplied by food consumption. Theoretical and empirical questions surrounding food production, consumption, and distribution are thus among the first and foremost energy and resource issues. As the human population grows, the capacity for humans to produce and effectively distribute the calories needed to sustain human life comes under intensified scrutiny and emerges at the center of the world food and resources debate.

The divergence of views concerning the salience of the world food problem as an environmental issue can be explained in part by the present pattern of world population growth. Food production and access to food tend to be more intractable problems in the less-developed countries (LDCs) where rapid population growth continues. However, the more-developed countries (MDCs) are by no means immune from the world food problem. Concern has been raised over the sustainability of agricultural systems in the United States and elsewhere in the industrialized world where farming requires large amounts of energy and can have detrimental effects on the land resource. Confidence in productivity increases in U.S. agriculture has yielded to worry over recent trends toward productivity stagnation, environmental degradation, and nutritive inadequacies. Even a society as affluent as the United States has not eliminated malnutrition and hunger for a substantial minority of American citizens. Conflicts over international food trade and distribution, the benefits and burdens of prevailing strategies of agricultural development, and the role of the transnational food corporations in the world economy further challenge the privileged position of the MDCs in terms of food. The food question is thus a global one. Food issues not only affect all nations directly or indirectly, but also go to the heart of relationships among societies in the context of the international economic and ecological order.

Since we cannot adequately deal with all aspects of the world food problem, this chapter focuses primarily on food production in the LDCs. The first section will be a basic overview of recent world trends in food production. The biophysical constraints operating to influence these production trends will then be discussed. The next section will address theoretical approaches to issues of population growth and hunger. Finally, the Green Revolution and its implications for the future of food production and distribution will be addressed, exploring the important roles that power and international inequality play in the distribution and production of food.

FOOD PRODUCTION IN HISTORICAL AND COMPARATIVE PERSPECTIVE

A review of the trends in food production over the last four decades reveals the success of global agriculture: Overall, the world has witnessed a steady growth in the production of most food crops. The overall rate of that growth, however, has been slowing over the past several decades. When coupled with world population growth projections, which predict an annual addition of 90 million people to the human population[1] and the increasing realization of environmental constraints, the dimensions of the stresses on the future global food system emerge. Of particular concern, and at the heart of much of the contemporary discourse about the future of food, are questions regarding the overall productive capacity of the world food system, the recent sharp increase in international food aid, and the food security dialog that focuses on the differentiation between meeting food production needs and equitably distributing that which is produced.

Productivity: Recent Trends and Future Projections

Since 1961 the world has witnessed steady growth in the production of most food crops, as seen in Figure 4.1. The optimism generated from these absolute trends must be tempered, however, by the consequences of rapid population growth for improvements in agricultural productivity. Even though many LDCs exhibit increases in total food production, in some cases rapid population growth dramatically reduces the impact of these increases on a per capita basis. Per capita trends, shown in Figure 4.2, reveal that in Africa and the former Soviet Union, for example, total food production has not increased rapidly enough to match the pace of population growth. In contrast, in Asia and Latin America, success is evident in both absolute and per capita terms.

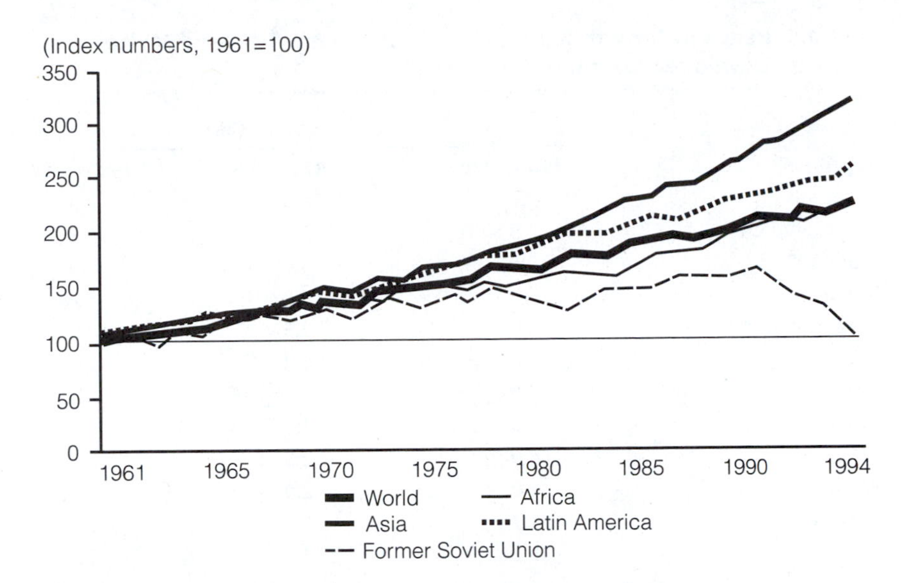

FIGURE 4.1 Trends in World Food Production, 1961–1994

SOURCE: World Resources Institute, 1996. Used by permission of World Resources Institute.

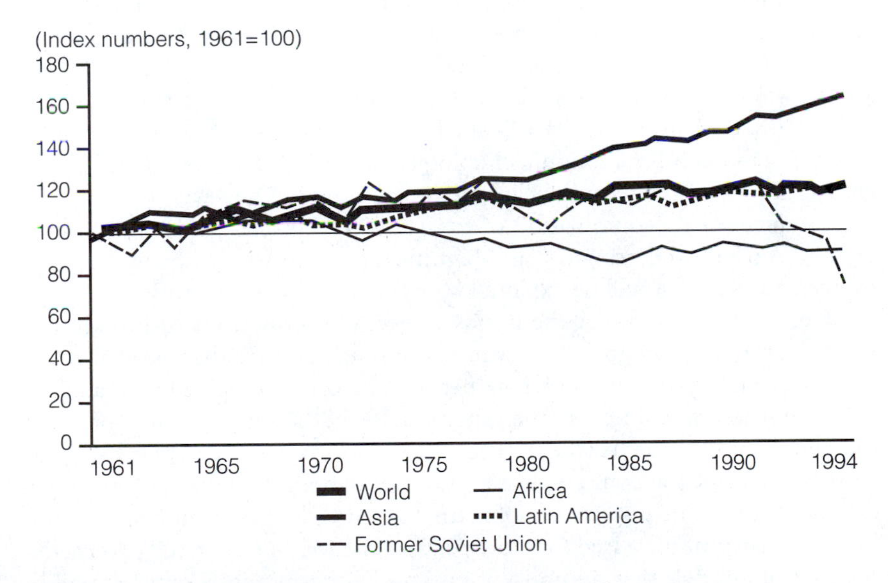

FIGURE 4.2 Trends in Per Capita Food Production, 1961–1994

SOURCE: World Resources Institute, 1996. Used by permission of World Resources Institute.

Table 4.1 Trends in Growth Rates in Yields of Wheat, Rice, and Maize in 93 Less Developed Countries, 1961–1992

	GROWTH (%)		
Crop Type	**1961–1970**	**1970–1980**	**1980–1992**
All cereals	2.8	2.6	2.1
Excluding China	1.5	2.1	1.9
China	6.0	3.7	2.9
Wheat	3.7	3.5	3.2
Excluding China	2.0	2.5	2.8
China	7.8	5.4	3.7
Rice	2.5	1.7	2.1
Excluding China	1.3	1.5	2.2
China	4.9	2.5	2.3
Maize	3.0	2.9	2.2
Excluding China	1.7	1.6	1.3
China	6.2	4.2	3.3

SOURCE: World Resources Institute, 1996. Used by permission of World Resources Institute.

Although overall growth has been steady, the actual *rate* of growth in food production has slowed considerably. Over the period between 1961 and 1992, annual growth in world agricultural production declined from 3 percent annually in the 1960s to 2.3 percent in the 1970s and 2 percent between 1980 and 1992.[2] Most of the initial growth reflected in the figures from the 1960s and early 1970s, evident in Table 4.1, is attributed to higher yields generated by the adoption of Green Revolution technologies in the LDCs discussed later in the chapter. Much of the recent decline in the growth rate, however, is considered a consequence of a combination of environmental, social, and technological factors. Agricultural growth places significant stresses on land and water resources, an issue that will be explored more fully in the next section.

Trends in yields mirror the trends in overall production: Although total yields increased, yield growth rates in developing countries have slowed.[3] Yet, many argue that yield increases, together with expanded cropland areas and improved efficiency, will remain the critical factor in the future growth of global food production. Whether the needed yield increases are indeed possible, given global environmental constraints, is a matter of considerable debate and statistical uncertainty. The importance of natural resource and environmental factors in the scenario of increased food production cannot be overstated; alternative production models that minimize inputs and environmental damage will be crucial to the success and sustainability of future food production.

Forecasts for future food productivity to meet increasing needs offer varying degrees of optimism. The Food and Agriculture Organization of the United Nations offers the optimistic projection that given substantial investments in agricultural research, production can accommodate the food demands of future human populations.[4] More pessimistic views, which suggest that human population growth is exacerbating environmental degradation and causing a reduction in productive cropland areas available per person, foresee declines in grain production and additional related problems such as widespread famine.[5] Later Brown and Hal Kane identify several factors they predict will undermine future food production,[6] including the limits to the biological productivity of rangelands and fisheries, an increasing scarcity of fresh water, the declining effectiveness of fertilizers, and the deterioration of social systems.[7]

Production versus Distribution: Food Security

Even if yields can keep pace with global population growth, increased production alone cannot guarantee food security for all of the earth's human inhabitants. As has been the case for decades, in 2000 a sufficient quantity of food was produced to provide needed calories to the global population, but due to inequitable distribution millions faced hunger and malnutrition. Production and availability, then, are quite different issues. Although sufficient quantities of food may be produced in the future, there are strong arguments that suggest that many people will be denied physical and economic access to the food produced.[8] Indeed, chronic undernutrition is already a long-term food security problem, which is particularly acute in sub-Saharan Africa and South Asia. The Food and Agriculture Organization (FAO), for example, reported that between 1990 and 1992, twenty-three African nations experienced low or critical food security levels.[9]

Despite impressive gains in agricultural production over the last four decades, then, hunger and chronic undernutrition remain critical problems for a considerable portion of the global human population. Hunger has a variety of significant adverse impacts on the economies of less-developed countries, just a few of which are increased human vulnerability to disease, reduced workforce productivity, disrupted physical and mental development in children,[10] and increased political and social instability.[11] Although our statistical understanding of the precise dimensions of protein-calorie malnutrition is incomplete due to widely divergent measurement techniques, definitions, and ideologies, numbers from UN agencies allow a general grasp of worldwide hunger. UNICEF estimates at least 200 million people starved to death or died from hunger-related disease over the past two decades.[12] The UN Population Fund reports that as many as one billion people were chronically undernourished in 1992.[13] Two-thirds of the world's hungry are in the densely populated countries of Asia, especially in India, Pakistan, Bangladesh, Indonesia, the Phillipines, and Kampuchea. Together with Zaire, Ethiopia, and Brazil, these countries comprise 70 percent of the world's hungry and undernourished.[14]

A related and important recent trend is the fall in prices of agricultural commodities over the past fifteen years.[15] This price trend has consequences wherever agricultural exports comprise a significant portion of a country's trade—the case in much of the LDCs. As prices fall, a given country must rely on exporting ever larger volumes of food in order to sustain the same earnings.

The pressures of external debts aggravate this situation, which will be further discussed in the final section of this chapter. Countries which increasingly focus on food production for export may fall short of meeting domestic food needs. When this is the case, governments have to buy from abroad or seek food aid, an increasingly common pattern. Given current trends, by 2010, the World Bank estimates that the LDCs could be importing 15 percent of their total grain consumption.[16]

Crisis and Food Aid

Over the last decade there has been a considerable increase in the frequency and severity of humanitarian crises (such as droughts, floods, wars, political turmoil, poor harvests) that require international food aid. The number of people affected by natural or political disasters rose from 44 million in 1985 to 175 million in 1993, and the number of people officially receiving protection and assistance from the United Nations rose from 1 million in 1970 to 17 million in 1993.[17] The result is a dramatic shift in aid priorities from development projects to programs that manage refugee needs and emergencies. The World Resources Institute reports that the global redistribution of food by public sector agencies amounted to a record 17 million metric tons in 1993: About 25 percent went to Somalia, Rwanda, and other countries in sub-Saharan Africa, while 41 percent went to Central Europe and the countries of the former Soviet Union.[18] Africa is by far the most seriously affected by food shortages; fifteen African countries face exceptional food emergencies.[19] Whereas in 1977 to 1979 cereal aid to sub-Saharan Africa was equivalent to 36 percent of the region's cereal imports, by 1987 to 1989 this figure increased dramatically to 68 percent.[20] Food aid, Tim Dyson notes, can serve both humanitarian and political objectives,[21] a point to be explored further in the final section of this chapter.

BIOPHYSICAL CONSTRAINTS ON FOOD PRODUCTION

Green plants require light, air, water, and soil for growth. Cultivated crops have additional energy requirements, including human labor and, often, chemical fertilizers, fossil fuel-based machinery, and irrigation. Presumably, human and biophysical constraints limit the extent to which food production on earth can be expanded. Failure to clearly define, recognize, and heed these limitations can result in irreversible damage to the productivity of the basic land resource.

Scientists commonly estimate, and disagree upon, the precise nature and parameters of the biophysical and human limits to food production. While some individuals and agencies provide optimistic predictions for the planet's capacity to increase food production, others claim that current rates of production are not sustainable and that further increases in production are unlikely, if not impossible, given the increasing severity of current and predicted future environmental degradation.

The goal of this section is not to evaluate the accuracy of these competing claims. Rather, we seek to identify and discuss the principal biophysical factor which could have a dramatic impact on the true capacity of global food production to expand and meet the nutritional needs of a growing planet. Thus the purpose of this section is to provide an overview of some key resource limits to growth in food production along with some reasoned estimates of the intractability of these limits. Biophysical constraints to expanding food production are increasingly important in the accurate estimation of overall food production capacity. Paul and Anne Ehrlich and Herman Daly identify eleven primary biophysical constraints to food production:[22]

1. potentially productive farmland is being lost to other uses, including urbanization, due to population pressures[23]
2. the increasingly limited supply of fresh water[24]
3. soil degradation and erosion[25]
4. the biological limits to yields[26]
5. limits to returns from the application of fertilizers[27]
6. the problems associated with chemical pest control[28]
7. loss of biodiversity resources of crops as well as their wild relatives[29]
8. possible decrease in yields as a result of increased ultraviolet B radiation[30]
9. air pollutants and their possible effect on crop yields[31]
10. the effects of climate change and rise in sea level on agricultural production[32]
11. "a general decline in the free services supplied to agriculture by natural ecosystems"[33]

Land

Food production can be increased by cultivating additional land, increasing the productivity of existing land, or both. Much attention has been directed to possibilities for bringing more land into production. The Food and Agriculture Organization (FAO) estimates that arable land expansion will account for 21 percent of production growth if we are to meet global food requirements in 2010. In addition to bringing more land into production, the FAO scenario predicts that increased yields will account for 66 percent of expansion, and increased cropping intensity for 13 percent.[34] The location and type of land the FAO views as available for cultivation is a point of discussion and controversy.

The FAO model suggests that the 760 million hectares currently under cultivation in the LDCs (excluding China) could increase by 12 percent, to 850 million hectares, by the year 2010.[35] Forests currently cover about one-half of the additional land needed to achieve this goal, however, and deforestation of these lands, while opening land for cultivation, would likely result in deleterious environmental effects, among them loss of biodiversity resources and carbon storage capacity, ecosystem functions from which humans benefit and upon which they depend.[36] Many argue, then, that in the worldwide potential for expanding cropland is limited by possibly extreme and significant environmental costs.[37]

Enthusiasm for bringing uncropped lands into production, then, should be tempered. The most productive agricultural land is already being cultivated. Uncultivated lands tend to be arid, too cold, too steep, under essential forests, or lacking in suitable soil structure. Yields on newly cultivated lands, as a consequence, will most likely be lower, since these lands are in general more marginal than those already in production.[38] Certain shortcomings of marginal land can be overcome with technological modifications: Irrigation can provide moisture; fertilizers can replenish or supplement inadequate fertility levels; and terraces can mitigate the effects of poor topography. Each of these remedies, however, requires energy, capital investment, or tremendous expenditures of labor. These remedies must also be balanced against losses in essential ecosystem functions.

Almost half of what is considered to be potentially arable cropland in sub-Saharan Africa and Latin America, for instance, is currently under forest cover or is located in protected areas.[39] These two regions, according to work by the FAO in Figure 4.3, will experience the most agricultural land expansion in this decade. Over 70 percent of potential cropland in these regions suffers from soil or terrain constraints.[40] Conversion from their current state to cropland, then, will entail high financial and ecological costs incurred both regionally and globally.

Much of the potentially arable land that is not under cultivation is located in the tropics, where soil conditions, among other things, limit the possibilities for long-term cultivation. On the surface, tropical soils appear attractive for agricultural expansion because they have abundant water and yield rich, biodiverse natural forests. Experience with farming tropical soils, however, suggests that severe environmental damage results from these endeavors. Tropical soils, once deforested, tend to become infertile and rapidly deteriorate, falling victim to erosion from heavy rains or extreme dessication from loss of humus and microbial life. Some tropical soils undergo deleterious chemical changes when they are deforested because of increases in soil temperatures. Farming on tropical soils, as a result, typically requires large amounts of manufactured fertilizer to maintain a minimal level of productivity. Efforts to clear and farm tropical moist forest land show that in many cases clearing is unsuitable for conventional farming and quickly degrades to unusable soil once cultivated.[41] In addition, the genetic and climatic implications of widespread tropical rain-forest loss constitute potentially serious threats that render these forests vital to important global processes.[42]

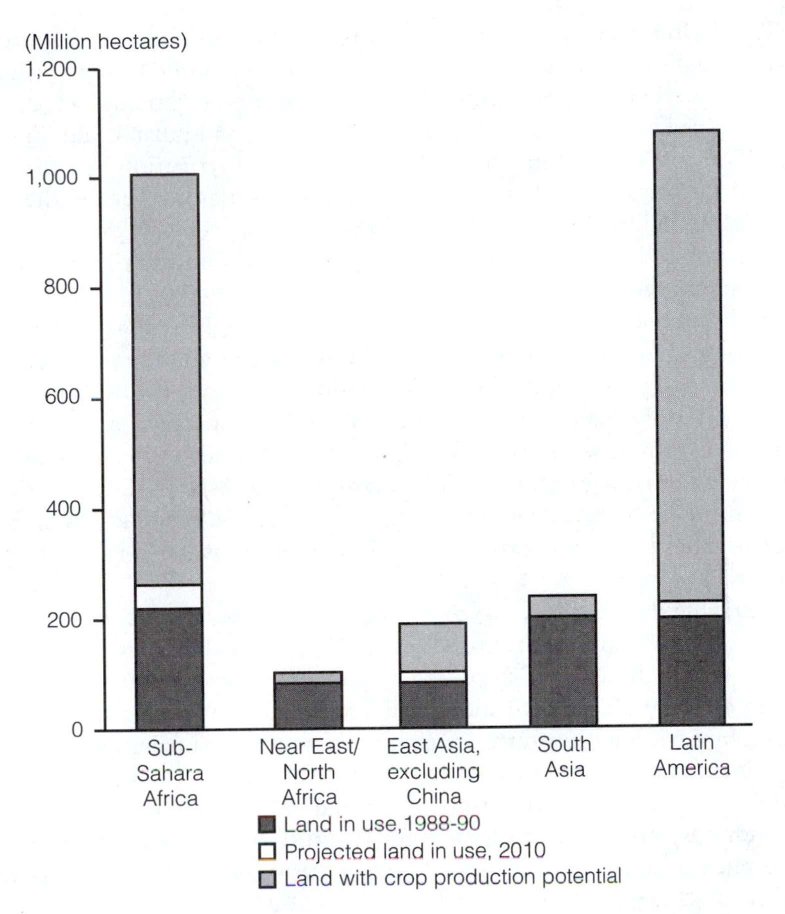

FIGURE 4.3 Projected Cropland Expansion in Less Developed Countries by 2010

SOURCE: World Resources Institute, 1996. Used by permission of World Resources Institute.

Water

Irrigation has been a critical component, over the past several decades, for achieving higher crop yields and multiple cropping; irrigation is thus seen as an essential part of future efforts to increase global food production. Expanded irrigation accounted for more than 50 percent of the total increases in worldwide food production from the mid-1960s to the mid-1980s; in the 1990s irrigation accounted for about one-third of total production.[43] The rate of expansion of irrigated land has been slowing consistently since the 1970s, but even at this slowed pace, the FAO predicts that more than half of the increment in crop production between the late 1990s and 2010 will come from irrigated land.[44] Farming, in 1989, accounted for an estimated 70 percent of worldwide water use.[45]

The declining rate at which cropland can be brought under irrigation can be attributed to a combination of ecological and economic factors. Salinization and waterlogging problems from certain irrigation techniques serve to reduce crop yields, and irrigation has negative impacts on human health in some cases. The cost of development and maintenance of irrigation is rising, as is competition for and conflict over water uses. Avoiding or reversing the siltation of dams and reservoirs is more and more costly.[46] Finally, the overall decline in real agricultural and food prices places limits on the possibilities for expanding irrigation.

Freshwater scarcity is also a crucial and increasingly important factor in irrigation expansion. Brown claims that water scarcity is already hindering growth in food production all over the world.[47] With a growing global population and expanded demand for irrigation and industrial water, the overextraction of groundwater resources is likely and is also already a critical problem in several regions, including much of South Asia.[48] At the same time that demand for freshwater is increasing, contamination by pollutants in freshwater supplies is also increasing, severely degrading water quality in many important sources.

Improving the efficiency of water already in use, some argue, is often more desirable than expanding irrigated lands. Often less than half of irrigation water is actually utilized by crops; the remainder is lost through seepage from unlined canals or evaporation and runoff from inefficient management.[49] Promoting more efficient water management practices and pricing water at its true value may serve to reduce wasted water resources.[50]

Because expanding irrigation is limited by many constraints, Nikos Alexandratos and the World Resources Institute suggest that investment in improvements to rain-fed cropland, which comprises about 84 percent of total cropped land, are critical to future food production.[51]

Energy

A future based on the use of chemical fertilizers, pesticides, and irrigation to increase agricultural productivity often is integral to an optimistic view about world food production. Each of these measures requires an increased use of energy. The manufacture of fertilizers and pesticides requires substantial amounts of energy,[52] and well irrigation likewise entails significant energy expenditures.

Since the 1970s, as we will discuss in Chapter 5, there has been a long upward trend in global commercial energy use; included in this trend is the rapidly rising consumption of energy in less-developed countries.[53] Financial resources are required in order to meet growing energy needs, creating problems where finances are scarce. Capital-intensive energy production can divert funds from other basic needs while at the same time increasing pollution levels and other energy-associated environmental degradation.[54]

Modern, technology-intensive methods of food production rely on large amounts of petroleum energy. Although at present petroleum appears to be in surplus, the price of petroleum remains much higher than before the oil crisis of the 1970s and constitutes a serious financial burden for many LDCs.[55] Prices of petroleum-based fertilizers and pesticides also remain high; the cost of increasing yields may thus be prohibitive for smaller producers in resource-limited LDCs.[56]

Fossil fuel combustion, at present, supplies about three-quarters of the world's energy.[57] The release of greenhouse gasses and the consequent threat of global warming associated with fossil fuel combustion has led to research on alternatives, including energy processed from plant matter, or biomass. The enthusiasm over biomass energy sources should be tempered, however, since replacing fossil fuels with biomass in the future energy budget would require dedicating the equivalent of approximately 15 percent of the land now in forests and 40 percent of that in croplands to biomass energy crop production.[58]

Augmenting traditional methods of agricultural production is possible by using crop rotations to replenish the soil. Also, improved, especially pest-resistent, plant varieties and limited uses of fertilizers and pesticides can increase agricultural productivity without requiring huge energy inputs. New approaches, such as agro-ecology, may represent alternatives to the current trajectory of energy-intensive production.[59]

Agroforestry, according to John Farrell and Miguel Altieri, is "the generic name used to describe an old and widely practiced land use system in which trees are combined spatially and/or temporally with agricultural crops and/or animals."[60] Agroforestry is often practiced on marginal land when agricultural inputs such as chemical fertilizers are unavailable to improve production outputs.

Norman Myers, who we discussed in Chapter 2, illustrates the use of agroforestry for tropical reforestation.[61] Using leguminous trees such as the acacia or ifilifil and plants such as vetch and clover, forest farmers intersperse their traditional grains, fruits, and nuts. The nitrogen enrichment from the legumes significantly improves agricultural output. Leguminous trees, according to Myers, improved the production of dryland rice in Indonesia by two and one-half times. As trees reach maturity, forest farmers move on, restoring the forest and farming more productively with agroforestry, a form of natural capitalism.

Environmental Degradation and Sustainability

Human history is replete with examples of societies whose agricultural production techniques resulted in environmental destruction and often human extinction.[62] Thus it is hardly novel that accelerating population pressure and demand for food are now leading to widespread degradation of ecosystems, as discussed in Chapter 3. What is new and significant, however, is the scope and rapidity of this process throughout the world and its effects on global productivity.[63]

Environmental destruction is the principal limit on sustainability. If production techniques irreversibly degrade the land resource base, immediate production increases cannot be sustained and productivity decreases will likely result unless abundant energy is available to compensate.

In addition to the biophysical constraints listed on page 113, other environmental problems and threats to sustainability include waterlogging and

BOX 4.1 Focus on the United States: Food on Campus

Our relationship with food is one of the fundamental ways we interact with the environment. Frances Moore Lappé's *Diet for a Small Planet,* originally published in 1971, made the case that vegetarianism was the socially and ecologically right lifestyle since a diet based on plants uses less resources and is less environmentally damaging than one based on meat eating. Vegans (vegetarians who don't use dairy products) agree and raise ethical concerns about the treatment of animals. A group called Vegan Action argues, "Animal agriculture takes a devastating toll on the earth. It is an inefficient way of producing food, since feed for farm animals requires land, water, fertilizer, and other resources that could otherwise have been used directly for producing human food." Students at the University of California-Berkeley who founded Vegan Action started a dorm food campaign in 1994 to have the dining hall serve vegan food. UC-Berkeley currently offers a vegan entrée at every meal.

Another way to eat more environmentally responsibly is to consume a diet that is produced close to home, in "local" food systems. Eating locally avoids tranportation-related environmental damage. Proponents of a bioregionalist outlook also argue that the future health of the planet requires that people develop an attachment to place. One way to do this is by purchasing foods at farmers' markets and getting to know farmers. A growing trend that supports local food systems is the movement for community supported agriculture (CSA). "In a CSA system, the farmer grows food for a group of shareholders (or subscribers) who pledge to buy a portion of the farm's crop that season. This arrangement gives growers up-front cash to finance their operation and higher prices for produce, since the middleman has been eliminated. Besides receiving a weekly box or bag of fresh, high-quality produce, shareholders also know that they're directly supporting a local farm. CSA also allow shareholders to re-establish a connection with the land that many people feel they have lost" (ATTRA). It is estimated that there are over 500 examples of CSAs throughout North America (Van En 1995).

On campuses, several student/food service/administration coalitions are working with local, sustainable farmers to purchase local foods. An initiative is taking place at the University of Wisconsin through its Center for Integrated Agricultural Systems. In addition to research, the Center has held a "Regional, Seasonal Foods Banquet." "The banquet brought together local chefs, . . . local farms, two local farm cooperatives, and the University dining hall staff and facilities. The chefs worked with the food service to prepare the meal. . . . The UW cooks and administrators were very impressed with the quality of the foods from local farms." Other colleges have planned similar events, such as Oberlin College's "All-Ohio Meal." Some schools are shifting their purchases toward local farms, often with organic products, including Bates College (Maine), Northland College (Wisconsin), College of St. Benedict (Minnesota), St. John's University (Minnesota), Iowa State Memorial Union (Iowa), and Hendrix College (Arkansas). A report that outlines the opportunities and barriers to getting sustainable agricultural products to college dining halls can be found online at the CIAS' website.

SOURCES: *www.attra.org* (ATTRA—Appropriate Technology Transfer for Rural Areas) *www.context.org* (Eating for Your Community by Robyn Van En. Fall 1995, p. 29. In *Context: A Quarterly of Human Sustainable Culture*) *www.wisc.edu/cias* (Center for Integrated Agricultural Systems) *www.vegan.org* (Vegan Action).

salinization of irrigated land, silting of irrigation canals, soil destruction in the tropics, overfishing, and overgrazing. Desertification is an ominous environmental consequence of the intensification of food production. Allan Durning estimates that desertification annually consumes 6 million hectares of land worldwide, an area the size of Switzerland.[64]

The obvious question is whether these environmental stresses are localized aberrations resulting from imprudent but largely correctable practices or substantial constraints that will irreversibly threaten sustainability and perhaps even reduce food yields from their present levels. Certainly if soil erosion and desertification—two processes that are basically irreversible in the time frame of a few generations—continue unabated, world agronomic potential must decline. Also, the intensification of agriculture resulting from rapid population growth accelerates the processes of environmental destruction. The world food crisis, thus, appears to be a significant environmental and energy crisis for both the human population and for the ecosystem on which food production depends.

POPULATION GROWTH AND HUNGER: THEORETICAL ISSUES

We know from Chapter 3 that the causes and consequences of population growth have been controversial issues ever since Malthus, in 1798, published the first edition of *An Essay on the Principle of Population*. In particular, Karl Marx's criticism of Malthusian theory of population in the 1860s—that labor exploitation, rather than population growth, was at the root of human misery and poverty—remains widely read and of considerable relevance over 150 years later. Malthusian theory of population, nonetheless, and more recent neo-Malthusian theory, have historically been the most widely accepted perspective on population in development circles, though perhaps more so among agricultural scientists than social scientists. There is scarcely an agricultural research and development planning document, for instance, that does not call attention to burgeoning world population growth and to the need for further agricultural research to keep pace with the relentlessly growing numbers of the world's hungry.

Table 4.2 provides a typology of theoretical perspectives on food production and hunger. The typology demonstrates that the major theoretical arguments on hunger are shaped by positions on two key issues. The first issue is whether increased agricultural production is the principal factor in alleviating hunger, and the second is whether population or nonpopulation factors (such as land tenure or the distribution of wealth and privilege) are major causal antecedents of hunger. The typology yields four categories, which we term (1) conservative non-Malthusianism, (2) radical non-Malthusian (political economy), (3) productionist neo-Malthusianism, and (4) ecological neo-Malthusianism. Table 4.2 also highlights illustrative examples for each theoretical perspective.

Table 4.2 A Typology of Theoretical Perspectives on Hunger in Developing Countries, with Contemporary Examples

ASSUMPTION REGARDING THE ROLE OF INCREASED FOOD PRODUCTION IN ALLEVIATING HUNGER	ASSUMPTION REGARDING THE ROLE OF POPULATION GROWTH	
	Non-Malthusian	**Neo-Malthusian**
Productionist Increased food production is central to alleviating hunger	**Conservative** Hayami and Ruttan (1985, 1987) Simon and Kahn (1984) Srinivasan (1988) Boserup (1965, 1981)	**Productionist–Neo-malthusian** Fornos (1991) Sadick (1990) Bongaarts (1996)
Nonproductionist Nonproduction factors are central in alleviating hunger	**Radical** Lappé and Collins (1977) McMichael and Raynolds 1994) Dyson (1996)	**Ecological Malthusian** Meadows et al. (1992) Meadows et al. (1972) Brown (1995) Ehrlich and Ehrlich (1990, 1991, 1995)

The two variants of neo-Malthusianism, as noted earlier, historically are the prominent views of population and hunger issues among scholars and policy analysts in international development circles. Increasingly, these two perspectives are becoming less distinguishable as scientists learn more about the ecological consequences of both population growth and modern energy-intensive agriculture. Scholars such as Bongaarts (1996), Fornos (1991), and Sadik (1990), nonetheless, place more emphasis on the need to reduce population growth, while others such as Catton (1980), Ehrlich and Ehrlich (1990, 1991, 1995), and Meadows et al. (1972, 1992) look beyond this issue to a need for a shift in values, as evident in the New Environmental Paradigm and more sustainable forms of energy consumption and production. But we stress that there are some major differences between Malthusian theory and that of contemporary neo-Malthusianism.

Malthusian theory of population, as discussed in Chapter 3, has three major postulates: that population has a natural tendency to increase faster than the means of subsistence; that population increases more or less rapidly according to the abundance of subsistence; and that as a result of these differential growth rates, human population growth has a natural tendency to induce poverty (for a comprehensive summary of Malthusian population and economic theories see Petersen[65] or our Chapter 3). Of these, modern neo-Malthusians would tend to agree only with the third; indeed, the most common argument among contemporary neo-Malthusians is that rapid population growth dictates that income, land, natural resources, and food be divided among more and more persons, thereby reinforcing poverty and hunger. There are, however, two other distinctive components of neo-Malthusianism. One, stressed mainly by productionist Malthusians, is that rapid population growth inhibits GNP and income growth by increasing the level of investments required to employ a growing labor force, and, by severely straining state budgets in providing education to a large number of children. The expense of international family planning programs is thus a necessary part of efforts to modernize productivity in the LDCs.[66] The second, stressed by ecological Malthusians, is that rapid population growth leads to poverty, underdevelopment, and malnutrition, not only because of the logic of more persons among whom to share food and income, but also because population growth undermines the natural resource base. Growing population is seen as leading to deforestation, soil erosion, desertification, and land degradation, which subsequently exacerbate poverty and lead to further population growth.[67]

While neo-Malthusian policies continue to be an important part of the efforts by donor nations to direct the development of recipient countries, it is probably fair to say that the persuasiveness of neo-Malthusianism has declined over the past decades. One reason, for instance, is the arguably un-Malthusian nature of the population problem in Africa over the past thirty years. Population growth rates in Africa in the 1960s, for example, were relatively modest—generally 1.5 to 2 per cent annually.[68] Subsistence farmers brought new land into production, which slowly enabled African countries to increase their food output and to be self-sufficient in staple foods (while some countries, such as Senegal and Nigeria, were major food exporters). After independence, however, the initiation of foreign-aid-financed development projects and the pursuit of economic development by African states led to significantly higher population growth rates and to declining per capita food production and income.[69] African development problems, thus, are not direct consequences of long-term population growth. Instead, rapid population growth is a relatively new phenomenon that, ironically, seems to have been caused by unsuccessful economic development efforts to reduce poverty.[70]

The prestige of neo-Malthusianism also suffers from continual criticism of limits-to-growth type models and projections.[71] In particular, these models have generally projected that population growth and resource scarcities, particularly of land and fossil fuels, perpetuate a decline in per capita food production. Observers from across the theoretical and political spectrum (e.g, from Lappé, 1986 to Simon and Khan, 1984) present theoretical arguments and convincing data that neo-Malthusian projections of declining per capita food production prove to be misleading. Indeed, a further reason for the declining persuasiveness of neo-Malthusianism is that this perspective not only has formidable critics on the left, following Marx's attacks on Malthus in the nineteenth century, but also increasingly vocal criticism on the right.

The scholars who subscribe to the non-Malthusian political economy position, as suggested in Table 4.2, insist that population growth tends not to be an autonomous cause of poverty and hunger, even though rising populations and population densities may play some role in exacerbating environmental

Table 4.3 Food Distribution in the Less-Developed Countries[a]

The maldistribution of food in developing regions occurs in five ways:

1. Food production increases since World War II have largely gone to the urban middle and upper classes. Much of the reason for this is the dramatic disparity between rural and urban incomes.[b] A further factor is the prevailing cheap food policies designed to stave off worker unrest and allow industrialists to keep wages low.
2. Considerable amounts of cereal grains are diverted into luxury products that the poor cannot afford. Lappé and Collins cite the example of the emergence of a Pakistani industry that processes badly needed corn into a sweetener primarily used for soft drinks.
3. Much cereal is devoted to the production of livestock—again, a product that the majority of the poor cannot afford. For example, two-thirds of Columbia's Green Revolution rice was fed to livestock or diverted into the production of beer. Considerable Colombian corn is fed to chickens.
4. Food grains are exported to generate currency to repay loans for agricultural inputs and industrial machinery.
5. Food is often dumped for lack of a market. Up to 65 percent of the fruits and vegetables produced by U.S. firms in Central America have been dumped or fed to livestock because of an oversupplied world market.

Lappé and Collins attribute these phenomena to the commercialization of food production and distribution and the irrational inequities of economic and political power that have resulted.

[a]Lappé and Collins, 1977, chapter 18.
[b]Lipton, 1974.

degradation and food shortages. In this view, instead, rapid population growth is contingent on poverty, inequality, and economic insecurity. As Lappé argues, "Hunger, the most dramatic symptom of pervasive poverty, and rapid population growth occur together because they have a common cause . . .: [the lack of] adequate land, jobs, education, health care, and old-age security [for] most people."[72]

Non-Malthusians typically insist that economic insecurity and poverty lead to high fertility rates because, for poor peasants and urban dwellers, having many children provides social security in old age and creates a domestic labor force for on-farm or off-farm work. Of interest is the differential in educational attainment between young women and men, which may give women seeking wage employment less opportunity to compete in the labor market for better-paying jobs.[74] Thus, fertility and population control will not tend to result in major gains in combating poverty and hunger unless policymakers first or simultaneously effectively deal with their more fundamental causes—economic dependency, landlessness and maldistribution of land, excessive reliance on capital-intensive industrialization and export-oriented growth, widespread unemployment, and educational opportunity. The principal evidence employed by non-Malthusian political economists is that population growth rates tend to decline in countries where high incomes or income redistribution result in a high level of economic security and a low level of abject poverty. Table 4.3, based upon the classic work, *Food First: Beyond the Myth of Scarcity,* elaborates on this important non-Malthusian argument.[73]

Conservative non-Malthusians tend to agree with the non-Malthusian political economists on one key point—that decreased fertility and population growth rates are governed by socioeconomic factors such as increased income, increased education, reduced child mortality, urbanization, and the enhancement of women's status. But conservative non-Malthusians theorize population, development, and poverty issues far differently than do their non-Malthusian political economy colleagues. Non-Malthusian political economists, for instance, tend to see Third World poverty and underdevelopment perpetuated by the rapid penetration of capitalism in the LDCs and by the subordination of these countries to dependency relationships with the MDCs.[74] Dependency refers most broadly to relationships between the less-developed and more developed countries in which the development of the Third World can be said to be strongly shaped or conditioned by that of the developed countries. The strong case for a dependency perspective is the argument that these relationships lead to the extraction of surplus wealth and resources from the Third World to the First, and hence are the cause of underdevelopment, poverty, and hunger in less-developed countries. The notion of dependency is, however, often used more generally to depict relationships consisting of Third World export-trade dependence, technological dependence, or debt-financial dependence. Conservative non-Malthusians argue that underdevelopment is the result of too little capitalism rather than too much—in particular, government policies that interfere with market forces.[75]

The most distinctive aspect of conservative non-Malthusianism, however, is their insistence on the notions that (1) there are no formidable ecological limitations to economic growth, and thus to income growth that will inevitable lead to reduced fertility, and (2) population growth can play a positive role in development. The first argument is buttressed by data showing that the relative costs of natural resources, food, and other primary materials have been in continuous decline over long periods of time. Evidence for the second argument, stemming from early historical work by Ester Boserup,[76] is based on comparative empirical work on the theory of induced innovation. As agricultural land/human labor ratios decrease, according to induced innovation theory, a society can adapt, either by bringing more arable land into production, or by biotechnological advances such as Green Revolution seeds and mechanical advances in irrigation technology. Hayami and Ruttan empirically trace increases in food production over time per unit of arable land in response to population growth in Japan and the United States, countries with vastly different cultural and agricultural environments.[77] While Hayami and Ruttan do recognize that rapid population growth can significantly impair the expansion of agricultural technology and other factors necessary for increased food production, their emphasis is on the creative market forces of capitalism induced by population growth, not managerial policies such as family planning programs.

Conservative non-Malthusians thus argue that population growth can contribute to economic development through increased population density, and hence through economies of agglomeration through improvements in transportation and communications that population density makes possible. Their reasoning runs exactly parallel to the process of Durkheimian dynamic density discussed in Chapter 2. These non-Malthusians envision population growth impelling economic expansion by encouraging more agricultural investment[78] and by increasing demand for goods (which results in expanded industrial production and greater economies of scale in industry). Further, population growth provides a larger labor force necessary for increasing output in agriculture and industry. Rapid population growth, according to conservative non-Malthusians, is not a major constraining factor in efforts to achieve economic development and reduce hunger (see, especially, Simon, 1981, for the most comprehensive statement of this perspective.)

BEYOND THE GREEN REVOLUTION: POWER, INEQUALITY, AND FOOD DISTRIBUTION

The controversy over the Green Revolution illustrates how various social scientists debate the issue of increasing agricultural productivity and efficiency. Laissez-faire or conservative social scientists were the principal architects of policies for diffusing the technologies and organizational forms that have become known as the Green Revolution. Managerial and radical social scientists have since become critics of the revolution on both sociological and ecological grounds.

The notion of a Green Revolution is used in two senses. The first pertains to the broad movement of the LDCs toward increased agricultural productivity and overall development, much along the lines advocated and hoped for by conservative students of the world food problem. The Green Revolution expression is also used in the more limited sense of the spread of high-yield seed varieties (HYVs) in less-developed countries. Conservative theorists generally assumed that the latter would lead to the former; however, in hindsight we see that this is not unambiguously the case. Here, our use of the expression "Green Revolution" will be restricted to the spread of the HYVs.

When the Rockefeller, and later the Ford, Foundation embarked on research and demonstration projects to develop higher yielding crop varieties in the 1950s and 1960s, there was surprisingly little opposition to the upcoming revolution on the part of social or biological scientists. There has been considerable criticism of the revolution's meager benefits, however, and massive social and ecological costs since the 1970s. Contemporary discourse includes an acknowledgement of the deleterious effects of the revolution and a variety of perspectives on how to balance future food needs, future food distribution, and the ecological integrity of the environment.

The Costs of High-Yield Production

The essential characteristic of Green Revolution seed varieties is their responsiveness to chemical fertilizers. The HYVs also require adequate water and pest control. Unlike mechanical technologies that are "landlord-biased" because they confer major benefits on large farmers and landowners, Green Revolution technologies are, at least in theory, neutral to scale. The technology thus offered the promise for all classes of farm operators to achieve greater productivity. Despite the theoretical neutrality of Green Revolution technologies, however, one particular class of farmers—large-scale farmers and landowners—typically male—became the principal beneficiaries of the revolution. The widespread adoption of HYVs, in addition, also impelled significant ecological costs.

Many governments in the LDCs, in collaboration with international funding agencies, promoted the Green Revolution production system through extension programs, subsidies for agricultural chemicals, tractors, irrigation, and credit.[79] Although the resulting yield increases associated with the adoption of HYVs have been dramatic, they are highly concentrated in a few ecologically advantaged regions of the developing world. Asia, and to a lesser degree Latin America, reaped the benefits of the new grain varieties while Africa experienced few gains.[80]

Even in regions that are said to have benefited from the Green Revolution, there are considerable criticisms. Vandana Shiva, in her description of the environmental and economic consequences of the adoption of Green Revolution technologies in the Punjab, insists that the processes of the Green Revolution led to a variety of social and environmental ills, including reduced genetic diversity, increased vulnerability to pests, soil erosion, water shortages, reduced soil fertility, micronutrient deficiencies, soil contamination, rural impoverishment, and increased social tension and conflicts.[81] The bulk of the benefits, she asserts, have been enjoyed by the agrochemical industry and large-scale producers. Her assertions are echoed by many contemporary critics of the aftermath of the Green Revolution.

Dissemination of Green Revolution HYVs brought a new dependence on fertilizers, irrigation, and pesticides. Ehrlich, Ehrlich, and Daily report a tenfold increase in worldwide chemical fertilizer use between 1950 and 1990, parallel with a nearly threefold increase in global grain production over that same period.[82] Replicating such a scenario to meet ever-increasing future food needs is considered unlikely by those who claim that fertilizers have reached a point of diminishing returns.[83] Dependence on fertilizers can also be harmful to natural ecosystems: Synthetic fertilizers are often contributors to the overall degradation of surface and groundwater supplies[84] and can damage forests and other natural ecosystems by disrupting nutrient cycling.[85] Irrigated land can lose fertility or can go out of production at an increasing rate due to salinization and waterlogging; Postel asserts that as much as a quarter of the world's irrigated land has been adversely affected by salinization.[86]

High-yielding varieties, because of their restricted genetic base and lack of native adaptations, are generally less resistant to diseases, insects, and drought than the local varieties they replaced. Disappointing yields often followed initial dramatic yield increases as pest infestations intensified. Increased use of pesticides was the logical result. While pesticides may enhance productivity, however, using chemicals to control insects, pests, weeds, and fungi can threaten human and ecological health.

As scientists learned more about the consequences of human exposure to pesticides, they became increasingly aware that continuing, long-term exposure to pesticide and chemical residues in food, water, and the air can be extremely hazardous, especially to children.[87] In the early 1980s, Gear estimated that approximately 10,000 people died each year in the LDCs due to pesticide poisoning and about 400,000 suffered acutely.[88] The deleterious effects of these chemicals are not restricted to human communities, since the chemical impact can be transferred throughout the food chain and thus potentially extend to all levels of natural ecosystems.[89] As a result of Green Revolution adaptations and practices, in the early 1990s about half of the pesticide poisonings of people and 80 percent of pesticide-related deaths occurred in the LDCs, though these nations accounted for only 15 to 20 percent of world pesticide consumption.[90]

Although the Green Revolution can be credited with increasing agricultural yields, many nutritionists regard the revolution as a protein disaster, since improved varieties generally increased production at the expense of lowered dietary protein content. In many regions, HYVs supplanted legume production—a particularly important source of protein in many regional diets and, as mentioned earlier in the discussion of agroforestry, a critical soil nitrogen fixer. The quality of human diets, as well as the capacity of the soil to reproduce production, was thus adversely affected.

The expansion and intensification of agriculture also meant that in many cases traditional crop rotations have been abandoned in exchange for year-round, continuous cropland use. In the Punjab, for example, the cessation of crop rotation, in combination with a dramatic reduction of cropland under legumes, resulted in soil erosion and depletion and greatly increased fertilizer dependence.[91]

Improved Varieties and Genetic Diversity

Critics note that widespread adoption of Green Revolution seed varieties also had the effect of reducing genetic diversity, both by replacing traditional native mixtures and rotations of crops with monocultures and by introducing wheat and rice varieties from a narrow genetic base.

Since their productivity tends to decrease with successive generations, HYVs often have to be purchased anew for each harvest,[92] commercializing what was previously a self-reproducing resource and transferring control over biological diversity from peasants to transnational agricultural corporations. Despite this growing dependency on transnationals, the process of the commercialization of seeds has been actively encouraged by powerful creditors such as the World Bank.

The agricultural monocultures that characterize the adoption of Green Revolution technologies are highly vulnerable from an ecological standpoint. By replacing native seed varieties, HYVs erode the genetic diversity of seeds used in agricultural production while historically minor diseases grew into significant ecological threats.[93] In addition, wild crop relatives, a further source of genetic variety, are being destroyed with the elimination and replacement of natural ecosystems.[94] Ironically, the maintenance of genetic diversity, which depends on the preservation of natural ecosystems and a diverse crop base, is essential for the continuation of research that facilitates the maintenance and improvement of high-yield agriculture itself. New genetic strains are constantly necessary for innovations that allow HYV seeds to succeed in dynamic conditions.[95]

Socioeconomic Effects of Green Revolution Technologies

While Green Revolution technologies increased yields, their adoption also tended to exacerbate rural income inequalities, promote class differences, and have a disproportionately adverse effect on women.[96] Agricultural mechanization often adversely affected rural peasants, restricting their access to land by rising land prices and curtailing their employment opportunities by mechanization.[97] Many observers, overall, argue that the commercialization of agriculture via the Green Revolution undermined the viability of household food production as a livelihood strategy among rural populations.[98] The process is closely linked, for many analysts, to rural proletarianization and a growing trend toward rural-to-urban migration throughout the LDCs.[99]

Women, in general, have been adversely affected by Green Revolution changes because previous tasks allocated by gender have been renegotiated in response to changes in ecological, social, and economic conditions.[100] The substantial role women have traditionally played in rural agricultural production in the LDCs was ignored in preliminary Green Revolution development schemes. Land, water, seeds, and technical training, in general, were offered to men while women were expected to continue their traditional tasks in newly cash-cropped fields.[101] The adverse environmental consequences of HYV adoption, such as desertification and the spread of soil erosion, disproportionately affected women, making their traditionally expected tasks such as water collection and fuel wood gathering more difficult and in some cases impossible.[102] Theorists argue that by virtue of their dependence on environmental integrity to successfully carry out their traditional survival tasks, the degradation of the environment associated with the Green Revolution resulted in intensified marginalization of rural women and a dramatic increase in their work burden. The differential impact of agricultural change on men and women and on women in different income classes has been gradually recognized as a significant problem by international development policymakers.[103]

Another initial problem with Green Revolution fertilization, pest control, and irrigation technologies is that they require farmers to have access to credit in order to purchase and realize the benefits of production with the HYVs. Credit systems in the LDCs have been notoriously biased against the small farmer: peasant, tenant, or woman.[104] Large farmers and landowners, on the other hand, have been able to finance the needed inputs with their own accumulated capital or by favored access to credit. Small farmers with restricted access to credit have tended to be denied the use of Green Revolution technology and have therefore become less competitive and frequently forced out of business.[105] Ecologically disadvantaged and land-poor rural masses, then, by and large have not benefited from Green Revolution advances in technology.[106]

Although mechanization was not conceptually an integral part of the Green Revolution, in practice farmers who took advantage of improved varieties were encouraged to mechanize and displace labor. This frequently occurred when investment in irrigation equipment was necessary. With a substantial investment tied up in wells and pumps, there is a clear incentive to cultivate more acres to spread out the costs. Large farmers thus became larger by evicting their tenants, firing laborers, and purchasing land from neighbors.[107]

Often commercial farmers in the LDCs can realize the potential profitability of HYVs only if governments subsidize production with price supports. Although this policy would seem beneficial for all farmers, the subsidy pertains only to high-yielding crops and not to the principal staples, such as millet and cassava, grown by the majority of small farmers. Government price supports, then, have benefited primarily large landowners cultivating Green Revolution crops.

Subsidies in the MDCs also affect the global picture. Food surpluses in North America and Europe, largely a result of government-subsidized production, are cheaper to export than to store; these are commonly exported as food aid, depressing international market prices and causing severe problems for LDCs whose economies are based on agriculture.[108]

A consequence of the MDCs dumping of cheap agricultural commodities is increased reliance on imports at the expense of local production in the LDCs. When importing food is less expensive than local production, governments neglect domestic production and direct scarce resources into higher priority uses.[109] Africa, in particular, witnessed the most significant deterioration in local food production,[110] leaving the nutritional well-being of Africa's population increasingly dependent on the agricultural policies of the MDCs and shifts in international market conditions.[111]

Power and Unequal Access

As traditional production relationships, such as the subsistence production of food by peasants, give way to the international market relationships of capitalism, access to an adequate diet comes to be determined by the class structure, division of privilege, and an individual's ability to command resources on the market. The person who does not have money, quite simply, does not eat. The commercialization of the world food system, world food researchers note, presents critical problems of food distribution and food security.

The draft policy statement for the World Food Summit states that enough food is produced every year to provide everyone, worldwide, with an adequate diet, but that 800 million people do not have access to the food produced.[112] The world food problem, states a document signed at the World Food Summit

(WFS), is a "failure of effective demand on the part of people with inadequate nutrition to grow as much as required to raise their consumption to levels compatible with the elimination of food insecurity and undernutrition."[113]

Thus the nature of the problem can be interpreted as an important, fundamental distributional inadequacy that results from economic and political inequalities. As food becomes a commodity to be bought and sold on the market, nations and social classes with higher levels of income and wealth have favored access to the crops. The wealthy MDCs are able to command so much grain that they can afford to allocate much of the crops to inefficient conversion into meat and dairy products. The impoverished classes of the LDCs, on the other hand, are disadvantaged by their lack of income, external debt, and the inability of their countries to purchase sufficient grain on the world market.

Export Dependence

A primary component of the imbalance between sufficient production and equal distribution of food is that many food-deficit countries must export much of the agricultural commodities they produce in order to earn the foreign exchange necessary to service their international debt.[114] The foreign debt more than quadrupled since the 1974 UN World Food Conference, totaling 2,068 trillion dollars in 1996.[115] Stabilization and structural adjustment policies, in much of the LDCs, required by international donor agencies reorient local economic activity toward debt repayment, which require the replacement of nationally oriented growth strategies with export-led development schemes.[116]

Primary export dependence, "the extent to which a country's exports consist disproportionately of minimally processed agricultural and mineral goods," Wimberley and Bello argue, has the effect of undermining economic growth by generating few backward and forward linkages to other local economic activity.[117] Obligations to creditors often necessitate export-oriented production. A case in point is the active encouragement of "nontraditional" exports, such as flowers and exotic foods, in the structural adjustment policies imposed by creditors.[118] Food imports, in the immediate sense, are required to augment the food supply in most food-deficit countries. Hence a larger proportion of a country's limited foreign exchange that has not been used to service the external debt is used for the purchase of needed food rather than for other, more broad development purposes.[119] The World Bank, as stated earlier, estimates that the LDCs could be importing as much as 15 percent of their grain consumption by 2010.[120]

Wimberley and Bello argue that primary export dependence has a particular importance for nutrition because of the detrimental impacts of agricultural primary exports on food consumption and hunger.[121] The conversion of farmland to export-oriented, cash-crop agriculture has the effect of limiting the production of food for local consumption. Hence, dependence on imported food, to the extent that the food is affordable by local residents, grows. Price fluctuations in global food markets thus affect the poor while they simultaneously lose their land to large-scale, more powerful commercial farmers.[122]

Capital-intensive, export-oriented agriculture, while replacing the possibility for subsistence, typically reduces opportunities for employment; thus, those who have lost their land may also have a limited capacity for purchasing food. Those displaced by the agricultural transition typically become marginally employed in unstable wage jobs, or migrate to urban areas to join the poor urban workforce.[123] Primary export dependence, in this interpretation, can be seen as a catalyst for hunger through export effects on the agricultural sector.[124]

International Food Distribution and Marketing

An intensely debated aspect of the world food question is whether the interests of the LDCs and their hungry populations are served by the international grain market. It can be conversely argued that free trade—or the removal of protectionist and other barriers to improving and exporting—enables nations to specialize in those commodities for which they have the greatest comparative advantage.

Greater integration of the LDCs into the world agricultural market runs counter to the present world grain trade, as world trade in agricultural commodities is dominated by multinational corporations. These grain traders have become rapidly integrated both vertically—through transportation, storage, and marketing of their commodities—and horizontally—through expanding territorial scope and dealing in several agricultural commodities. Their superior access to information on grain yields, demand, and supplies enables them to be virtually independent of even the most powerful nations.

Multinational corporations can be understood to aggravate hunger in the LDCs by reducing effective demand for food through a combination of social, political, and economic effects. The low incomes promoted by transnational corporate investment, via increased inequalities and slowed economic growth, render many of the poor in developing countries unable to purchase the food they need.[125] Multinationals are said to further undermine nutrition with intensive advertising practices that encourage the displacement of cheap, high-quality foods for more expensive and less nutritious ones.[126]

Control of the movement of industrial agriculture, and indeed major policymaking, is increasingly dominated by transnational corporations and banks.[127] Through lobbying, financing candidates, public relations, control of the media via advertising budgets, and effecting the outcome of elections, global corporations are able to influence national and multilateral policymaking.

Moving beyond the Green Revolution

The world food problem encompasses questions of both production and distribution. The inequitable distribution of production assets, unemployment, and underemployment are central to the problem of hunger in the LDCs. Moving beyond the Green Revolution depends not only on raising global production, but on reducing distortions in the structure of the world food market.[128] The World Food Summit concludes that the best way to relieve world hunger is to increase production in food-deficit countries.[129] the same conclusion was drawn almost a decade earlier by the World Commission on Environment and Development.[130]

For many LDCs, such a shift, if increased production is possible, can succeed only insofar as the natural resource base is secure. Ecological change, then, is as essential as the structural change necessary to reverse the global problem of food maldistribution. Practicing sustainable, yet productive and efficient agricultural management emerges as a central challenge for the decades beyond the Green Revolution.[131]

Emerging initiatives to develop sustainable agriculture are challenging the predominant agricultural development patterns and technologies through agro-ecological practices, particularly in the area of pest management. Integrated pest management, or IPM, relies on biological and other natural or cultural methods of pest control within an ecological framework.[132] In this alternative method, chemical pesticides are used only as a last resort, and energy input requirements are dramatically reduced. Contrary to the expectations of many researchers and conventional farmers, IPM proves to be profitable—in some cases, even more profitable than conventional methods.[133] Among the most successful implementations of IPM on a large scale, to date, is the use of IPM in Indonesian rice production. There, between 1987 and 1990, the volume of pesticides used in rice production fell by over 50 percent while yields increased 15 percent, to the benefit of both farmers and the government.[134] Critical to the Indonesian success story, and to most successful IPM programs, are the human dimensions of such efforts: particularly the institutional, socioeconomic, and policy factors that support IPM and other agro-ecological methods like intercropping and cover crops. Eliminating agro-chemical subsidies and building state and community empowerment programs are identified as important elements in the success of these agro-ecological methods.[135]

For other students of the world food problem, innovations in technology will continue to rise to meet the challenges of food production and distribution in this new century. Recent advances in agricultural biotechnology may provide new methods for producing food and boosting productivity and yields in the context of a new industry largely free of contemporary ecological constraints.[136] Exemplary biotechnological advances include the in vitro production of tomato pulp, orange juice, and apple sauce; genetically engineered plants resistant to viruses or insects; and altering the reproductive cells of fish, poultry, sheep, and pigs. Propelled by the demands of the market, advances in biotechnology may bring high-tech solutions to seemingly intractable food production problems. For those who base their optimism in such advances, truly sustainable agriculture in a contemporary context requires biotechnology.

In either production scenario, the problem of equitable distribution of food remains an increasingly complex and intensifying dilemma. In the global economy, the world's several million hungry face considerable obstacles to food security. As the export-oriented model of development replaces the nationally oriented model, local agricultural production and national food security will likely be increasingly threatened. The growing corporate-dominated free-market regime, which requires purchasing power in order to perpetuate sufficient production and equal distribution, may insure that dependency in the LDCs will deepen over the next several decades.[137]

SUMMING UP: THE POTENTIAL OF GLOBAL AGRICULTURE

World food researchers conclude that the only lasting solution for hungry nations is for them to reintroduce and improve the productivity of their own indigenous agricultures. The distortions of the world grain trade and the frustrations involved in establishing a food security system make this observation particularly germane. Dramatic increases in international food aid over the last decade and somber projections for future food security in many regions further underscore this critical need.

Recent opinion concludes that there may well be significant unrealized potential for traditional agricultural practices in the developing world, but this potential cannot be reached within the confines of Green Revolution programs. Although one must not overlook the necessity for and potential of changing social structural arrangements that hinder food production and distribution, new hope for the productivity of traditional agricultural systems has come from suggestions that agricultural researchers attempt to blend the strengths of new technologies with those of traditional agriculture. An important step would seem to be the redirection of the priorities for plant researchers away from developing a few strains of cereals with wide applicability to selecting new varieties particularly suited to specific growing areas.

Regardless of strategies, the necessity of clearly defining and addressing the ecological implications of agricultural practices—both traditional and technocratic—has become clear over the past decade. Biophysical constraints to food production must be acknowledged and understood if world food production is to approach sustainability, an imperative made more salient as the human population continues to grow.

The world food problem encompasses production, distribution, and complex ecological concerns. The dual challenges of adequate and ecologically sustainable food production emerge at the center of humankind's contemporary struggle to maintain human life without destroying the sustaining power of the earth.

CITATIONS AND NOTES

1. World Resources Institute, 1996, p. 225.
2. Alexandratos, 1995, pp. 38–44.
3. World Resources Institute, 1996, p. 227.
4. World Resources Institute, 1996, p. 228.
5. Brown, 1991b, pp. 13–15; Ehrlich and Ehrlich, 1990, p. 16.
6. Brown and Kane, 1994.
7. World Resources Institute, 1996.
8. Dyson, 1994.
9. FAO, 1995, pp. 12–13; World Resources Institute, 1996.
10. Warnock, 1987, pp. 1–5.
11. Ehrlich, Ehrlich, and Daily, 1993, p. 5.
12. United Nations Childrens' Fund, 1992.
13. United Nations Population Fund, 1992.
14. Buttel and Raynolds, 1989, p. 328.
15. World Resources Institute, 1992, p. 225.
16. Alexandratos, 1995, p. 8; Mitchell and Ingco, 1993.
17. Webb, 1995, p. 2; World Resources Institute, 1996, p. 227.
18. Webb, 1995, p. 1.
19. FAO, 1994, p. 11.
20. Dyson, 1994, p. 371.
21. Dyson, 1994.
22. Ehrlich, Ehrlich, and Daily, 1993.
23. Brown et al., 1990; World Resources Institute, 1992.
24. Postel, 1990; Falkenmark and Widstrand, 1992.
25. Brown and Wolfe, 1984; Oldeman, Van Engelen, and Pulles, 1990; Aber and Melello, 1991; World Resources Institute, 1992.
26. Bugbee and Monje, 1992; Walsh, 1991.
27. Brown, 1991; Brown et al., 1990; Smil, 1994.
28. Francis, Flora, and King, 1990; Ehrlich and Ehrlich, 1990.
29. National Research Council, 1991; Plunkett et al., 1987.
30. Worrest and Grant, 1989.
31. World Resources Institute, 1986.
32. Parry, 1990; Schneider, 1989.
33. Ehrlich et al., 1993, p. 7.
34. Alexandratos, 1995, p. 170; World Resources Institute, 1996, p. 230.
35. Alexandratos, 1995, p. 170; World Resources Institute, 1996.
36. World Resources Institute, 1996, p. 232.
37. Dyson, 1994, p. 380.
38. Crosson and Anderson, 1992, p. 20.
39. Alexandratos, 1995, p. 152; World Resources Institute, 1996.
40. Alexandratos, 1995, p. 155; World Resources Institute, 1996.
41. Carruthers, 1994; Ehrlich, 1988; Sanchez, 1976; Tivy, 1990.
42. Houghton, 1992; Myers, 1992; Wilson, 1989.
43. World Resources Institute, 1996, p. 231.
44. Alexandratos, 1995, p. 160; World Resources Institute, 1996.
45. Postel, 1990, p. 6.
46. World Resources Institute, 1996, p. 231.
47. Brown, 1995.
48. World Resources Institute, 1996, p. ??.
49. Gleik, 1993, p. 6; World Resources Institute, 1996.
50. Smil, 1994, p. 270.
51. Alexandratos, 1995; World Resources Institute, 1996.
52. Pimentel and Pimentel, 1979; Steinhart and Steinhart, 1974.
53. World Resources Institute, 1996, p. 273.
54. World Resources Institute, 1996, p. 273.
55. McLaughlin, 1996, p. 52.
56. McLaughlin, 1996, p. 52.
57. International Energy Agency, 1996; World Energy Commission, 1993.
58. Hall, Mynick, and Williams, 1991; World Resources Institute, 1996.
59. Thrupp, 1996; Altieri, 1995.

60. Farrell and Altieri, 1995, p. 217.

61. Myers, 1992.

62. Catton, 1980; Burch, 1971; Dale, 1955.

63. Catton, 1980; Brown, 1987; Eckholm, 1976.

64. Durning, 1989, p. 6.

65. Petersen, 1979.

66. Sadik, 1990; Fornos, 1991.

67. Brown, 1995; Meadows, Meadows, and Randers, 1992.

68. Eicher, 1986.

69. Paulino, 1987.

70. Stokes, 1995.

71. Meadows, Meadows, and Rander, 1992; Meadows et al., 1972; Barney, 1980.

72. Lappé, 1986, p. 25.

73. Lappé and Collins, 1978.

74. Simon, 1981.

75. Boserup, 1965.

76. Hayami and Ruttan, 1985, 1987.

77. Boserup (1985) is the classic analysis of this point, though it should be recognized that this argument may not be generalizable beyond the rice economies of Asia (Bray, 1986).

78. Buttel and Raynolds, 1989; Griffin, 1974.

79. McMichael and Raynolds, 1994.

80. Shiva, 1991, p. 57.

81. Ehrlich, Ehrlich, and Daily, 1993, p. 11.

82. Brown, 1991; Brown et al., 1990; Walsh, 1991.

83. World Resources Institute, 1992.

84. Smil, 1991.

85. Postel, 1990.

86. WCED, 1987, p. 126; Wargo, 1996.

87. Gear, 1983.

88. WCED, 1987, p. 126.

89. Conway and Pretty, 1991.

90. Shiva, 1991, p. 59.

91. Shiva, 1991, p. 59.

92. Shiva, 1996, p. 16.

93. Hoyt, 1988; Vaughan and Chang, 1993.

94. Ehrlich, Ehrlich, and Daily, 1993, p. 9.

95. Knudsen and Nash, 1990, pp. 84–94; Griffin, 1974; Dankelman and Davidson, 1988; Jiggins, 1986.

96. Pearse, 1980; Raikes, 1988.

97. McMichael and Raynolds, 1994; Barkin, 1987.

98. McMichael and Raynolds, 1994.

99. Dankelman and Davidson, 1988, p. 12.

100. Dankelman and Davidson, 1988.

101. Dankelman and Davidson, 1988; Agarwal, 1986, 1992.

102. Jiggins, 1986.

103. Freebairn, 1973; Griffin, 1974; Dankelman and Davidson, 1988.

104. Havens and Flinn, 1975.

105. WCED, 1987, p. 124.

106. Perelman, 1976.

107. WCED, 1987, p. 123; Freidmann, 1993; Wessel, 1983; McMichael and Raynolds, 1994.

108. McMichael and Raynolds, 1994.

109. Bradley and Carter, 1989, p. 104.

110. McMichael and Raynolds, 1994; Raikes, 1988; Sanderson, 1989.

111. McLaughlin, 1996, p. 51.

112. WFS, 1996.

113. McLaughlin, 1996, p. 52.

114. World Debt Tables, 1996, p. 31.

115. Wood, 1986; McMichael and Raynolds, 1994.

116. Wimberley and Bello, 1992, p. 900.

117. Freidmann, 1993, p. 50.

118. McLaughlin, 1996, p. 53.

119. Mitchell and Ingko, 1993; Alexandratos, 1995.

120. Wimberley and Bello, 1992.

121. Wimberley and Bello, 1992.

122. Wimberley and Bello, 1992.

123. Barkin, Batt, and DeWalt, 1990; Lappé and Collins, 1978.

124. Wimberley, 1991, p. 412.

125. Wimberley, 1991, p. 412.

126. Ritchie, 1996; McLaughlin, 1996, p. 54.

127. WCED, 1987, p. 128.

128. WFS, 1996.

129. WCED, 1987, p. 128.

130. World Resources Institute, 1996, pp. 233–35; WCED, 1987, pp. 133–38.

131. Thrupp, 1996; Moore, 1995; Vanden Bosch, 1978; Altieri, 1995.

132. Thrupp, 1996, p. 7.

133. World Bank, 1996; Thrupp, 1996, p. 7.

134. Thrupp, 1996.

135. Kennedy, 1993, pp. 70–81; Kloppenburg, 1988; Busch et al., 1991.

136. McMichael and Raynolds, 1994.

5

Energy and the Environment

The Reemerging Energy Crisis

By August 2010, people in twenty nations in the Northern hemisphere experience another scorching summer. The word scorching is important. With rising average temperatures each year for at least the past three decades, the summer really is a hot one. Experiencing a scorcher in 2010 is not the same as experiencing a scorching summer ten years ago.

Corn, traditionally grown in the U.S. Midwest, now grows in the normally cooler Canadian central plains. Soil productivity for corn, unfortunately, is lower in this part of Canada than in the United States. So, when the temperature in 2010 heats up, drought and crop damage really put a dent into North American food production. Food prices pinch the pocketbooks of even the wealthiest consumers in the United States. The media is buzzing about the cost of food.

To add fuel to the fire, Elisabeth—a Class 5 hurricane—roars up the Chesapeake Bay in September, slamming into the nation's Capitol and then blows northeast into the heart of Philadelphia. Estimates of property damage alone far exceed those of Ann in New Orleans last summer, the worst in U.S. history until now. Two weeks later, a cyclone strikes Tsingtao, southeast of Beijing, displacing millions living in the shantytown surrounding Kiachow Bay. Because of this brutal storm, U.N. relief officials estimate as many as one-half million Chinese may have lost their lives.

The U.N. Intergovernmental Panel on Climate Change (Panel), with unprecedented speed, assembles a conference with over 100 leading climatologists in Geneva. On October 14 the group holds a press conference about their deliberations, broadcast live on CNN. The Panel scientists label Elisabeth, Ann, the cyclone in China, and other recent storms as unnatural, anthropogenic disasters connected with the buildup of greenhouse gases in the atmosphere, specifically carbon from motor vehicles and the consumption of coal in China and India. What the popular environmental writer Bill McKibben calls the "end of nature" because of human-induced alterations to the planet is now officially recognized by the world's leading scientists. They urge national political officials worldwide to take action by moving their citizens away from coastal areas and by initiating aggressive action to reduce greenhouse gas emissions as soon as possible. Stocks on the Tokyo and Wall Street markets, shortly thereafter, plummet.

A special session of the U.N. General Assembly convenes on December 5 to consider the Panel's announcement. U.N. officials invite the president of the United States and the youngest prime minister in the history of China to attend the meeting as the representatives of the two countries most responsible for carbon emissions. The Chinese prime minister endorses the policies presented to the Assembly by Amory Lovins, the director of the Rocky Mountain Institute, where hundreds of ways to gain energy efficiency such as solar technology are being developed. The president of the United States remains noncommittal.

The ripple effects of the Assembly gathering, in spite of the United States' neutrality, are unprecedented. Through round-the-clock negotiations at the United Nations, country after country, especially the more-developed countries, begins adopting what Lovins calls paths to sustainable development through renewable energy consumption. Germany and Canada commit their nations to ending coal consumption for generating electricity within a decade. India and China, representing more than one-third of the world's population, adopt plans to cut coal consumption by one-half within five years.

By 2011 the Worldwatch Institute, a U.S-based environmental think tank, begins reporting stepped-up efforts to adopt new energy policies in countries all over the world. Multilateral lending agencies, such as the World Bank, start funding multimillion dollar loan programs such as solar-powered lanterns for home lighting in India and rooftop photovoltaic units for electrifying houses in the Dominican Republic. The Belgian firm, Solel, actively pursues the development of solar hot water systems for housing projects throughout the less-developed counties. In Europe, a carbon tax, gauged to the amount of carbon emission for particular makes of cars, subsidizes buyers who choose more fuel-efficient vehicles.

Whether such a scenario becomes a reality, of course, is a matter of speculation. What we can do is examine trends in energy consumption throughout the world; discuss how leading international energy consortia envision the world fuel mix now and in the next ten to twenty years; and show how the energy crisis will reemerge, in part, because of increasing scientific evidence about the seriousness of carbon-based fuel consumption. Because the United States plays a large role in the consumption of fossil fuels, especially petroleum, we will also trace the historical growth of American petroleum dependence. Then, in a closely related section, we will discuss the Persian Gulf crisis—a war fought over the security of increasingly vital Middle Eastern oil supplies. Finally, we will discuss recent work on ecological modernization, including work by Hunter and Amory Lovins, mentioned earlier. They suggest that societies can enjoy high standards of living more efficiently without contributing to global climate change and natural resource scarcity.

WORLD PRIMARY ENERGY DEMAND

To understand conventional views about the flow of energy in societies, we draw upon two recent appraisals of fuel supply and demand for the world and for regions of the world: work by the International Energy Agency (Agency) and work by the World Energy Commission (Commission). Formed in 1976 in the wake of the first international oil shock, the Agency serves as the energy information and policymaker for the Organization of Economic Cooperation and Development (Organization). Representing mostly the more-developed countries, the Agency includes twenty-six countries such as Australia, the European Union, Japan, Portugal, Turkey, and the United States.

The Commission, a much older organization founded in 1924, represents diverse energy organizations from a variety of 100 more- and less-developed countries. Members include major international lending agencies such as the World Bank, national utilities in France and Zimbabwe, national energy research institutes in Sweden and Germany, and former energy ministers from the United States, Korea, and China. The Commission's $5 million research project, initiated in 1989, led to the publication of *Energy for Tomorrow's World* (1993).

Global Fuel Mixes: The Present and the Foreseeable Future

According to the Agency and the Commission, the world's nations are rapidly converging on a standard, modern fuel mix of coal, oil, and natural gas. Just over 77 percent of the world primary fuel mix, as one can see in Table 5.1, consists of fossil fuels. Ninety percent of the world primary energy mix will be composed of fossil fuels by the year 2010 as more and more less-developed countries move away from traditional renewables such as wood and animal dung.[1]

The Commission and the Agency agree that an important structural shift in world energy consumption is occurring. Both groups believe that this shift in energy consumption will be complete by 2010. Currently about 70 percent of all commercial energy is consumed by the more-developed countries, one-quarter of the world's population. Commercial energy consumption in the less-developed countries, however, is growing much more rapidly than in the more-developed countries. At this rate, by 2010 no less than 40 percent of world commercial energy will be consumed by the less-developed countries. The less-developed countries consumed only 28 percent of world commercial energy in 1993. The Commission expects less-developed countries' energy

Table 5.1 Energy Consumption, 1970s and 1990s: United States and World Total

	PERCENT USED			
Energy Supply	**United States (1972)**	**World (1970)**	**United States (1994)**	**World (1993)**
Coal	17.2	32.1	22.9	25.1
Petroleum	41.9	42.8	40.7	37.6
Natural gas	35.9	18.5	25.0	24.1
Hydroelectricity	4.2	6.2	3.4	6.8
Nuclear energy	0.8	0.4	8.0	6.4

SOURCE: *U.S. Bureau of the Census, 1997,* pp. 584 and 589.

consumption to reach 50 percent of all the world's commercial energy by 2020. Declining rates of energy consumption in the growing service economies of the more-developed countries and rapid population growth and industrial development in the less-developed countries account for this structural shift.

While neither the Commission nor the Agency foresee a departure from the commercial fuel mix depicted in Table 5.1, there most certainly will be a growth in total commercial energy demand. The imperative question for energy analysts, then, is how much that growth will be? According to Agency estimates, growth of total world commercial energy consumption will rise from a staggering 8,000 million tons of oil equivalents in 1997 to somewhere between 10,900 and 11,800 million tons of oil equivalents in 2010. The difference in their estimates largely depends on whether industrial and household energy dependent technology becomes more efficient. Nevertheless, even in the more conservative estimate of the Agency, world commercial energy consumption will increase 36 percent by 2010.

The Commission works with hypothetical cases to project world commercial energy consumption between 1990 and 2020. We discuss three of their major scenarios here. In the high-energy scenario, world commercial energy consumption grows to 16,000 million tons of oil equivalents, or by 82 percent between 1990 and 2020. Their own preferred scenario, the one they consider most likely, estimates that world energy consumption will reach 13,500 million tons of oil equivalents by 2020, a 50 percent growth in world commercial energy consumption.

Their third scenario involves a "massive drive to raise energy efficiency." Growing international concern of the oil-importing countries for the long-term availability and price of oil drives this scenario. Here the world demand for commercial energy grows by 28 percent or nearly 12,000 million tons of oil equivalents. This scenario employs an increase in the use of natural gas, a rapid expansion in energy from renewable sources such as solar energy and small-scale hydroelectric generation, and a modest increase in the use of nuclear power, assuming that safe waste storage methods will be found. This scenario, however, is not the one the Commission identifies as most likely. Thus, for an organization controlled by large national utility companies, international development banks, and energy research institutes, a 50 percent increase in world commercial energy consumption between 1990 and 2020 most likely will occur.

Neither the Agency nor the Commission take issue with the argument that the world commercial fuel mix is unlikely to change in the next two decades. As the Agency report states, "the projections for the world energy fuel do not indicate a significant shift among different fuel types over the outlook period."[2] In addition, both groups foresee a substantial growth in world energy consumption. The lowest Agency estimate for the period between 1990 and 2010 is 36 percent. If, for the sake of comparison, we crudely assume that the demand for world commercial energy is linear—with equal increments of growing energy demand in each decade—with the pattern continuing for another ten years, world energy consumption possibly will grow by 54 percent, a change comparable to the Commission's preferred scenario.

Except for the Commission's "massive drive scenario," neither group envisions renewable sources of commercial energy playing a major role in the near future commercial fuel mix. This is not to say that either group discounts the importance of renewable fuel sources. In the language of the Agency, "renewable energy sources (here defined as wind, wave, solar, and geothermal power), albeit registering the highest growth rates among all fuel types (1990–2010), are projected to account for only around 1 percent of total primary energy in 2010."[3] With already massive volumes of energy consumption and growth, fuel sources now playing such a minor role in world commercial energy production will require many years of high rates of growth before the renewables can make a significant contribution to the world fuel mix.

The Commission and Agency agree about the potential for renewable energy sources in the future. The Commission, for reasons to be discussed in the next section on oil, anticipates a demand for renewables, but not until a time beyond the period of 1990–2020. As with the Agency, the Commission states, "Even given clear and widespread public policy support, the new renewables (such as solar and wind energy production) will take many decades to develop and diffuse to the point where they significantly substitute for fossil fuels."[4]

Key players in the use of these renewables, according to the Commission, will be the less-developed countries without the foreign exchange needed to purchase fossil fuel imports; without the wealth needed to develop nuclear power; in locations too far from natural gas to make transporting this fuel source economically feasible. The local availability of solar, geothermal, or biomass in these countries will make renewable fuel sources most likely.

Neither the Agency nor the Commission make a strong case for nuclear power playing a major role in the future world fuel mix. The Agency recognizes that nuclear power now contributes to the commercial fuel mix in South

Africa, China, South Korea, Taiwan, Pakistan, Argentina, Brazil, and extensively in parts of Central and Eastern Europe. They also recognize, however, few plans to expand nuclear power production in the more-developed countries in the next ten years, the countries using 80 percent of the nuclear power produced in the world.

Because nuclear power does not contribute to the buildup of carbon in the atmosphere, the Commission remains somewhat optimistic about its utilization. The future of nuclear power, however, depends upon finding solutions to significant technological and administrative problems. According to the Commission, nuclear power "will first require . . . technical safety in operation, management skills, effective international inspection, and safe long-term disposal of waste."[5] For many of the less-developed countries, where the growing demand for commercial energy is most pronounced, the costs and managerial problems associated with the use of nuclear power make its widespread use unlikely.

Of the energy-related problems facing societies throughout the world, the greatest is meeting the acute and rapidly growing demand for energy in the LDCs. Concluding analysis by the Commission is telling: "Over half the world's current population do not now have access to commercial energy . . . and use has actually gone down . . . reflecting poverty. With the population increases occurring in the developing world . . . even the maintenance of energy per capita availability will be a challenge."[6] For the Agency and the Commission, overcoming the barriers of energy poverty throughout the world without creating ecologically damaging emissions is a top priority.

For the Commission, the group more explicit about international energy policy, one finds a commitment to the managerial paradigm with "the wider spread of the market system within a framework of effective government support and stimulation."[7] Thus, decision making by leaders in the less-developed countries, working together with multilateral lending agencies, will provide the key to expanding the supply of traditional commercial fossil fuels. These leaders are also likely to promote a shift to nonfossil fuels in the next two decades.

While energy-related challenges to the less-developed countriess are formidable, to say the very least, more-developed countries face significant energy-related challenges as well. One, shared by the less-developed countries, is the risk of facing climate-related disasters. Leading energy analysts foresee no significant change in the fuel mix of the more-developed countries until after 2020. In fact, as less-developed countries make the transition to the modern fuel mix, the entire world takes an even greater risk of coping with the disasters believed to be associated with global climate change.

While there remains much uncertainty about the prospect of global climate change, there is somewhat more certainty about a second problem: the prospect of non-Organization for Petroleum Exporting Countries' petroleum reserves being depleted. To understand this problem we will first discuss how the world level petroleum glut of the 1990s came to be. Then we will turn to a discussion of why energy analysts anticipate that the glut will become a windfall for the Organization for Petroleum Exporting Countries once again.

World Crude Petroleum: Past, Present, and Future

In the 1970s a number of societies, especially in the United States and Western Europe, became well acquainted with the sting of a vital natural resource scarcity. In the course of two sudden reductions in the supply of crude petroleum production first in 1973 and again in 1978, these more-developed countries learned they were not exempt from the political/ecological limits of natural resource scarcity. The causes of those shocks and their ramifications for subsequent world petroleum production are the subject of this section. Because the United States has the world's highest per capita consumption of oil, emphasis will be placed on American petroleum dependency. We will then turn to what energy analysts cautiously regard as the future of world petroleum in the years ahead.

Oil Shock, 1973 Throughout the twentieth century, the price per barrel of unrefined petroleum remained surprisingly low and stable. Figure 5.1 suggests that the price (in 1993 dollars) fluctuated slightly around ten dollars per barrel for decades. Oil prices actually declined even more by the time of the first dramatic shock in 1973. Once that shock occurred, the possibility for a return to cheap oil seemed inconceivable.

Both shocks reflect the political and military volatility in the Middle East in the 1970s and earlier as well. Prompted by Israel's participation in the Suez Crisis and their victory in the 1967 Six Day War, Egypt's President Anwar Sadat aspired to end the military and political stalemate between Israel and his nation. By 1973 Egypt spent 20 percent of its GNP annually on defense, thereby limiting Egypt's prospects for national economic development.[8] With the cooperation of Syria's President Hafez al-Assad, the two countries embarked on an invasion of Israel, on October 6, 1973, taking Israel by surprise on the holiest of Jewish holidays, Yom Kippur.

At the same time, the Organization of Arab Petroleum Exporting Countries, including the very oil rich Iraq, Kuwait, Saudi Arabia and the United Arab Emirates, showed support for Egypt and Syria in waging war against Israel.[9] The Arab countries did so by increasing the price of crude oil by 5 to 10 percent monthly in retaliation for Western support for Israel in the war. In October oil sold for as little as $5.40 per barrel. With the increases oil prices almost tripled to $16.00 per barrel.[10] The Arab exporting countries also decided to place an embargo on oil to countries financing Israel's war effort, especially the United States. Prior to the embargo, the United States had been importing 1.2 million barrels of Arab oil per day; by February Arab oil producers reduced exports for the United States to a trickle: 18,000 barrels per day.[11]

The Arab retaliation with higher oil prices and embargoes provoked a frantic, worldwide search for new sources of oil and alternative energy sources. In the United States President Ford proposed to build 200 nuclear power plants, 150 coal-fired electric generators, 30 major oil refineries, and 20 major synfuel plants for coal liquefaction. In addition, in 1975, Congress, after eight years of debate, approved the Trans-Alaska Pipeline System for

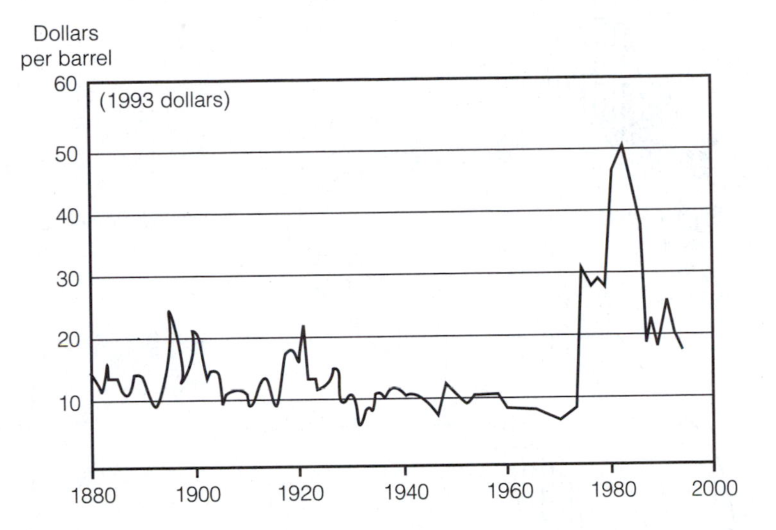

FIGURE 5.1 The Real Price of Oil, 1880–1993

SOURCE: From *Power Surge: Guide to the Coming Energy Revolution,* by Christopher Flavin and Nicholas Lenssen. Copyright © 1994 by Worldwatch Institute. Used by permission of W. W. Norton & Company, Inc.

transporting crude petroleum 798 miles from Prudhoe Bay on Alaska's North Slope south to the port of Valdez at the eastern edge of Prince Willliam Sound.[12] Figure 5.2 maps the route of the pipeline. By the end of the 1970s, the pipeline carried two million barrels of oil per day, a quarter of the total U.S. domestic crude petroleum.[13]

Congress also passed new legislation to raise fuel efficiency standards for automobiles. By 1985 the average fuel efficiency of an American-made car doubled, in response to the new standard, from thirteen to twenty-seven miles per gallon of gasoline. This legislation had a major impact on the world's oil balance in as much as one out of seven barrels of oil consumed daily, worldwide, fueled American cars.[14] By 1990 the average new American car traveled twenty-nine miles per gallon of gas.[15]

Internationally, non-Organization for Petroleum Exporting Countries, both the more- and less-developed, increased oil production through existing or newly discovered sources. The former Soviet Union produced 12.6 million barrels per day, mainly for internal and East European markets. Mexico developed oil export capacity with discoveries in the southern state of Tabasco and offshore in the Bay of Campeche. North Sea oil became a major revenue source for Norway and the United Kingdom. China and India also became sources of oil in international markets, as suggested in Figure 5.3.

Oil Shock, 1978 In the midst of these significant efforts to reduce international dependence on oil from the Organization for Petroleum Exporting Countries, a second worldwide oil shock occurred in 1978. In this case oil

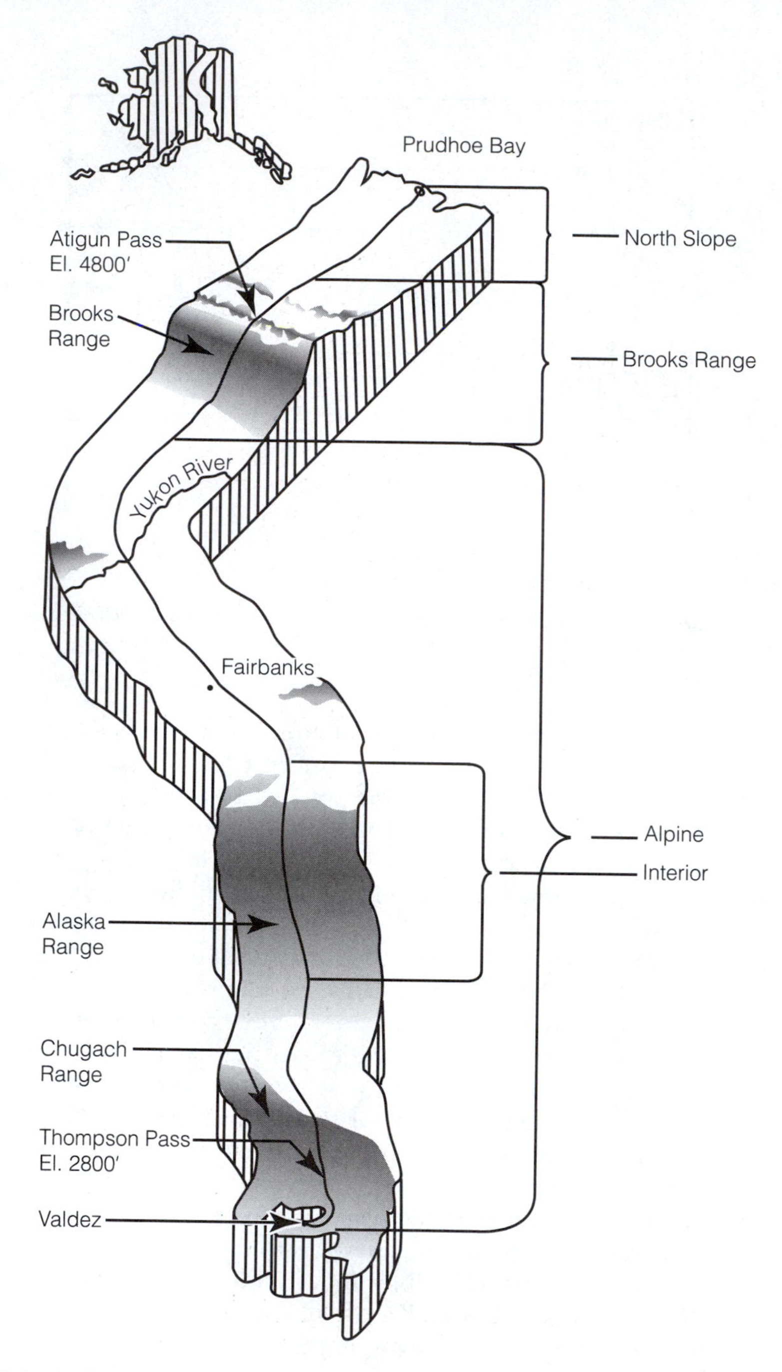

FIGURE 5.2 The Trans-Alaska Pipeline (TAPS)

SOURCE: Coates, 1991. Used by permission of the Alyeska Pipeline Service Co.

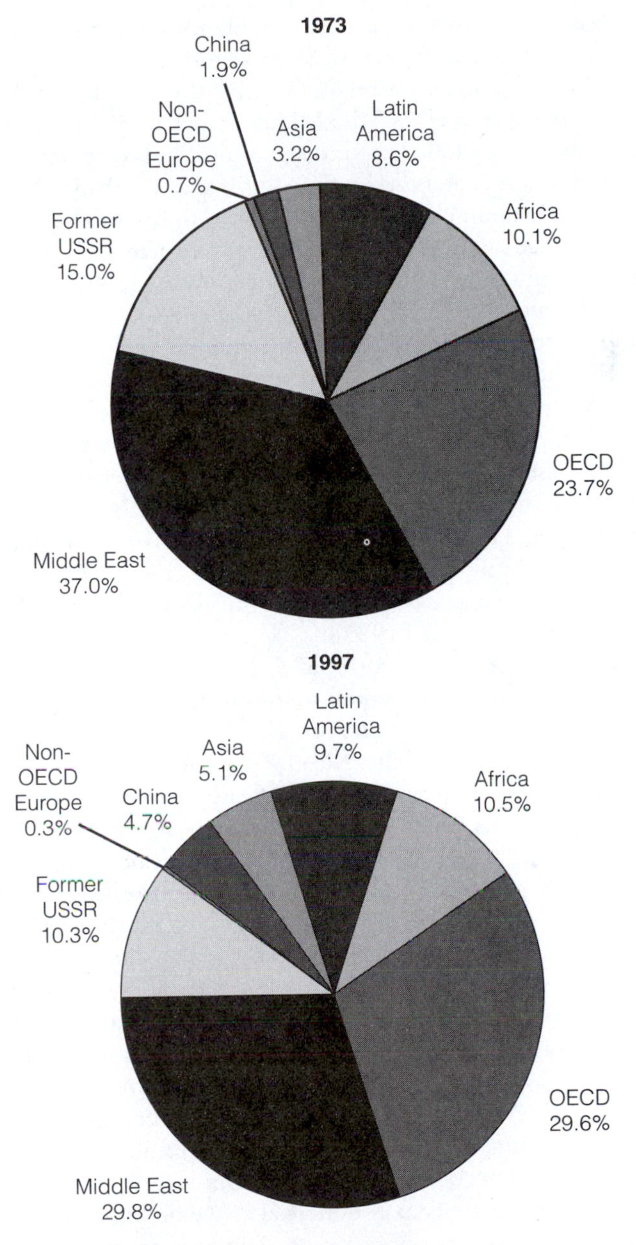

FIGURE 5.3 The 1973 and 1997 World Regional Shares of Crude Oil Production

SOURCE: U.S. Energy Information Administration, 1998.

prices increased because of a fundamentalist religious uprising against the pro-U.S. Shah of Iran. Iran had been a world oil producer, second only to Saudi Arabia until this fundamentalist revolution. By the mid-1970s the sharp increases in oil revenue throughout the Middle East led to closely related social changes. The changes included the growth of demand for employment in the Iranian oil production centers and stepped up rural to urban migration. Iranian cities such as Teheran experienced a massive influx of rural migrants, inundating an outdated rail-based public transit system. Electric companies could not keep pace with the growing demand for power. Public discontent with the Shah and his efforts to modernize Iran grew as electric power outages disrupted both industrial production and domestic life.[16] A growing number of Iranian fundamentalists called for a return to a traditional Islamic theocracy, the kind endorsed by the exiled Shiite religious leader, the Ayatollah Ruhalla Khomeni. To the Ayatollah, the Shah, the lifestyle represented by the United States, and the Western world, was both foreign and reprehensible.

As Shiite fundamentalism grew, Iran's oil industry became enmeshed in the power struggle. The Shiite's principal target was the giant Oil Service Company of Iran, crippled by striking workers in late 1977. Iranian oil production, according to oil analysts, declined from 5.5 million barrels per day to no production at all for a time in 1978.[17]

The Shah left Iran in January 1979, allegedly for a vacation. Street demonstrations and celebration by the Shiites followed the Shah's departure. In February Kohmeini returned from exile, bringing the Moslem theocratic regime into power. The Shah died in exile. Meanwhile, the world would have to adjust to a second energy crisis with prices increasing from $13.00 per barrel in 1979 to $34.00 per barrel in 1981.[18]

The search for alternative sources of oil and for greater efficiency in transportation, electrical generation, and manufacturing processes would prove to be a short-term victory for conservative human exemptionalists. Driven by the market price of oil, the major oil producers found significant quantities of crude petroleum outside the Organization for Petroleum Exporting Countries. The effect of these discoveries was an oil glut. However, the years immediately following the two shocks caused significant structural changes in the United States and other major industrial powers as well. For the first time since World War II, Americans experienced a shortage of a vital natural resource with long waiting lines at gas stations, gas rationing, a ban on lights for commercial Christmas decorations in 1973, lower settings for home thermostats, and mandatory blackouts in commercial office buildings at night.[19]

Mandatory and voluntary conservation, however, were only a part of the aftermath from the shocks. Equally important, the increase in energy prices cut into the profit margins of manufacturers everywhere, but more so in the United States, especially the heavy manufacturers of rubber products, automobiles, and metals. The shocks affected U.S. manufacturers more adversely than, say, German, Japanese, or newly emerging South Korean manufacturers. The differences stemmed from the age of industrial production technology in U.S. basic industries compared with the age of manufacturers in parts of Europe or Japan. Having rebuilt their basic manufacturing capacity after World War II, their newer, more energy efficient production methods meant their profit margins declined less than the older, less efficient U.S. producers.

Increasing foreign competition through the 1970s forced U.S. manufacturers to search for more efficient modes of production. Manufacturers closed plants and downsized their workforces.[20] For working-class America, the period before less costly oil reemerged was a time of heavy-handed suffering. As the U.S. Office of Technology reports, "from 1979 to 1984, 11.5 million American workers lost jobs because of plant shutdowns or relocations, rising productivity, or shrinking output . . . . In 1984, 1.3 million of these displaced workers were still unemployed, joining the millions of others seeking work during a recovery that has not yet pushed the civilian unemployment rate below 7 percent."[21]

The social and economic stress in the aftermath of the shocks, however, was not shared evenly nationwide. It was concentrated in the Old Industrial Belt from Boston south to Baltimore and west to St. Louis and north to Chicago and Duluth. For residents in this multiregional area, the relationship between energy and society was all too clear.

Future Prospects for the Crude Petroleum Supply There is considerable agreement among energy agencies such as the Agency and petroleum geologists about the role of petroleum in the world's energy mix from now until 2020. As long as oil prices remain relatively low, these energy policy analysts argue that oil will continue to play a major role in the world energy mix, reducing the likelihood of substantial change in the mix with the addition of renewable or other energy sources.

By 2010, according to the Agency, the world will consume about 95 million barrels of oil daily, a 30 percent increase over 1995.[22] While this rapid growth in oil consumption represents a 2 percent annual growth in the consumption of refined petroleum, the statistic is misleading. Most growth in oil consumption will come from the less-developed countries where growth in oil consumption now is as much as 4 percent annually. Rapid population growth, urbanization, and industrialization bring new demand for automobiles, mass transportation, and oil to generate electricity for new buildings. The Agency estimates that the demand for oil will grow in the more-developed countries by about 1 percent annually, mainly because of growing transportation activities such as long-distance commuting and commercial freight hauling.

Both the Agency and petroleum analysts expect a return to dependence on Middle Eastern oil by 2010. According to the petroleum analysts, "After the turn of the century, (the Western Europe) oil supply is expected to pass its peak, perhaps after reaching 47 million barrels per day, principally from the North Sea."[23] The United States' domestic supply of oil has been declining since the early 1970s, and Canada's domestic oil supply is expected to begin declining early in this decade. Petroleum analysts estimate that 60 percent of Europe's oil supply will depend on the Middle East by 2010. With some North

American oil still flowing from Alaska and the U.S. Gulf coast, their dependence on the Middle East will grow from 40 percent in 2000 to 45 or 50 percent in 2010.[24]

These estimates are significant because they are based on an extension of a surprisingly accurate projection method developed by M. King Hubbert in 1956. Hubbert successfully predicted the beginning of the decline in U.S. domestic oil production in 1970. Hubbert pioneered the notion of the production cycle, or, a postulated series through which the production of a nonrenewable resource tends to go. The first phase is increasingly rapid growth in the rate of production as demand increases, as the industry achieves greater efficiency in energy production, and as per unit costs of the resource fall. The second stage involves a leveling off of production as the resource becomes scarcer and the price begins to rise. The final stage is a continuous decline in the rate of production. This decline is assumed to result as scarcity, increasing prices, and the declining quality of the resource overwhelm advances in the technology of extraction.

Hubbert's approach has the advantage of recognizing that a resource is never completely or suddenly exhausted. Instead, the highest-grade supplies are exploited first, and later production exploits progressively less concentrated and accessible supplies. The model emphasizes two points in the production cycle. The first is the year of peak production, and the second is the point at which 90 percent of the ultimately recoverable reserves are exhausted.

Hubbert's work predicted that the U.S. domestic oil production cycle would peak in 1970 and that 90 percent of the domestic supply of oil would be consumed by the year 2000. Both predictions proved to be accurate. More recent work in 1998 on the future of the world's consumption of oil employs the same methodology and is also consistent with the predictions of the Agency and the Commission. Thus, there is reason to believe that a resurgence in the role of the Middle East in world oil production is likely. The socioeconomic and environmental effects this may have on international relations, energy prices, and climate change remain as key issues in the study of energy and society.

THE BIRTH OF AMERICAN PETROLEUM DEPENDENCY

Of course, there is no single explanation about how societies such as the United States come to be dependent on a politically unstable region for a resource as vitally important as oil, escape that dependence, only to return to the dependency once again. A look at the competing views about the social forces creating this situation in the United States, however, provides insights into the complexities and unintended consequences of relationships between environment, energy, and society in the context of petroleum dependence.

The Growth of American Petroleum Dependency

Consumers in the United States use one-seventh of all refined petroleum in the world, especially in their cars, trucks, and aircraft. The current popularity of the sport utility vehicle, featured in Box 5.1, definitely is in the spirit of that American trend. To understand how this form of petroleum dependence evolved, one has to look at two major developments in American society between the Depression and the end of World War II. The first development is intra-class competition between the railroad-oriented coal-electricity-steel coalition and the newer auto-oil-rubber coalition. Between 1925 and 1945 the auto-oil-rubber coalition would literally drive the electric trolleys into obsolescence.[25] The second major development is the realtor-banker-home-builder coalition that emerged after World War II to meet an acute demand for affordable housing by the growing middle and working classes. The story of these social changes that played important roles in the growth of American petroleum dependence is a double-edged one, with conservative and radical accounts.

The Auto-Oil-Rubber Coalition Both the conservative and radical accounts of how the auto-oil-rubber coalition became a powerful force that shaped American energy consumption begins with the history of growth of urban mass transit, developing in tandem with the expanding industrial city of the late nineteenth century. First horse-drawn trams and later the electric street railway companies served the needs of the growing working and middle classes in their daily urban journey to work. At their peak in 1917, American street railway transit companies controlled by wealthy families such as the Yerkes in Chicago, the Belmont and Dougherty families in New York City, and the Mellon family in Pittsburgh maintained over 45,000 miles of steel tracks for the streetcars. By 1923 the street railway companies nationally provided 13.5 billion rides in a single year. By the 1950s ridership plummeted to 4 billion rides annually even with decades of urban growth.

For conservative transportation analysts, the demise of the street railway transit companies was an artifact of a changing market. New consumer demand for a supply of faster, affordable, and more flexible mass transit service grew as cities and the early streetcar suburbs decentralized beyond a comfortable walking distance to the electric railway lines. One transportation analyst observes, "The technological development of the motor bus after the middle 1920s produced a vehicle which was better suited to the more diffuse and changing passenger demands of later times."[26] Newer diesel-powered buses running on rubber tires developed more flexible routing patterns without the costly maintenance problems presented by the steel tracks for the railcars. They also could be chartered for a variety of uses in off-peak travel times.

Radical scholars add another dimension to the story of the demise of electric street railways in the United States. As one transportation scholar put it, the shift away from the electric street railways "was not altogether due to consumer choice expressed in a competitive market."[27] Alfred P. Sloan, the

BOX 5.1 Focus on the United States: SUVs and Climate Change

Sport Utility Vehicles (SUVs) have become incredibly popular. The authors of The Ultimate Poseur Sport Utility Page ("The only SUV satire page on the web!") point out the irony of this trend. "People laughed at their parents for driving behemoth gas guzzlers in the 50s, 60s, and 70s, yet today's SUVs are even bigger and more ridiculous." From an environmental perspective, a serious problem with SUVs is their low gas mileage. The popularity of SUVs in spite of their environmental consequences has led a number of environmental groups to campaign against the SUV. Friends of the Earth calls theirs the "Roadhog Reduction Campaign." They point out another paradox of SUVs: "Television advertisements present them as a way to return to nature, yet they actually accelerate existing environmental problems." Consumer choices are linked to cultural trends. The humorous "Poseur" website makes fun of SUV drivers who don't use the vehicles for their off-road capabilities, but who instead drive them to the mall. The authors decry the SUV culture: "Millions of lemmings buy SUVs for no reason other than to fit a trendy image and 'look tough'."

The U.S. Department of Energy and U.S. Environmental Protection Agency acknowledge that burning fossil fuels creates global warming. They argue, "One of the most important things you can do to reduce global warming pollution is to buy a vehicle with higher fuel economy." A third of carbon dioxide emissions in the United States comes from cars, and overall carbon dioxide emissions are increasing.

A huge range of options exists in terms of auto fuel economy. For example, Honda has recently marketed the "Insight," a small, two seater, which gets 61 miles per gallon in the city and 68 on the highway. The Insight is a hybrid—its internal combustion motor is boosted by an electric motor that does not need to be charged like electric cars. Contrast the Insight with three of the least fuel efficient vehicles, all SUVs—the Land Rover Range Rover, the Lincoln Navigator, and the GMC Yukon. These vehicles get only 12 mpg in the city and 15 or 16 mpg on the highway. Federal standards for fuel economy (corporate average fuel economy—"CAFE standards") are set at 27.5 mpg for cars, but only 20.7 mpg for SUVs because they fall under the category of light trucks.

New car and truck sales in 1992 totaled 12.9 million; in 1997, sales increased to 15.4 million. During this period, car sales showed little change (8.1 million to 8.3 million). The increase was largely due to new truck sales (the SUV category). Sales jumped from 4.4 million to 7.1 million.

How might we reduce our contributions to global warming? Some support changes in the government's CAFE standards (a managerialist action). The state of California has required producers to reduce SUV emissions by 75 percent by the year 2004. Others support voluntary changes in consumer choices and habits (a conservative action). For example, a number of groups, such as the Bicycle Civil Liberties Union, support "car free days" in which drivers are encouraged to walk, bicycle, and use public transportation. Producers, too, are working on changes, seeing a potential market for "green" products. For example, Ford Motor Company has announced its work to produce a less-polluting SUV and other manufacturers are following suit. In addition, technological advances are being explored by industry and are receiving some support from the government for the creation of less polluting alternatives, such as fuel cells.

SOURCES: *poseur .4x4.org, www.fueleconomy.gov, www.suv.org,* Statistical Abstract of the United States online, Tables 1028 and 3407, *www.bclu.org.*

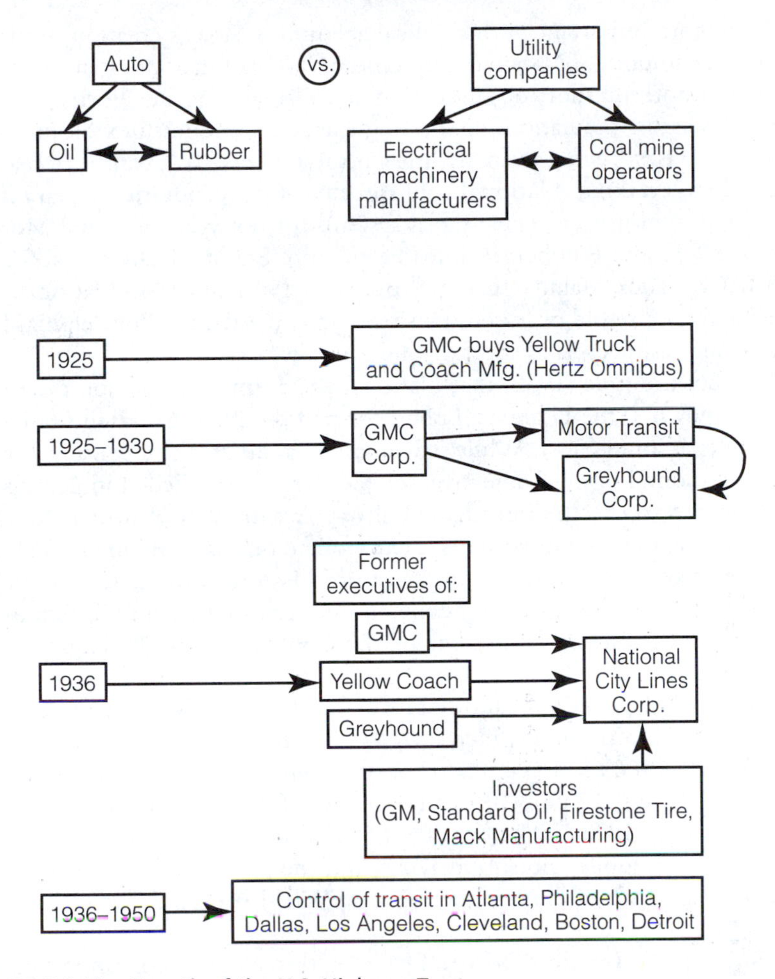

FIGURE 5.4 The Growth of the U.S. Highway Trust

president of General Motors Corporation, and Pierre DuPont, a controlling stockholder in Sloan's growing corporate giant, orchestrated a new treadmill of production fueled by oil. Figure 5.4 identifies key events in the General Motors' takeover of the electric street railways from coast to coast. In 1925, under the corporate leadership of Sloan, General Motors purchased the Yellow Truck and Coach Manufacturing Company. Through this purchase General Motors gained control of the patent and production rights for the Hertz Omnibus.[28] This diesel-powered, rubber tire vehicle would be used by a General Motors subsidiary, the Motor Transit Corporation, to produce buses for the Greyhound Corporation. Greyhound provided inexpensive interurban transportation services that would compete with the older electric (and some steam) railway systems, whose energy came from coal.

Even more important in this radical account is Sloan's creation, with the help of Greyhound, of National City Lines (National) in 1936. National's objective was nothing short of purchasing and dismantling the electric railway companies across the nation, converting the aging and less flexible streetcar railways to buses, paving over the lines in the process. To finance such a vast financial undertaking, National sold stocks to the corporations guaranteed sole-supplier status to the bus transit systems for ten years: General Motors, Firestone Tire and Rubber, Phillips Petroleum, Standard Oil of California, and MACK Truck Manufacturing.[29] Between 1936 and 1950 National purchased and then resold or leased transit services in Atlanta, Philadelphia, Dallas, Los Angeles, Cleveland, Boston, and Detroit.

At about the time (1932) that National was formed, Sloan and others, to become known as the Highway Lobby, formed the National Highway Users' Conference (Conference). While it would take a quarter of a century for the group to accomplish their objective, the Conference succeeded in getting the U.S. Congress to pass the Interstate Highway Act in 1956. Within a decade a 40,000 mile system of limited access metropolitan beltways and interstate highways subsidized the major treadmills of production in the auto-oil-rubber coalition and served the growing demand for automobiles and oil. Figure 5.5 identifies the key interest groups behind the Conference and the Highway Act they produced.

Between 1945 and 1950 automobile ownership grew by 3 million cars annually. Ownership continued increasing by 2 million cars annually throughout the 1950s. By 1966 the average number of passengers per route mile on urban buses (38,500) was 12 percent of the average of 300,000 per route mile on street railways in 1923.[30] The era of American petroleum dependence, commuting by automobile, pleasure driving, and the family vacation by car now defined a major part of the new, post–World War II American lifestyle.

The Suburban Housing Boom Sharp increases in rates of family formation and childbearing accompanied the return of veterans after World War II. Both the Depression and war, however, suppressed the growth of the home-building industry and the demand for housing. The existing housing stock was largely expensive, custom-built housing for the upper and upper-middle class.[31] Demographers estimate that in 1947 at least 6 million urban families searched or planned to search for affordable housing.[32] Yet, because of more than a decade of suppressed demand and the tradition of building custom-designed housing for middle and upper-middle income housing, the building industry was not in a position to meet this unprecedented demand in the new era of post–World War II prosperity.

Beginning in 1945 large residential house-building companies led the massive suburbanization of cities. Exemplary large builders include Levitt and Sons on Long Island, New York, John Mowbray outside Baltimore, Maurice Fishman in Cuyahoga County surrounding Cleveland, Don Sholz in what would become suburban Toledo, J. D. Nichols outside Kansas City, and Dell

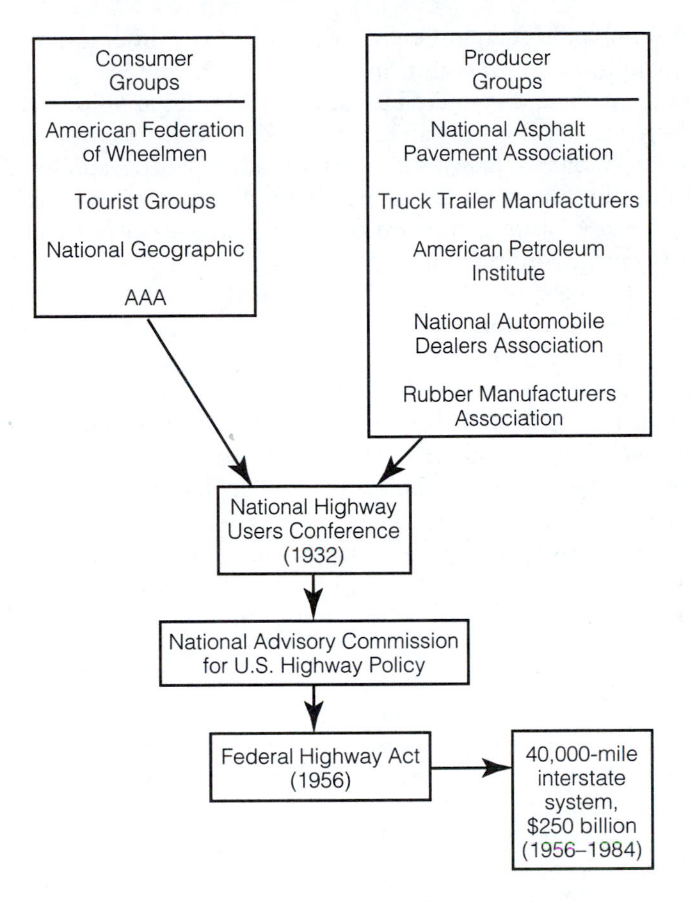

FIGURE 5.5 The Growth of the Highway Users Conference and the Interstate System

Webb in suburban Phoenix. One housing analyst defines large builders, in the context of the late 1940s, as companies capable of producing at least 100 houses in a single year with a minimum of 100 workers and annual sales of one million dollars or more.[33]

Levitt and Sons was a path-breaking large homebuilder in Nassau County, Long Island. Having learned to mass produce housing on military bases during World War II, Levitt converted potato farms into suburbs nearly overnight.[34] As one housing scholar observes, "By 1948 Levitt was completing more than 35 houses per day and 150 houses per week and rapidly selling the low-cost ($7,999) product."[35] Between 1947 and 1950 Levitt built his first of three Levittowns with 17,000 Cape Cod style houses, each with two bedrooms, a bath, and equipped with an 8 inch black-and-white television, a washing machine, four fruit trees, and a signed agreement to mow the lawn regularly.[36]

With companies such as Levitt's, what had been an industry producing nearly 2.8 million housing units in the 1930s, tripled to nearly 7.5 million units in the 1940s, and then doubled again to 15 million units in the halcyon years of American suburbanization, the 1950s.[37] Conservative, market-oriented accounts for this massive change in the spatial organization of American communities and the petroleum dependence that accompanied suburbanization vary, but they share a common argument that the builders met a pent-up consumer demand for housing.

For human ecologists, the post–World War II suburbanization of cities resumed the pre-Depression decentralization of the urban community. Decentralization was a response to industrialization, growing urban population density, and the development of an affordable automobile used for shopping and the journey to work. Urban population decentralization began in 1921 when the number of registered motor vehicles increased from 8,000 to 10,500,000, or, from one automobile per 10,000 Americans to one for every ten persons.[38]

Other conservative analysts emphasize the consumer demand interpretation of the suburban trend. They insist that suburbanization reflects deeply held American cultural values articulated in Jeffersonian thought. Thomas Jefferson's aversion to city life, with large urban scale, complexity, and high population density, deflated the individual's sense of independence and commitment to their community. For Jefferson, the ideal community arrangement was the independent farm family, the yeoman farmer, who played an essential role in protecting the democratic way of life. American rural life, for Jefferson, instilled a sense of rugged individualism and independence. This individualism, in turn, played an important role in the vitality of small-town, grassroots government, the basic unit of a democratic society.

While 17,000 identical homes on 1,400 acres in Levittown is hardly a representation of the yeoman farmer at the edge of the American frontier, some scholars insist on such a connection. Jeff Hadden and Joseph Barton link residence in the early, affluent streetcar suburbs such as Evanston, Illinois, Shaker Heights, Ohio, and Llewellyn Park, New Jersey, to the attainment of the Jeffersonian ideal.[39] These early suburbs set the standard of residential prestige for middle- and working-class families to emulate a century later. As another suburban analyst observes, "the tract suburbanite settles for a little open space and fresh air, removed from the smoke and noise and dirt of the metropolis."[40] Only after World War II and the development of affordable, single-family suburban housing could the average American household move a step closer to the Jeffersonian ideal in a modern, metropolitan context. To win the American dream, however, home buyers had to become commuters, something to be explained in the next section.

Not all scholars agree that suburbanization, with its inherent demand for the automobile, was an innocent supply-and-demand phenomenon. For Barry Checkoway, "Suburbanization prevailed because of decisions (by) large opera-

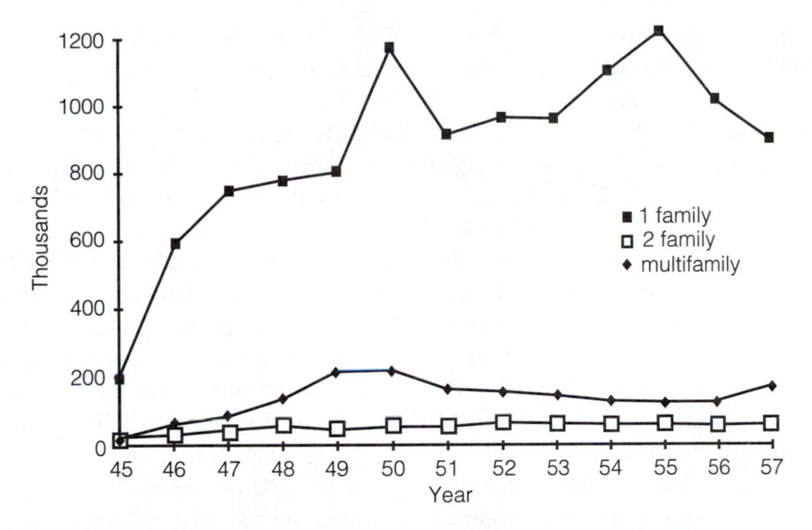

FIGURE 5.6 U.S. Housing Starts, 1945–1957, by Residential Type

SOURCE: U.S. Census, 1975.

tors (home builders) and powerful economic institutions supported by federal government programs, and ordinary consumers had little choice in the spatial arrangement of housing that resulted from these decisions."[41] The Housing Act of 1934, legislation whose effects would be visible in two decades, is crucial to this change in urban spatial organization and American petroleum dependency.

The Housing Act, said to be the most important legislation about American residential patterns in the twentieth century, created the Federal Housing Administration (Administration), the managerial agent charged to oversee and administer a home loan guarantee program.[42] The Administration thus protected banks against mortgage defaults by home buyers. In effect this opened the doors of banks and other lending agencies to middle- and working- class families for the first time in U.S. history. The Administration's program increased home ownership from 39 percent of adult households in 1937 to 57 percent as early as 1950.[43] Administration-approved home loans, however, had to be for new, single-family units located in areas with low population density. Figure 5.6 illustrates the pattern of home ownership created in the built environment of the post–World War II American suburbs.

The architecture of this legislation is understandable, perhaps in part through the Jeffersonian ideal about American settlement, but more—in the radical account—through the workings of capitalism. Two highly prevalent groups of capitalists are especially influential in shaping this legislation: the National Association of Real Estate Boards (Board) and the National Association of Homebuilders (Homebuilders). Support for the Board comes from the hundreds of members of the U.S. Savings and Loan League, the U.S. Chamber of Commerce, the American Bankers' Association, and the Mortgage

Bankers' Association. The Building Products Group, the National Retail Lumber Dealers' Association, the Association of General Contractors, the National Clay Producers' Association, and the Producers' Council also support the Board. In 1942 the Homebuilders and the Board joined forces, creating the well-financed Realtors' Washington Committee. This Homebuilder-Board coalition lobbied, and continues lobbying, to expand the total Administration lending powers since the end of World War II.

Two other federal housing policies promised a secure and prosperous future to this banker-builder-real estate coalition. First, the Veterans Administration (Veterans) initiated their own home loan guarantee program for returning war veterans, beginning in 1946.[44] Second, Internal Revenue Service regulations continue promoting private home ownership and promise windfall profits to the coalition. The Internal Revenue Service does so by allowing mortgage holders to deduct the interest paid on their home loans from their federal income taxes. This Internal Revenue Service tax policy, in effect, creates the illusion of ownership and the reality of debt among American homeowners, a situation that has not changed since the beginning of the post–World War II era.[45] The Internal Revenue Service's tax policy, in combination with the Administration's and Veteran's loan guarantees, perpetuates the dispersal of millions of homeowners to the suburban periphery of cities and towns. With the new house, the car, and the fuel to travel, suburbs and petroleum dependence went hand-in-hand.

Looking Back

In Chapter 2 we noted that no single theory or story telling why a society acts the way it does or why social changes occur can provide a complete explanation for a social custom or trend. The growth American petroleum dependence certainly illustrates the multifaceted character of sociological change.

Petroleum dependence in the United States reflects both a cultural development, the growth of a collective consciousness about the appropriate spatial arrangement of communities, and the political economy of American energy policies. Petroleum dependence is a by-product of the development of two cultural values. First, the dependency reflects Americans' desire for autonomous ease of movement across space with the petroleum-fired automobile. The dependency also reflects the pursuit of the American dream: ownership of a home in a small suburban town. The combination of the two values makes for probably the most mobile society in the world, energized by petroleum.

Yet, we also know that American petroleum dependence reflects the actions of powerful capitalist actors such General Motor's Alfred Sloane and Pierre DuPont. Working with others in the auto-oil-rubber coalition, these capitalists managed to turn the resources of the federal government into an agent working to secure their financial future through the construction and maintenance of the Interstate Highway Program. The emergence of the car,

National City Lines, and the Highway Users' Conference represents concerted efforts by capitalists to realize their own financial self-interests. Large home-builders and their work to shape the Housing Act plays an equally formative role in shaping the spatial organization of post–World War II America. Without the deliberate, protracted efforts of these capitalists, Americans would not have realized their dreams and debts; nor would they have their voracious appetite for energy, especially oil.

THE UNITED STATES IN THE PERSIAN GULF CRISIS: A WAR FOR OIL?

The world oil market, in the summer of 1990, looked like the market in 1973. The demand for oil was growing annually by nearly 1 percent. The Organization for Petroleum Exporting Countries controlled 78 percent of the world's proven oil reserve, much of the reserve in the Persian Gulf.[46] The United States imported oil at a higher rate than ever. Unlike a decade earlier, however, no known vast new sources of oil existed outside the Organization for Petroleum Exporting Countries for the producers to tap.

"On August 2, 1990, Iraqi tanks had sped the 80 miles to Kuwait City to begin the occupation of Kuwait and the deployment of the first 100,000 troops—many more than was needed to conquer a country with a military force of less than 30,000."[47] What was Iraq's justification for invading Kuwait? Kuwait "was guilty of economic aggression against Iraq—a perception that was widely shared in the Arab world."[48] Yet, Iraq's deployment of so many troops led observers to sense that the aggressor was fully prepared to invade and seize the vast Saudi Arabian oil fields. Iraq did so by stationing military troops along the Saudi border. Iraq's strategic positioning in Kuwait can be seen in Figure 5.7. A single command and move from Iraq's General Saddam Hussein would put Iraq in control of 40 percent of the world's remaining oil reserves.[49]

The United States responded, at first, with an international call for a complete trade embargo against Iraq. As Hussein continued to build his military forces in Kuwait, however, the United States, along with a smaller supporting military coalition of thirty-five nations, felt that more aggressive action was essential. On November 29, 1990, the United Nations Security Council established a resolution authorizing the use of military force against Iraq, if Hussein did not withdraw from Kuwait by January 15, 1991. On January 12, 1991 the U.S. Congress granted President Bush approval to use military force in the Persian Gulf. As the deadline for withdrawal of Iraqi troops approached, war seemed inevitable.

On February 16, 1991, Iraq fled Kuwait City, forty-two days after the start of the offensive. Kuwait City was liberated on the following day, and

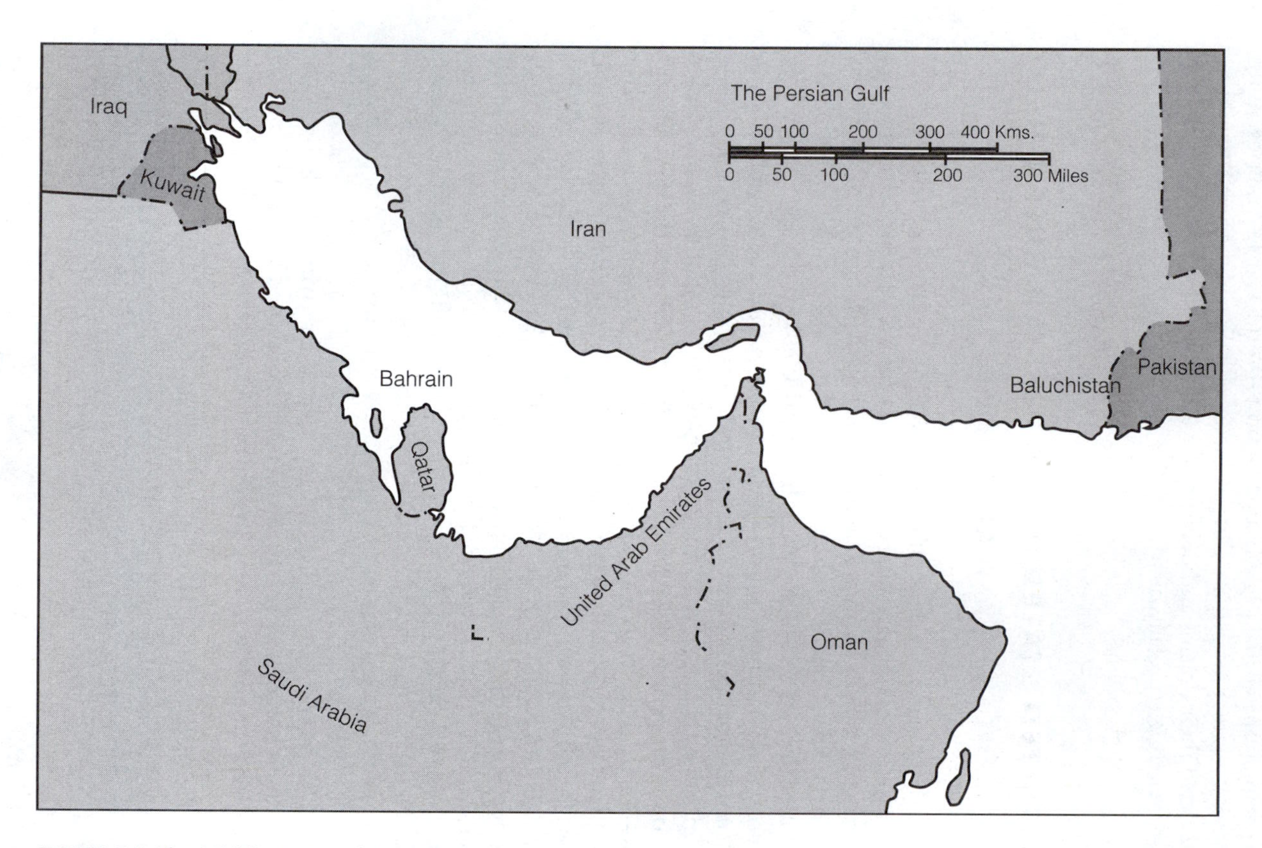

FIGURE 5.7 The Middle East and Persian Gulf

President Bush and his advisors agreed to put a halt to the war. What was the impact of the U.S.-led coalition forces? In June 1991, the United States estimated that more than 100,000 Iraqi soldiers tragically died in combat; 300,000 were wounded; 150,000 had deserted the war; and 60,000 had been taken prisoner. The United States' casualties were far fewer. In total 148 were killed in combat, including 11 women soldiers; 458 were wounded; and 121 were killed in non-hostile acts such as transportation accidents. According to W. J. Perry, a U.S. Secretary of Defense after the war, "These losses were so lopsided—roughly a thousand to one—that there is virtually no historical precedent."[50]

Hussein and his troops' retreat from the war-torn Kuwait was far from passive. Kuwait Oil Company officials, by May, estimated that Hussein's troops exploded 732 Kuwaiti oil wells.[51] Six million barrels of oil per day soaked the Kuwait desert and gushed into the Persian Gulf for months. By November the oil-soaked intertidal beaches covering 400 miles of Kuwait and Saudi Arabia choked highly sensitive marine estuaries. A thick, oily black cloud of soot, sulfur dioxide, nitrous oxide, carbon dioxide, and ozone was so encompassing that the communities over 400 miles down wind of the blown wells had their coolest summer on record.

The tragic nightmare of the war and Hussein's retreat, unfortunately, did not end with one of the world's most extensive air and water pollution incident—an episode not contained until teams of oil fire fighters capped the last blown well in November 1991. Besides roaring fires radiating 2000 degree heat and gushing oil wells, Hussein abandoned land mines, undetonated missiles, and bombs. T. M. Hawley's account of the war's aftermath, *Against the Fires of Hell,* claims that more than 1,300 civilians died or suffered injuries from the debris, including the deaths of 50 ordnance demolition experts. How much explosive debris remains in Kuwait's shifting desert sand remains unknown. The hidden debris will change people's perception of the Kuwaiti desert for decades.

While the final story about the war may be a lopsided victory for the United States and U.S. allies, the tragic conflict was a costly one, nevertheless, involving thousands of lives, $60 billion in war expenses, massive ecological damage to Kuwait and the Persian Gulf, and a recession in the United States linked to higher cost energy supplies during and after the conflict.[52] We cannot know whether this war is a prelude to future conflicts as Middle East oil becomes central to many countries energy mix in the decade ahead. We can, however, look at what radical, managerial, and conservative policy analysts say about this oil-related international conflict, exploring interrelationships between energy and society in the process.

The Radical Perspective on the Gulf War

In the radical perspective on the war, U.S. military intervention in Iraq reflects state dependence on capital in the form of the major oil producers. The state depends on major oil producers for the portion of capital earned in the world exchange of oil, for a steady supply of oil to so many other sectors of the economy, and for an important source of state revenues. Was the Persian Gulf War a war over oil? The radical perspective suggests the war certainly was over a fundamental part of capitalism—oil, "the lifeblood of capitalism."

The central role of a stable supply of oil in the ongoing expansion of global capitalism remains at the heart of the radical analysis of the war. Radical analysts such as James O'Connor insist that, unlike any other natural resource or commodity exchanged in the global economy, oil is capitalism's lifeblood. Oil is essential to military mobility and national economic development—the chief concerns of capital, labor, and state civilian and military officials. As O'Connor writes, "without oil there is no productivity, no markets, no profits, no petrodollars to help to fuel the world's financial system, and no Pentagon. There is no capitalism as we know it."[53] Thus, the intervention in Iraq could have occurred even if U.S. foreign oil imports did not increase from 12 percent in 1970 to 50 percent at the time of the war.[54] In the Marxian instrumentalist account, the state "seeks to promote the global spread of its multinational capital . . . by taking on a key role as the guardians of the conditions of capital accumulation."[55]

Not only is oil the basis of modern capitalism; oil also gives capital the ability to free itself from dependence on the working class. Oil serves as "a powerful weapon against the working class movement" by "liberating capital from its dependence on human labor."[56] The development of plastics, for example, enables capital to be less dependent on strongly unionized metal workers.

Michael Tanzer, in an earlier analysis of the first of the oil shocks, explains the dependence of "home governments" on powerful oil capital. In some instances such as the United Kingdom, the dependence on oil is direct, as the British government has substantial ownership in the British Petroleum Corporation (BP). Governments have a direct interest in a ready and safe supply of oil to energize their military power. Oil is also an important source of voter support in the sense that there is a direct relation between incumbents' political legitimacy and growing manufacturing and consumer sectors nurtured by abundant, inexpensive oil. National oil income, especially in the United States, the home base of major oil producers, plays a vital role with the nation's balance-of-payments in international trade. Finally, there are important interlocking directorates between oil corporations and federal agencies. Exxon, for example, could let a top executive go to serve as a Minister on Foreign Affairs. The Minister, in this capacity, works to maintain good diplomatic relations with major foreign sources of Exxon oil.

In applying eco-Marxism to Iraq's aggression against Kuwait, O'Connor argues that Iraq was forced into a unique capital accumulation strategy in comparison with the other Organization for Petroleum Exporting Countries. For at least the past two decades, most oil exporting countries, ironically, pursued an "abundant supply approach," expanding consumption and reinvesting profits in other phases of petroleum production or other sectors in the global capitalist system.

Iraq, earlier, incurred financial debt and a slower record of expanding their production capacity. A territorial dispute with Iran between 1980 and 1988 perpetuated Iraq's debt and oil production problems. To expand Iraq's petrodollar income, Hussein wanted to restrict the world oil supply. Having helped finance Iraq's war effort against Iran, Kuwait was especially aggressive with the abundant supply approach. Kuwait, accordingly, exceeded their Organization for Petroleum Exporting Countries' quota for several years before Iraq's short-lived seizure of Kuwait's oil. By seizing Kuwait, Iraq would cancel its debt to Kuwait; gain the ability to reduce the world supply of oil; gain control of superior production and refining capacity; and obtain the revenue needed for an even stronger military with the buildup of weapons of mass destruction.

For eco-Marxian scholars such as O'Connor, the Persian Gulf War is another in a series of conflicts that reflect a world capitalist system driven by profit and the treadmill of production rather than basic human needs. Nothing short of socially revolutionary forces, for scholars such as O'Connor, can reverse this treadmill of production involving evermore inputs from the global environment to maintain the growth of profits, employment, and consumption.

O'Connor captures the magnitude of the challenge: "Given the international structure of the ecological crisis and social inequality, especially important is the need to combine or sublate social ecology with issues of economic and social justice."[57] In other words, workers, environmentalists, feminists, and peace activists from both the more- and less-developed countries need to join together and work to achieve social and environmental justice. For O'Connor and other eco-Marxists, only such a vast coalition can be an effective transformative agent, a "red/green coalition" committed to redirecting the world capitalist system responsible for tragedies such as the Persian Gulf War.

Key targets for such transformative agents in the more-developed countries would include the policies or growth paths of the oil-auto-rubber coalition. Schnaiberg, in an early analysis of the first oil shock, suggests a policy where oil sales taxes going into the Highway Trust Fund could be diverted from road building projects to subsidized rail and mass transit systems.[58] This restructuring of U.S. transportation policy would be progressive both in terms of the labor intensity of the proposal and the ecological shift away from the high per capita energy consumption due to our reliance on automobiles. Schnaiberg radically envisions the first oil shock as a result of more-developed countries' vulnerability to political retaliation. More-developed countries' commitments to capital intensive growth in the treadmill of production perpetuates that vulnerability. Schnaiberg's policy analysis, interestingly, is a managerial one. The analysis focuses on the Highway Trust and new ways of managing state transportation policy.

O'Connor's work shows eclecticism similar to that of Schnaiberg. In redirecting the path of global capitalism, O'Connor insists that "highest priority [needs to be given] to dismantling the national security state, now an anomaly even on its own terms."[59] Demilitarization will free up vast resources for more progressive efforts at improving employment opportunities and for mass transit, environmental restoration, better education, and affordable housing. Demilitarization will also eliminate a major structural source of imperialistic action by countries such as the United States to protect the world flow of oil for energizing military action. Such a transformation, for O'Connor, represents "an America committed to peaceful solution to regional and international political conflicts and equitable and rational solutions to the plethora of economic and social problems on the home front."[60]

For Tanzer, the radical transformation of world capitalism must occur both in more- and less-developed countries, a vast, global restructuring consistent with the thinking of O'Connor and Schnaiberg. In more-developed countries such as the United States, there has been "virtually complete failure in the last two decades either to reduce demand sharply by energy conservation or to expand such renewable energy resources as solar power that has heightened the dependence . . . on oil."[61]

The less-developed countries' short-term success of the first oil shock, according to Tanzer, represents a natural experiment in diverting the international flow of wealth from the more-developed counties to the less-developed countries. Whether the less-developed countries can overcome the divide that drove oil prices to record lows in the 1990s remains a critical issue for the potential radical transformation in the distribution of wealth in the world capitalist system.

The Managerial Perspective on the Gulf War

While the war reflects the state as an agent of U.S.-based transnational oil corporations, the confrontation also reflects a significant political action by the Bush administration, late in the president's first term, close to the next presidential election. For students of international relations such as Simon Bromley, the action, which put hundreds of thousands of men and women on both sides of the conflict in danger, was an effort to reassert U.S. global hegemony.[62] Success with the war would not only secure a vitally important natural resource, but also internationally demonstrate U.S. commitment to maintain order among nations in the Middle East.

The timing of the managerial effort in the Gulf was not coincidental. The war occurred in the aftermath of Mikhail Gorbachev's effort to end the Cold War through *detente.* Gorbachev was making unprecedented efforts to improve relations between his government and Western nations such as the United States. Had Gorbachev's effort with detente not existed, President Bush probably would not have initiated the military action against Iraq. The president would have risked an East-West confrontation. The Soviets most likely would come to the assistance of Iraq, whom they assisted in the Iran-Iraq war. As the Persian Gulf crisis unfolded, however, the Soviets supported the UN Security Council's resolution to use NATO forces in the Gulf.

Without the threat of retaliation, President Bush, a former World War II pilot, could not have tackled the "Vietnam Syndrome"—American reluctance to support war efforts in the aftermath of Vietnam.[63] A swift defeat of Iraq's aggression against Kuwait would open doors to future opportunities for the U.S. military bureaucracy to act as a global police force. Expanding such opportunities would provide the rationale for continued expansion of defense-related contracts, a powerful force in building the president's constituencies in Congress and in manufacturing sectors associated with surveillance, aerospace, and arms manufacturing.

In the effort to extend global hegemony, the United States went to war to destroy Iraq's military power. President Bush knew well in advance that Iraq would invade Kuwait. In their meeting on July 25, 1990, U.S. Ambassador April Glaspie implied to Saddam Hussein that in the event that Kuwait and Iraq confronted each other, the United States would not get involved. Greer argues that the Bush administration ignored Iraq's offers to surrender, and would not allow Iraq to initiate a retreat, without being attacked.[64]

Managerial decision making by past administrations also played a role in the war. Thus, energy policy analysts such as Christopher Flavin at the Worldwatch Institute, and Amory and Hunter Lovins at the Rocky Mountain Institute, find the Bush administration's response to the Gulf crisis reflective of a failure by the state to develop an energy management policy, which greatly enhanced U.S. vulnerability to a politically unstable region.[65] Flavin blames the Reagan administration for eliminating government programs aimed at lowering oil dependency in the United States. Domestic oil production in the United States was on the decline during the first half of the 1980s, but that trend turned around after 1985 with an expansion in the supply of oil outside the Middle East and the fall of oil prices. Oil consumption has increased ever since.

We do not want to present radical or managerial perspectives on the war, or U.S. petroleum dependency in general, as competing frameworks. As O'Connor observes, "There was a political logic running more or less parallel with the economic logic, and Bush's political motives (not so) peacefully co-existed with U.S. economic motives."[66] The words "more or less parallel," however, are important in that transnational oil companies might have gained had Hussein succeeded in raising the price of oil.

In the next section we will see that conservatives argue fear of another oil shock was not the cause of the war. Hussein, even if he gained control of the entire Gulf region, would need to sell oil at a relatively low price, as it would remain the key exportable resource in the Gulf. Thus, the argument that the war was driven by an administration that wanted to demonstrate internationally its hegemonic world power gains the upper hand.

Conservative Perspectives on the Gulf War

For the participants in the Cato Institute-sponsored conference, "America in the Gulf: Vital Interests or Pointless Entanglement?," vital U.S. interests were not at stake in this international entanglement. Four arguments suggest the conservative verdict that the war was largely pointless:

1. U.S. petroleum supplies would not be threatened by Iraq's invasion of Kuwait. Oil would remain the major source of Iraq's foreign exchange. Moreover, while Hussein might try to increase the price of oil, he could not exceed twenty dollars per barrel. Leaders of the Organization of Petroleum Exporting Countries know that oil costing more than this amount invites energy producers and consumers to shift to alternatives such as plentiful, clean-burning natural gas.[67]
2. Thwarting Saddam Hussein's international aggression would not bring stability to a historically politically unstable region.
3. The war set a precedent for the United States to act as a global police force. The commitment was an even more encompassing defense-related goal than the one the United States had during the Cold War. In the wake of the Gulf War, any new international disorder could be a cause for U.S. military intervention.
4. U.S. effort to maintain international stability, in the Middle East or elsewhere, undermines the strength of the nation in the global economy. This militarism diverts research and development efforts away from economic and technological innovations that could enhance the United States in international trade and growth.

One conference participant argues that the deployment of U.S. military power against Hussein "was akin to using a sledgehammer to kill a gnat."[68] Iraq's military performance was so anemic that other countries with a really vital interest in the takeover of Kuwait, such as other Arab Middle Eastern countries, certainly had the capacity to contain and reverse the aggression.

Writing in *Atlantic Monthly,* Christopher Layne (1991), a foreign policy analyst for Cato, presents an encompassing conservative analysis of the war. As an alternative to the U.S. military containment action in the future, Layne argues for a balancing strategy analogous to military pluralism that would be in every nation's best interest. Amounting to a kind of Durkheimian leveling out of the global division of military labor, the strategy involves burden-shifting, as opposed to burden-sharing. Other nations could build their military strength with U.S. loans and technical assistance.

Layne argues that the result of such efforts would be U.S.-supported military counter-coalitions that would actually reduce the possibility of the kind of aggression presented by Iraq or other aggressors. Since all nations have an interest in international stability to facilitate trade, the United States should play the role of a facilitator of burden-shifting to help other countries play their part. Accordingly, "the less the United States does militarily, and the less others expect it to do, the more other nations will do to help themselves."[69]

The reverse side of Layne's argument focuses on the question of what happens if the United States continues a foreign policy commitment to the "bound to lead" mentality? Continued U.S. military action, for Layne, creates a self-fulfilling policy. Other nations will let the United States serve as their global police force. As Layne characterizes the problem, "As long as other nations believe that the United States will do the hard work of defending their interests, they are tempted to set back and let it do so."[70] Thus, while other countries invested to enhance their international competitiveness in the global economy, the United States risked human lives at substantial cost to their own economy.

James Phillips, senior policy analyst for the Heritage Foundation, another prestigious institution-builder of the conservative paradigm, also writes policy briefings on the Gulf War crisis and the future implications of the crisis. What makes Phillips unique, though, is the way he molds and crafts his argument with both conservative, free market economic logic, and managerial military policy.

The crux of Phillip's argument is that "America's first line of defense against future oil shortages is the free market, not the armed forces."[71] Since the end of the Cold War, U.S.-Middle East foreign policy changed. The United States no longer has to fear Soviet alliances with anti-Western Persian Gulf states. Thus, countries such as the United States need not be concerned over access to oil within the region. Without a military power in the foreseeable future that can deny the West access to Persian Gulf oil, our interest becomes one of protecting the global free market. Any short-term interruptions in the supply of Persian Gulf oil will lead to higher oil prices worldwide. Higher oil prices, in turn, provide an incentive for greater oil production elsewhere, greater conservation, more efficient consumption, and the use of alternative energy sources such as coal, nuclear power, or natural gas.

Unlike Layne, who favors military defense burden-sharing outside the United States to improve the international economic competitiveness, Phillips advocates bilateral military agreements with Persian Gulf countries friendly to the United States such as Saudi Arabia, Kuwait, or Bahrain. These agreements will enable the United States to store the weaponry needed for rapid deployment if a country such as Iraq threatens a sweeping seizure of oil throughout the Gulf. While oil would still be sold internationally after such a seizure, as others in this discussion recognize, the profits from the sale of oil could be used for an extensive military buildup threatening Western national security interests. For Phillips, guaranteed access to U.S. military equipment in the Gulf represents a strong backup to maintaining a world free market for oil.

Looking Back

Was the Persian Gulf War an international conflict over oil? To answer this important question, we return to an observation made by O'Connor. He writes, "There was a political logic running more or less parallel with the economic logic, and Bush's political motives peacefully coexisted with U.S. economic motives."[72]

The economic motives, of course, involve keeping the very lifeblood of international capitalism—petroleum—freely flowing. That interest was pronounced in the United States after over sixty-five years of growth of the auto-oil-rubber coalition. As both the International Energy Agency and the World Energy Commission point out, however, Europe too was heading toward a point where Middle Eastern oil would be a critical part of the European fuel mix. Thus, the decision to use military force against Iraq was made by the UN Security Council, not just the United States.

O'Connor's "political logic" does refer to efforts by the Bush administration both to demonstrate, internationally, the renewed U.S. military power and to demonstrate U.S. intolerance for Hussein's aggression against a neighboring oil-rich country. The political logic of the war, however, also included international consensus that Kuwait had the right to manage their own country; that many countries had the right to a free, secure supply of oil; and that countries wanted to ensure international free trade. The war in the Persian Gulf was a conflict over oil, over Kuwait's national sovereignty, over global capitalism, and over the management of the international balance of power.

ECOLOGICAL MODERNIZATION THEORY: CLAIMS AND CRITICS

According to some modern social theorists and environmental policy analysts, the impending scarcity of oil, the carbon buildup in the atmosphere, and the potential for global climate change are among the leading ecological problems now facing the world. These problems do not speak well for Western cultural traditions such as the expansion of capitalism, modern transportation systems, and the internationalization of industrial development. These theory and policy analysts foresee the twenty-first century as the period of ecological modernization or the re-rationalization of societies into more ecologically sustainable forms.

Ecological modernization theory, developed by scholars such as Maurie Cohen and Raymond Murphy, identifies the directions that society can progress to protect the environment and continue to expand economic production.[73] Ecological modernization theorists discuss emerging "win-win" social changes that include: (1) cleaner, more efficient technologies and production practices; (2) social and ecological planning; (3) new accounting methods that cause firms to internalize the ecological costs of their production; and (4) government regulations that encourage these changes such as ecological tax reform. Research and writing by Amory and Hunter Lovins as well as Ernst von Weizsacher contributes to the literature on ecological modernization, so we will discuss their main work next.

The Hard and Soft Energy Paths

Physicist Amory Lovins first presented his ideas about ecological modernization (without using that term) in an article, "Energy strategy: The road not taken?" in a 1976 issue of *Foreign Affairs.* Central to this work is a comparison

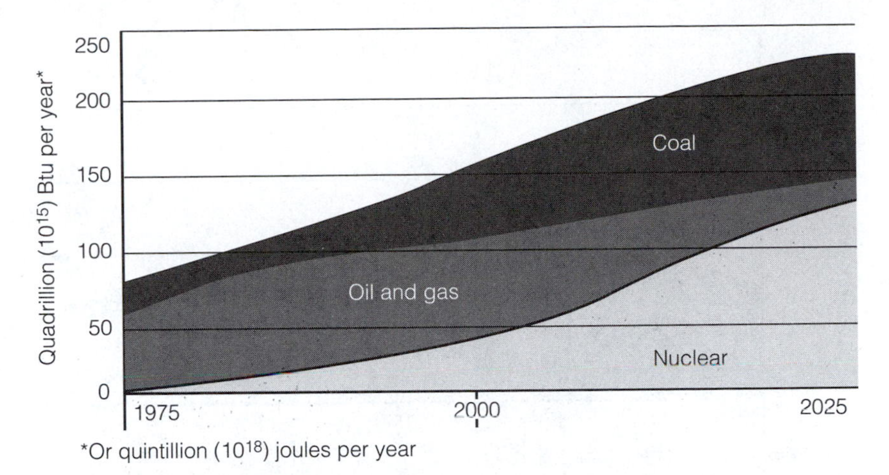

FIGURE 5.8 Lovin's Characterization of the Hard Energy Path: Schematic Future for U.S. Gross Primary Energy Use

SOURCE: Used with permission of Foreign Affairs.

of hard and soft energy paths, with the soft path involving more efficient renewable energy sources, paradoxically "the road (possibly) not taken" in the future of societies such as the United States.

Lovins defines the hard energy path as one that "relies on rapid expansion of high technologies to increase supplies of energy, especially in the form of electricity."[74] As one can see in Figure 5.8, the hard path involves a 300 percent increase in U.S. energy use between 1975 and 2025. Nuclear power and coal become the primary sources of energy, compensating for diminishing sources of oil and natural gas after the year 2000.

Lovins, as a physicist, argues that the pursuit of the hard energy path involves immense inefficiencies with only one-half the energy produced reaching the consumer. One-third of the energy in coal is lost if the mineral is converted into a more transportable, versatile liquid form. Two-thirds of the energy in coal, oil, or natural gas is lost in the production of electricity. There is considerable heat loss as the fuel is burned to spin the turbine generating electricity, and electricity itself is lost as the energy travels through wires for miles.

In addition to substantial problems with efficiency, the hard path, according to Lovins, involves the development of large, centralized bureaucracies to administer the production and the sale of energy. While the economic costs of administering these organizations would be considerable, cost problems are only one of several Lovins identifies as this vast energy production system unfolds. In his own words, "In an electrical world, your lifeline comes not from an understandable neighborhood technology run by people you know who are at your own social level, but rather from an alien, remote, and perhaps humiliatingly uncontrollable technology run by a faraway, bureaucratized, technical elite who have probably never heard of you."[75] Social problems involving conflict between democratic values and the inordinate

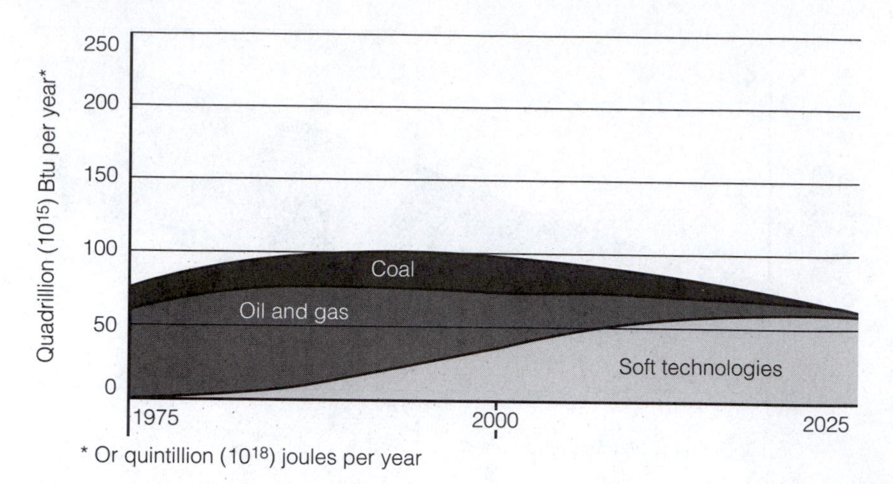

FIGURE 5.9 Lovins's Characterization of the Soft Energy Path: Alternative Schematic Future for U.S. Gross Primary Energy Use

SOURCE: Used with permission of Foreign Affairs.

power of technocratic elites controlling complex nuclear and fossil-fuel fired power generators thus become intertwined with thermodynamic inefficiency on the hard path.

Lovins's user-friendly, sustainable alternative is the soft energy path shows in Figure 5.9. He characterizes this energy mix in two ways. First, the soft path consists of energy flows that are always there such as the sun and wind and vegetation. Second, these forms of energy are diverse with each designed for maximum effectiveness. Home heating, for example, would often be energized by the sun rather than electricity produced by a centralized, coal-fired electrical generating station. The latter, for Lovins, is "like cutting butter with a chainsaw." Alternative, renewable energy sources can provide more efficient, dependable power more appropriately matched in scale and quality to their end uses. Because the energy source is direct, as with solar power heating a house, it is more efficient. With more efficient soft energy the United States could maintain a high standard of living with one-third of the energy required by the hard path.

Factor Four

Most recently, Lovins and his wife, Hunter, working in cooperation with Germany's Ernest von Weizsacher, director of the Wuppertal Institute, published *Factor Four: Doubling Wealth, Halving Resource Use.* In the spirit of ecological modernization, their work focuses on waging a worldwide "efficiency revolution." Historically, they note, production efficiency improved through changes in labor practices: industrialization, automation, and robotics. For von Weizsacher and the Lovins, the new focus of the production efficiency revolution will be gains in the use of natural resources, notably energy. To wage this revolution, they propose harnessing the power of markets through price adjustments to create incentives for technological innovation.

This ecological modernization in fact, is already underway. The U.S. National Association of Regulatory Utility Commissions, for example, changed the rules of operation for state utility commission in 1979. Since that time utility companies demonstrating improvements in the efficiency of electricity use among their customers can take a larger percentage of their sales in profits. This gives utility executives a market-based incentive to produce power more efficiently. The ruling encourages companies such as the vast Pacific Gas and Electric Company to close their building program and concentrate on generating "negawatts" or reduced electricity consumption through a multimillion dollar consumer education and rebate program. Customers buying the most efficient industrial equipment, new buildings, or household appliances receive electricity at lower rates. Then the utility sells the surplus power to new customers, avoiding the costly investment necessary to expand production capacity with new generating stations.

The authors of *Factor Four* cast a wide net, focusing on how the efficiency revolution applies to transportation, design and building methods, natural resources, agriculture, and already mentioned energy production systems. Common to these ways of using energy and natural resources more efficiently is the argument that "in many cases saving resources could cost less than buying and using them."[76] Examples include the Morro Bay, California homebuilding program. Here builders must demonstrate that they reduced water consumption by twice what their next new homes will consume by free installation of water efficient plumbing in already existing homes. Other examples include the use of more costly fluorescent lamps that last ten times longer than incandescent lamps; laptop computers that use 1 percent of the electricity consumed by larger desktop units; and more efficient air conditioning, in part through "superwindows" made to emit light, not heat.

Amory and Hunter Lovins, along with Paul Hawkin, identify Hypercar, Inc. as a voice of the efficiency revolution. The company, incorporated in 1998, consists of automotive engineers, designers, financiers, and an administrative staff based in England. An American branch of the company is located in Basalt, Colorado. Hypercar, Inc. raises private capital, funds research, builds automotive systems and subsystems, produces prototype model cars, and sells technology expertise to car manufacturers worldwide. The hypercar, of course, is the centerpiece of the company's efforts. Capable of making a coast-to-coast trip in the United States on a single tank of fuel, the car represents an excellent example of ecological modernization. The vehicle is shown in Figure 5.10.

The hypercar achieves revolutionary fuel efficiency with a streamlined, "slippery" body that is ultra-light and a hybrid electric drive train. Hypercar engineers make the car light in weight using strong, safe carbon fiber material. A small liquid-fueled engine energizes electric motors mounted adjacent to each tire. "Hypercars and their cousins," according to Lovins, Lovins, and Hawken, "could ultimately save as much oil as OPEC now sells."[77] Without

FIGURE 5.10 An Exemplary Hypercar

the large bank of batteries required in a fully electric vehicle, the hypercar also circumvents the problem of managing the waste buildup of aging batteries that could leak acid into the ground or water.

Critics of Ecological Modernization

While proponents of ecological modernization find numerous examples where increased efficiency and environmental protection lead to profits with improvements in environmental quality, there are leading critics of this perspective as well. The two main lines of criticism are environmental economics, and policy analysis.[78] The environmental economics critic questions the very existence of such win-win situations and the cost of environmental protection. The main line of argumentation is that if opportunities for increased energy efficiency and economic profit exist in tandem, corporate managers probably already seized them. In other words, ecological modernization is like finding a ten dollar bill on the ground. You rarely find the money because someone already seized the prize. Finding energy efficiency opportunities, in a competitive market economy, cannot be otherwise.

Other critics argue that while some win-win situations in the recent past do exist, such as the 3M Company's $1 billion emission reduction that saved them $500 million, most of the accomplishments readily put in place in Western industries have already happened.[79] They note that industrial improvements in efficiency and waste reduction reflect the fact that the percentage of U.S. gross domestic production for meeting environmental regulations went from 0.88 percent in 1972 to a projected 2.42 percent or $200 billion in the year 2000.

They also claim the burden of environmental costs also falls disproportionately on the chemical and petroleum industries. These industries now face serious problems with overproduction and declining market value for their output. To expect significantly more expenditure on environmental quality controls, therefore, would not be realistic.

The policy analysis critique insists that gains made through technological improvements do little to address the core issues of household consumption in the more-developed countries and population growth in the less-developed countries. Instead, ecological modernization and eco-efficiency shift the ways we assess technology, but they do not address the underlying causes of environmental disruption. The substitution of new chemicals for the chlorofluorocarbon in aerosol spray cans, for example, addresses problems with the depletion of the earth's ozone layer; the substitution does not address the growing waste management problem associated with the disposal of the cans, especially their nonbiodegradable plastic parts.[80]

These are important limitations to the logic of ecological modernization. Work of the kind by the Lovins and von Weitzsacher, nonetheless, suggests that major policy analysts see innovative efficiency processes as central to twenty-first century environmental and energy policies. That *Factor Four* is a report to the Club of Rome, a major policymaker for European transnational companies such as Fiat, Volkswagen, and Olivetti, suggests that market-driven environmental policies will play a central role in mainstream environmental policy for years to come.

SUMMARY

Benefitting from the comprehensive work on the International Energy Agency and the World Energy Commission, we can now see a world of nations moving through the first two decades of the twenty-first century taking two kinds of international risk. The first is planetary in scope: the risk of global climate change. With an estimated 50 to 80 percent increase in commercial energy demand to be met with carbon-based fossil fuels such as oil, coal, and natural gas, the build up of carbon in the atmosphere is unavoidable. Whether this carbon build up leads to the kinds of natural disasters such as the hypothetical one at the onset of the chapter or when this development might happen are questions with much less certainty among scientists. The second risk is social and political: the growing dependence on petroleum in the more-developed countries, and, increasingly, the less-developed countries on Middle Eastern oil. What will be the outcome of taking these global risks? Was the Persian Gulf War a precursor for U.S. international policing to keep the lifeblood of capitalism flowing, or will international defense efforts lead to burden-sharing where a number of the more-developed countries build their military capabilities? Burden sharing would create international checks on aggressive action to control a natural resource by a single state in an era of growing international resource scarcity.

Obviously, only the future holds the answer to these fundamental questions in environmental sociology. This chapter, however, uncovered at least two fundamental viewpoints about the consequences of taking these ecological and political risks. Conservatives such as Layne and Phillips see the expansion of free markets and the unfettered international flow of capital as essential to taking these risks. They optimistically endorse the expansion of global free markets in ways that promise to benefit a world population that will increase by 20 to 30 percent by 2020, accelerating the demographic transition in the process.

The growing scarcity of oil, in the conservative approach to these risks, will create the incentives necessary for the process of ecological modernization to occur through soft energy paths with hypercars, increased renewable energy production in the less-developed countries, more thermodynamically efficient buildings, and more extensive use of mass transportation systems. For conservatives, the redistribution of wealth into oil rich nations of the Middle East will provide new markets for a wide range of high technology products from the more-developed countries such as pharmaceutical, electronics, and biomechanical products.

The chapter also has discussed environmental sociologists such as Schnaiberg and O'Connor who blend radical and managerial paradigms in thinking about energy policy. The growing political tensions associated with energy scarcity and the increasing frequency of natural disasters, for these radical scholars, are outcomes of the very processes that conservatives embrace. Easing these global ecological and political tensions will require the emergence of transformative agents such as international coalitions of workers, environmentalists, feminists, and peace activists. These coalitions, activated by the global environmental problems discussed in this chapter, are essential for shifting the managerial policies of the state into a wide variety of forms of ecological modernization. Schnaiberg and other radical environmental sociologists make this argument knowing full well that multinational oil and other energy producers are becoming even more powerful in a world of growing energy scarcity.

This chapter implicitly addresses several of the basic questions raised by environmental sociologists and identified in Chapter 1. What is the importance of population size and population growth for the emergence of an explosive demand for the conventional fuel mix? While it certainly is not the sole source of rising energy demand, the 4 percent annual growth in the less-developed countries' commercial energy demand predicted by the World Energy Commission and the International Energy Agency, twice the world average, does have an important demographic root.

Must the United States and other more-developed countries depend on the less-developed countries for sources of raw materials to ensure the economic survival of the more-developed countries? Focusing just on the United States' growing Middle Eastern oil dependency, the answer clearly would be yes and within ten years. Spatially, the United States population now is widely decentralized. Cultural traditions based in Jeffersonian values are one reason for the U.S. decentralized, fuel-intensive lifestyle. Beyond these values, moreover, this chapter uncovered powerful social forces causing petroleum dependency such as the Highway Lobby. Over a ninety year period in the twentieth century, the Lobby worked to ensure highly lucrative financial futures for oil, rubber, and automotive producers as well as bankers and the real estate establishment. The now expanding Interstate Highway system literally built oil dependence into the geography of the American landscape.

Will resource scarcity such as the growing shortage of oil foster worldwide struggles over access to limited resources as we saw in the case of Brazil and the Mendes tragedy? Will the Persian Gulf War—a struggle over access to a limited natural resource—reoccur? For conservatives, the answer would be no. No matter who controls oil, according to the conservative mentality, the resource will be sold at an affordable price. Exportation is a matter of economic self-interest for nations with a supply of oil. But more was at stake in the war than oil, including President Bush's managerial effort to reassert U.S. military strength and supremacy in the eyes of the world. But the United States also protected or worked to protect stability in access to a natural resource of vital interest to powerful transnationals based in a number of the more-developed countries. Radicals, in that sense, have a convincing argument that capitalism has and will drive worldwide struggles over access to limited natural resources in the years to come.

CITATIONS AND NOTES

1. International Energy Agency, 1996.
2. International Energy Agency, 1996, p. 21.
3. International Energy Agency, 1996, p. 25.
4. World Energy Commission, 1993, p. 94.
5. World Energy Commission, 1993, p. 28.
6. World Energy Commission, 1993, p. 226.
7. World Energy Commission, 1993, p. 229.
8. Yergin, 1991, p. 593.
9. Non-Arab members of the Organization for Petroleum Exporting Countries include Iran, Venezuela, Indonesia, Nigeria, and Ecuador.
10. Flavin and Lennsen, 1994, p. 36.
11. Sampson, 1975, p. 313.
12. Coates, 1991.
13. Yergin, 1991, p. 666.
14. Yergin, 1991, p. 661.
15. Lovins and Lovins, 1991.
16. Luciani, 1989.
17. Luciani, 1989; Adelman, 1995.
18. Luciani, 1989, p. 125.
19. Harper, 1996.
20. Bluestone and Harrison, 1982; Harrison and Bluestone, 1988; Wilson, 1996.
21. U.S. Office of Technology Assessment, 1986, p. 3.
22. International Energy Agency, 1996.
23. Campbell and Leherrere, 1998, p. 81.
24. Campbell and Leherrere, 1998.
25. Yago, 1980.
26. Dewees, 1970, p. 59.
27. Yago, 1980, p. 314.
28. Yago, 1980.
29. Yago, 1980.
30. Dewees, 1970.
31. Kelly, 1993.
32. Hauser and Jaffee, 1947.
33. Maisel, 1953.

34. Gans, 1967.
35. Checkoway, 1984, p. 157.
36. Kelly, 1993.
37. Checkoway, 1984, p. 154.
38. Hawley, 1971, p. 146.
39. Hadden and Barton, 1973.
40. Donaldson, 1969, p. 55.
41. Checkoway, 1984, p. 168.
42. Mollenkopf, 1983.
43. Mollenkopf, 1983, p. 70.
44. Mollenkopf, 1983.
45. Checkoway, 1984.
46. World Resource Institute, 1994, p. 167.
47. Pfiffner, 1993, p. 24.
48. O'Connor, 1998, p. 217.
49. O'Connor, 1998.
50. Perry, 1991, p. 66.
51. Hawley, 1992, p. 14.
52. Moore, 1993.
53. O'Conner, 1998, p. 215.
54. Tanzer, 1991, p. 12.
55. Bromley, 1991, p. 56.
56. O'Connor, 1998, p. 215.
57. O'Connor, 1998, p. 224.
58. Schnaiberg, 1975.
59. O'Connor, 1998, p. 224.
60. O'Connor, 1998, p. 224.
61. Tanzer, 1991, p. 13.
62. Bromley, 1991.
63. O'Connor, 1998; Mayer, 1991.
64. Greer, 1991.
65. Flavin, 1990; Lovins and Lovins, 1991.
66. O'Connor, 1998, p. 219.
67. Henderson, 1991.
68. Carpenter, 1991, p. 2.
69. Layne, 1991, p. 81.
70. Layne, 1991, p. 80.
71. Phillips, 1992, p. 1.
72. O'Connor, 1998, p. 219.
73. Murphy, 1994; Cohen, 1997.
74. Lovins, 1976, p. 65.
75. Lovins, 1976, p. 92.
76. von Weizsacker et al., 1997, p. 146.
77. Lovins, Lovins, and Hawken, 1999, p. 151.
78. Cohen, 1997.
79. Walley and Whitehead, 1994, p. 49.
80. Redclift, 1996.

6

The Environmental Movement

Historic Roots and Current Trends

In mid-1998, the United States Forest Service granted approval to Vail Mountain Ski resort to expand Vail's ski area into backcountry areas that some environmentalists believed to be the habitat for the endangered lynx. Environmental groups fought to prevent this expansion that would require clearing forest for ski trails, roads, and a restaurant. In June, a group of environmentalists, including the Colorado Environmental Coalition, the Wilderness Society, and the Southern Rockies Ecosystem Project, filed a lawsuit in federal court to challenge the Forest Service's approval. A spokesperson for the Coalition argued that the project would ". . . sacrifice the public's land and Eagle County's quality of life to help fatten Vail Associates' bottom line. VA made more money than ever in a below average snow year in 1997–98. Why do they need even more terrain? It's sheer greed."[1]

When the environmental groups lost the lawsuit in October 1998, another environmental group, the Earth Liberation Front (ELF), set a series of fires at the Vail ski resort "on behalf of the lynx" that caused $12 million in damages. ELF stated, "Putting profits ahead of Colorado's wildlife will not be tolerated. This action is just a warning. We will be back if this greedy corporation continues to trespass into wild and unroaded areas. For your safety and convenience, we strongly advise skiers to choose other destinations until Vail cancels its inexcusable plans for expansion."[2] Members from the Coalition group condemned ELF's tactics, saying: "[The arson] is a heinous act. One of our contentions was and still is that the Forest Service has not followed the law in approving this. . . . If we're going to insist that the Forest Service has not followed the law, we're not going to go out and break a different set of laws."[3]

The Vail case raises a number of issues regarding the environmental movement: Are the primary goals of the movement to preserve ecological communities or to protect human economic capacity? What are the most effective tactics to attain these ends? Should the destruction of property or legal battles be the dominant course for environmentalists? Do all environmentalists believe in the same means and ends? These are questions that environmentalists and conservationists have struggled with since the Progressive conservation movement of the late 1800s. The decisions that these and other environmental organizations make about the best ways to solve environmental problems reflect key questions of this book: What are the causes of environmental problems? What are the best ways to protect the environment—changing individuals' behavior, modifying the legal system, or reorganizing social and political structures?

In the early twenty-first century, one has difficulty speaking of "the" environmental movement. Within the United States alone, organizations as diverse as ELF, the Sierra Club, and the Center for Health, Environment, and Justice identify themselves as environmental organizations; however, their interests, composition, tactics, ideologies, and goals differ considerably. If we consider environmentalists outside the United States, such as Chico Mendes and the Brazilian rubber tappers, and the women of India's Chipko movement, one can see that the environmental movement is not a unified whole; the movement represents a spectrum of political interests from the radical, to the managerial, and the conservative.

Despite the movement's diversity, or perhaps because of that diversity, the environmental movement in the United States has significantly altered individuals' lives and social structures. Environmental issues, for example, frame a number of daily decisions—to recycle our cans or toss them in the trash? to purchase recycled or less-expensive virgin paper for our printers? to drive to school using fossil fuels or to walk? In political life, politicians find it important to present themselves as environmentalists. During the 1988 presidential campaign, George Bush proclaimed, "I am an environmentalist"; 1990s Vice President Al Gore authored the bestseller, *The Earth in Balance;* and when Newt Gingrich and the predominantly Republican 104th Congress, who referred to themselves as "browns" (as opposed to environmentalist "greens"), attempted to gut environmental regulations, their efforts did not generate support.[4] In economic life, corporations have had to respond to the government's environmental regulations and to their customers' environmental attitudes. Large corporations with poor environmental records such as Dow Chemical, Chevron, and Waste Management, Inc. hire environmental managers to help them comply with laws and employ public relations firms to portray them as good environmentalists. In intellectual life, the environmental movement spurred the emergence of what we now call environmental sociology.[5] While these examples point to the apparent success of the movement, environmental problems persist. The existence of toxic pollution, acid rain, and tropical deforestation suggest that despite positive environmental accomplishments, the environmental movement continues to face many challenges.

In this chapter, the conceptual tools of social movement theorists will aid us in the examination of the diversity of the environmental movement from its origins in the progressive conservation era to the movement's present status. In this chapter, we analyze (1) social movement concepts, (2) the origins of U.S. environmentalism, (3) the emergence of the contemporary environmental movement, (4) the trends in the organization, composition, tactics, and ideologies of contemporary environmental movement groups, and (5) the outcomes of the environmental movement.

SOCIAL MOVEMENT CONCEPTS

A social movement is "a collectivity acting with some degree of organization and continuity outside of institutional channels for the purpose of promoting or resisting change in the group, society, or world order of which it is a part."[6] In our discussion, we will focus on environmental movement *organizations* as key components of the broader environmental movement.

Mayer Zald and John McCarthy distinguish between social movement organizations and social movement industries. A social movement organization "is a complex, or formal, organization that identifies its goals with the preferences of a social movement . . . and attempts to implement those goals."[7] This is relevant to our discussion because "the" environmental movement in the United States comprises diverse actors belonging to groups whose ideologies differ, yet they all operate under the umbrella of the environmental movement. Diversity within social movements is relatively common; "social movements are rarely unified affairs."[8] This observation leads to the conceptualization of what Zald and McCarthy call a social movement industry. A social movement industry is composed of "All [social movement organizations] that have as their goal the attainment of the broadest preferences of a social movement."[9] Organizations within a social movement industry compete for members, resources, and ideological legitimacy. Each organization wishes their view of the causes and solutions to problems (in this case, environmental problems) to be the legitimate and accepted analysis of the problem.

The organizations in the environmental movement vary in their ideologies, tactics, and goals. Conservative environmental organizations, for example, advocate changes in individuals' behaviors; managerial groups for policy reform; and radicals for social-economic restructuring. The idea of a social movement industry reminds us that not all environmental groups agree on what the environmental problem is or what should be done about the problem.

THE ORIGINS OF U.S. ENVIRONMENTALISM

Contemporary environmental concern can be traced back to seminal events in the 1960s and 1970s—the publication of Rachel Carson's (1962) *Silent Spring,* the first Earth Day (1970), and the United Nations Conference on the Human Environment in Stockholm (1972). While these events reflected an increased environmental awareness in the early 1970s, this was not the first wave of U.S. environmentalism. Americans expressed environmental concern at the turn of the century during the progressive conservation movement. In other parts of the world, this concern was expressed even earlier.[10]

Environmentalism in the early twenty-first century, 1970s, and late 1800s share a number of connections. In this section, we will focus on the progressive conservation movement of the late 1800s and lay the groundwork for showing how four critical elements of the movement carried forth into the contemporary environmental movement. Namely, this era and the contemporary movement are marked by (1) environmental crises that led to the emergence and mobilization of organized environmental groups, (2) lack of agreement within the movements of a single ideology, (3) relative success in gaining legislative changes, and (4) inequitable distributions of environmentalism's negative effects.

Emergence and Mobilization around Crises

"Environmental crises" spurred the emergence of the progressive conservation movement in the late 1800s just as they played a role in the reemergence of an environmental movement in the 1970s. At the turn of the century, one reason for the popularity of progressive conservation was the destructive environmental practices wrought by the captains of industry as they exploited the nation's natural resources for private profit after the Civil War.[11] For example, devastating environmental catastrophes turned public sentiment against logging. The residual bark, branches, and other "slash" that remained after harvesting large stands of trees was a form of pollution that led to a number of devastating fires. Approximately 1,500 people died and 1,300,000 acres of land burned in the Peshtigo, Wisconsin, fire in 1871. A second fire in 1894 at Hinkley, Minnesota, killed 400 people. Other environmental disasters, such as the famous 1889 Johnstown, Pennsylvania, flood, partly reflected clearcutting—a process that does not allow a forest to retain ground and surface water. Such wanton environmental destruction caused broad-based public support for efforts to curb the abuses of private ownership of resources. From the late 1960s to the present, environmental crises such as oil spills, nuclear contamination, toxic waste dumping, and rapid deforestation continue to rouse public concern and provide a basis of support for the environmental movement.

Ideologies, Actors, and Organizations

The progressive conservation movement, like the contemporary movement, was also characterized by disagreement about what should be done about environmental crises. During the progressive conservation movement, preservationists and conservationists debated how resources should be protected and for what ends. Symbolized by John Muir, the first president of the Sierra Club (1892), preservationists favored the protection of undeveloped habitats so the public could gain historic, scientific, and recreational values from natural areas. These areas could undergo ecological succession with minimal human interference. Gifford Pinchot, who would later become the chief of the U.S. Division of Forestry (1898) and the chief of the Forest Service (1905), advocated conservation based on three principles—"development (the use of existing resources for the present generation), the prevention of waste, and the development of natural resources for the many, not the few."[12] Pinchot had a close alliance with the reform-oriented President Theodore Roosevelt who worked with technical resource experts to make conservation, or the "wise use" of resources, a key in his domestic policy.[13] The idea of wise use was to provide the "greatest good for the greatest number." Muir's biocentric philosophy advocated maintaining wilderness in its natural state as a value and goal in itself; Pinchot's anthropocentric philosophy advocated a scientific, rational management of natural resources, specifically forests, for the benefit of humans.

Despite ideological differences, the members of both preservationist and conservationist groups shared an upper-class or upper-middle-class status, were concerned about the rapid destruction of the public domain and private forests, and sought to use the legal and political power of the state to protect forest lands from resource exploitation.

The conservation versus preservation debate persists to this day in conflicts over how national parks and protected areas should be managed. For example, environmentalists (with a preservationist ideology) and industrialists (with a conservationist ideology) are debating whether or not the Arctic National Wildlife Reserve should be opened for oil exploration. Environmentalists from the Wilderness Society oppose exploration, citing concern over species preservation. Industrialists and some Alaskan Congressmen argue for exploration. To support their stance they cite the human benefits of exploration: employment opportunities, profits, and tax revenues. The two sides have different value systems, one biocentric[14] and the other anthropocentric, one expressing the New Environmental Paradigm, the other the Dominant Western Worldview (see Chapter 2). This debate mirrors other ongoing controversies such as the spotted owl controversy in the Pacific Northwest and the debate over whether or not gold should be mined on the government-owned land surrounding Yellowstone National Park. In each of these situations, the value of human benefits (often economic) have been constructed as being at odds with the protection of wilderness habitat. Thus, the Pinchot-Muir debate lives on.

Relative Success

Despite disagreements among the major actors of the progressive conservation movement, their efforts led to the creation of a number of influential organizations and the establishment of legislation and public institutions designed to protect the environment that exist to this day. Private organizations such as the Sierra Club (established 1892) and Audubon Society (1905), for example, are still influential players in the environmental movement a century later, each currently with over half a million members. Governmental structures such as the U.S. Forest Service (established 1905) and the National Park Service (1916) originated in this period, and they continue to manage public lands today.

The earliest legislative landmark in U.S. conservation was the 1864 act of Congress that transferred Yosemite Valley and the Mariposa Redwood Grove to the state of California for public use.[15] This was the first of a number of legislative changes. Shortly thereafter, in 1872, Congress established the first national park, Yellowstone National Park. The aim of protection was both for public enjoyment and scientific research.[16] In 1891, the U.S. Congress passed the Forest Reserve Act, which allowed the president to create forest reserves. The follow-up to the Forest Reserve Act, the 1897 Forest Management Act, was passed to ensure the management of the reserves. The management, regulation, and rational development that the Act advocated reflected Pinchot's utilitarian ideology.[17]

Women of the middle and upper classes were influential in these achievements. One of the successful female leaders of this time, Mrs. Lovell White, was instrumental in having the federal government set aside lands to protect the redwoods of California.[18] Women of this period ". . . not only brought hundreds of local natural areas under legal protection, but also promoted legislation aimed at halting pollution, reforesting watersheds, and preserving endangered species."[19] Upper-class women participated in progressive conservation because of a "trilogy" of values: womanhood or nurturing relationships with others, including nature; commitment to the home that would be more protective for the family with clean air and water; and the protection of children whose future would be protected by natural resource conservation.[20]

By the time that Roosevelt took office in 1901, more than 46 million acres had been placed in reserve, and by the end of his tenure in 1909, more than 150 million acres were included in 159 national forests.[21] These organizations, agencies, and environmental laws have formed the basis of the contemporary environmental movement's work.

Distribution of Impacts

In addition to these specific outcomes, the progressive era had far-reaching consequences for society-environment interactions. The ultimate legacy of the progressive conservation movement was the increased involvement of the government in the economy. The government worked to secure a more stable pattern of economic growth, thus forging one of the links in what Schnaiberg terms "the growth coalition."[22] Although progressive conservation has been depicted as a "grassroots" movement that opposed the power and privilege of large corporations, historian Samuel Hays reminds us that "conservation neither arose from a broad popular outcry, nor centered its fire primarily upon the private corporation. Moreover, corporations often supported conservation policies, while the 'people' just as frequently opposed them."[23] The establishment of organizations such as the Forest Service and the Park Service to promote land reclamation and irrigation actually helped large businesses. At least until after World War II, most of the 180 million acres of mature, public national forests in the West remained intact, thus protecting the market value of privately owned forests for capitalists in the timber industry.[24] The National Park Service helped newly emerging railroads such as the Northern Pacific gain travelers by luring the leisure class away from Europe into posh, rustic hotels beginning with those built by the Park Service in Yellowstone National Park. Thus, progressivism not only contributed to the concentration of wealth, but also set in motion the trajectory of government-augmented economic growth and consequent ecological destruction.

The activists of the progressive era espoused managerial ideologies. They sought to reform the political system so that the federal government could make better decisions regarding natural resources. Specifically, they institutionalized a link between scientific expertise, such as forestry, and the legal and financial power of the state to help rationalize the use of the nation's resources. At that time, a critique regarding the distribution of the costs and benefits of conservation measures was not well developed. However, in the 1980s and 1990s, a critique of the inequitable distribution of the negative effects of environmental policy has emerged, contributing to a radical end to the ideological spectrum of environmentalism. The critique has two elements. First, radicals argue that legislation is not improving the environment. Second, radicals argue that the impacts of legislation are felt most by the poor. "Some environmental policy initiatives unintentionally intensified or displaced problems they were designed to cure."[25] The reason for this is that many environmental laws cut into corporations' profit margins. To maintain profit levels, corporations often use anti-environmental measures, such as lowering costs through ecologically destructive technologies, through more dangerous means of attaining natural resources (for example, offshore drilling), and by moving their operations abroad with environmentalism's "consequent export of environmental degradation and health problems to the Third World."[26] In sum, radicals conclude that the unintended consequences of environmental legislation is that environmental quality is worse now than it was at the beginning of the contemporary environmental movement in the 1960s.[27] Others, like biologist Barry Commoner, agree.[28]

Placing decision-making authority in experts' hands was a goal that managerial actors in the progressive era eagerly pursued. The radical critique of environmentalism questions the ability of the government and other experts to solve environmental problems. Trust in experts has been decreasing through a series of delegitimating events, such as Love Canal, discussed in Chapter 1, and the near meltdown of the Three Mile Island nuclear reactor in 1979. In these cases, citizens lost faith in the government, they questioned the ties between government and industry, and they questioned whether the government was truly for the people. Presently, many grassroots environmental organizations are fighting to keep decision making local, not centralized in a federal authority.

The next section discusses the emergence of the contemporary environmental movement in the late 1960s. We skip over a large section of history from the progressive era to the 1960s. However, a number of significant events took place during this period—the construction of large dams such as the Hoover, the loss of topsoil in the Dustbowl, and growing pollution (smog) in urban areas such as Los Angeles, to name a few. We jump forward because of the notable increase in activism that took place in the late sixties and because most social movement scholarship on environmentalism focuses on the period since this time. An exception is Harry Potter's work that challenges conventional sociological thinking that routinely marks the beginning of the contemporary period of environmental awareness with Earth Day. Potter shows that scientists and scientific organizations were concerned with environmental issues prior to Earth Day and that they promoted legislation to protect wilderness and alleviate pollution.[29] For example, the Wilderness Act, passed in Congress in 1964, played a pivotal role in the growth of public interest in the environment. For more than a decade after Congress passed the Act, people in hundreds of communities participated in public hearings to decide upon protecting roadless areas in national forests and parks.[30] In addition, since the 1950s, there has been significant media attention focused on issues such as endangered species and air pollution that likely contributed to public awareness of environmental issues.[31]

THE EMERGENCE OF THE CONTEMPORARY ENVIRONMENTAL MOVEMENT

Social movement scholars show particular interest in the origins of social movements. What led to the emergence of the civil rights movement? the women's movement? Why did the environmental movement reemerge in the United States in the late 1960s? Researchers suggest a variety of answers to these questions that operate at three levels:

1. Broad macro-level explanations focus on historical changes and discontinuities in social, economic, and demographic structures.[32]
2. Meso-level explanations discuss the importance of political opportunities and organizational resources[33] and isolate specific trigger events such as oil spills and nuclear power plant disasters.[34]
3. Micro-level explanations focus on how individuals and groups interpret or "frame" situations as problematic.[35]

Table 6.1 identifies these three main frameworks for explaining the roots of the movement. The explanations differ in scale and are not mutually exclusive. Together, the range of theoretical explanations suggests that a number of conditions are necessary for a social movement to emerge. Multicausal, multilayered explanations are necessary to understand movement emergence. Each of these three levels will be used to explain the emergence of the contemporary environmental movement.

Table 6.1 The Origins of Contemporary Environmentalism

Macro Level	Broad socioeconomic processes including: • Economic transformations: post–World War II affluence • Technological changes: increased use of chemicals and synthetic products • Demographic shifts: shift from agricultural to industrial occupations
Meso Level	Changes in organizational resources and political opportunity structure including: • Increases in elite support and sponsorship to environmental organizations and events such as congressional support for Earth Day 1970 • Changes in political opportunity structure; for example, educated and activist culture acquired political skills to operate within system Events that "strain" society including: • Santa Barbara oil spill • Love Canal • Three Mile Island
Micro Level	Reinterpretation of understanding of environment including: • Carson's framing of pesticides that shifts their interpretation from a *benefit* to society to a *danger* to society • Media attention to issues such as toxics at Love Canal that instill concerns over environmental issues in everyday citizens

Macro Level

According to macro-level explanations, changes in the United States' social, political, and economic structures since World War II have set the stage for environmentalism. One thesis that receives much attention is Ronald Inglehart's "post-material" thesis. Inglehart argues that broad-scale changes reshape society through generational change. As post–World War II birth cohorts reached adulthood, they brought new values and attitudes into society. They were the first to be raised in suburbs, never to experience the Great Depression or a worldwide war. Instead, they witnessed economic prosperity, rising educational levels, and the power of media. Inglehart suggests that these historical structural changes (what he calls "source level changes") are the cause of a shift from "materialist" to "post-materialist" values. He argues that macro historical changes manifest themselves in changes in individuals' values, beliefs, and actions. A materialist thinker concerns him or herself with strong military defense forces, maintaining political order, a stable economy, economic growth, and lowering prices.[36] The younger generation of post-materialists find greater importance in the concepts of free speech, and of greater input on the job, in the community, and in the government. A description of the Green movement in Germany suggests that "post-materialists stress quality of life issues over economic growth, civil liberties over order, participation in decision making over hierarchy, direct over indirect democracy and decentralized over centralized decision making styles."[37]

The important aspect of Inglehart's thesis for social movements is that the ultimate consequences of the structural changes are new forms of political participation. Source level changes equip people with the skills that are needed to pursue goals linked to post-material values. Specifically, Inglehart argues that post-materialist thinking leads to political action whereby: (1) issues regarding economic needs decrease while the importance of lifestyle issues increases, including the environment, gender roles, and public participation in decision making at all levels; (2) political groups form on the basis of post-material interests, such as environmental concerns, rather than economic or class interests; (3) support for national institutions declines; and (4) demands for political participation increase.[38]

A number of researchers have tested Inglehart's thesis empirically as his argument relates to the environment. Some find support for Inglehart's thesis that "postmaterialists tend to be more concerned about the environment than are respondents who identify as materialist"[39] A common critique of Inglehart's thesis, however, is that his argument only accounts for one type of environmentalist: environmentalists who are educated, economically secure, and belong to national environmental organizations. In a cross-national analysis of opinion polls, Steven Brechin and Willett Kempton found that Inglehart's thesis does not adequately explain global environmentalism.[40] In the past twenty years, many of the activists in the environmental movement have not been the recipients of post–World War II affluence and education, nor do they possess post-material values. Instead, these actors are the recipients of post–World War II pollution and degradation.

An international survey conducted in 1992 on the "Health of the Planet" examines twenty-four nations, including more- and less-developed nations. The survey evidence does not support Inglehart's thesis, either. Individuals in rich and poor countries are concerned about the environment and support environmental protection.[41] Figures 6.1 and 6.2 show that citizens in both industrialized and developing nations have personal concern about the environment and believe that environmental problems in their country are very serious.

Thus, there are environmentalists that do not fit into Inglehart's scheme. For example, his scheme does not explain blue-collar environmentalists who are mobilized by crisis events that affect their material interests, such as the residents of Love Canal who reacted to an immediate problem that threatened their economic and physical well-being.[42] Inglehart's later work takes these criticisms into account.[43] Using evidence from forty-three nations, he found that objective concerns based in real problems such as air pollution coupled with a post-materialist outlook are equally important in explaining support for environmental protection. Values alone cannot explain the existence of

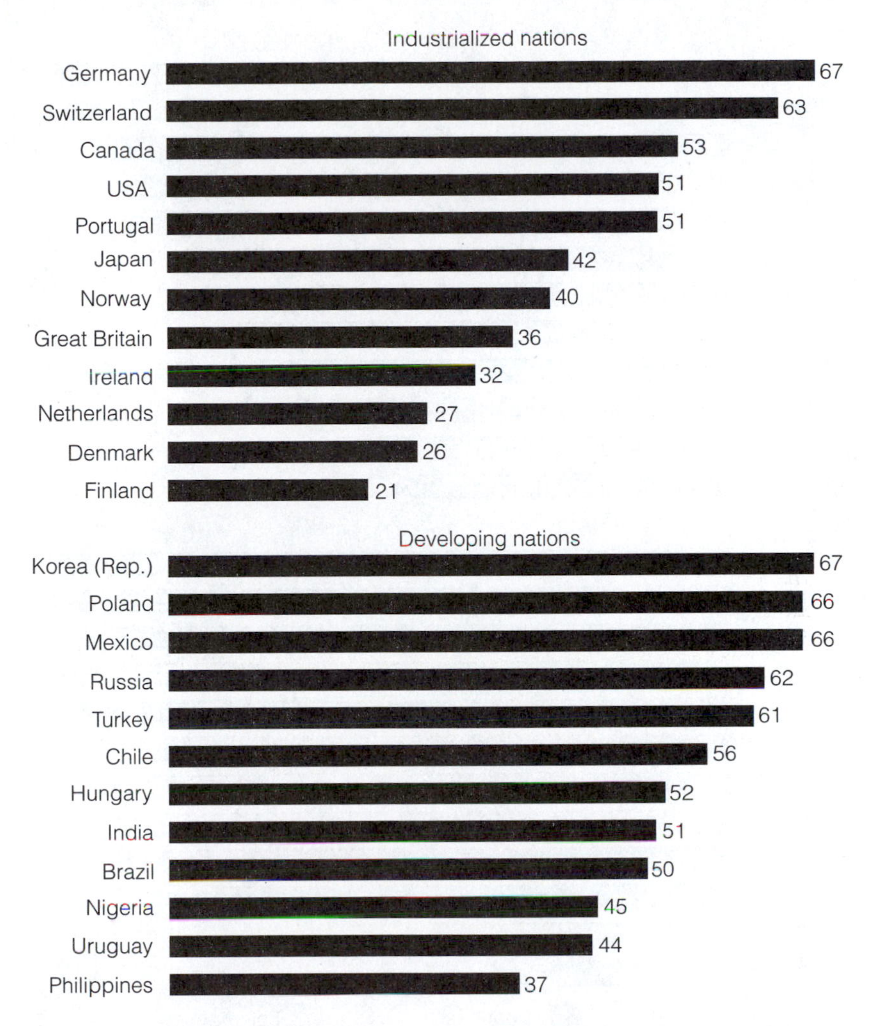

FIGURE 6.1 Seriousness of Environmental Issues—Percent Who Say "Very Serious"

SOURCE: Dunlap, Gallup, and Gallup, 1992. Used by permission of George H. Gallup International Institute.

poor, black environmentalists in the United States or Third World environmentalists who lack formal education. Love Canal is an oft-cited case that is worth explaining.

You may recall from Chapter 1 that citizens in the New York neighborhood of Love Canal organized in response to a collective grievance. Residents suffered from what they perceived was a disproportionately high rate of birth defects, still births, cancer, and other serious child health problems. A study conducted by the state that suggested that toxic contamination from the Hooker Chemical Company could be the cause for these serious child health problems confirmed the residents' fears. At the time, the Niagara Falls School

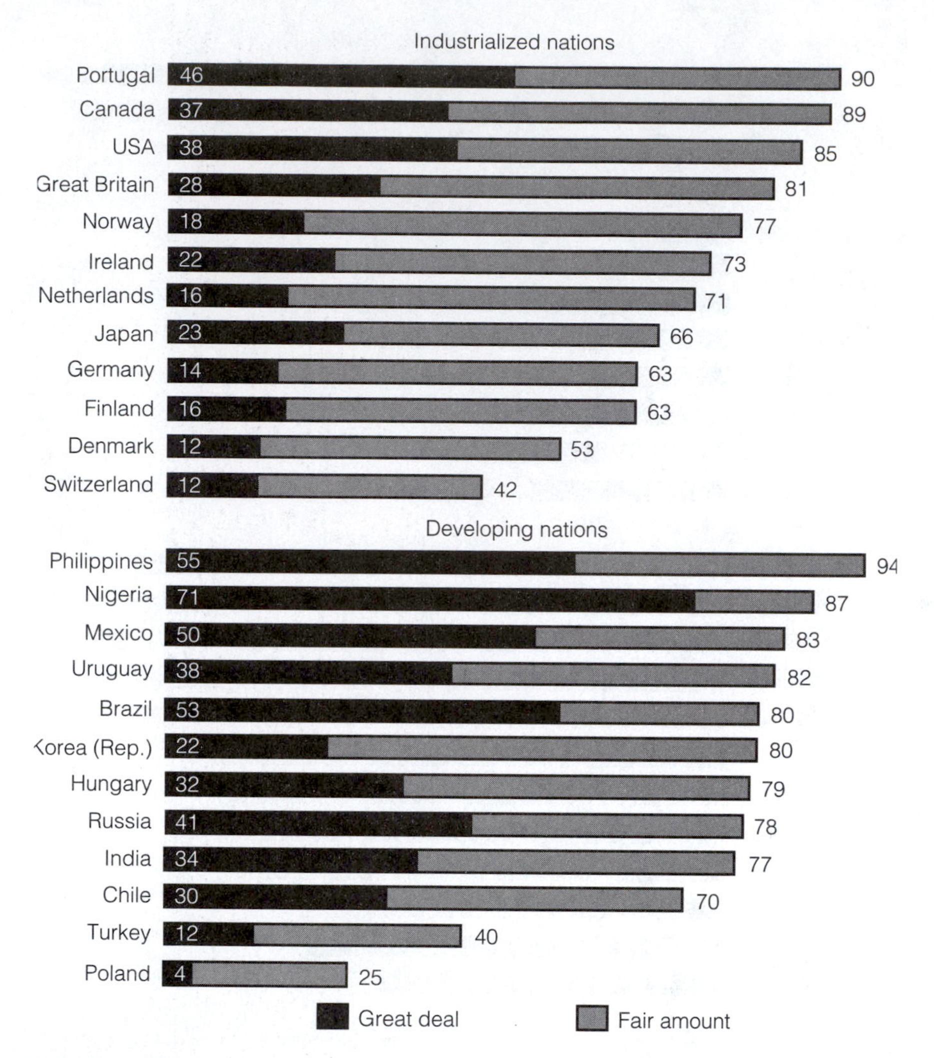

FIGURE 6.2 Personal Concern about the Environment—Percent Who Say a "Great Deal" or "Fair Amount"

SOURCE: Dunlap, Gallup, and Gallup, 1992. Used by permission of George H. Gallup International Institute.

Board owned the land that the chemicals were dumped on. Six hundred families organized an action group called the Love Canal Homeowners Association and began a campaign to have the state buy their homes and relocate their families. The local newspaper, merchants, supermarkets, labor unions, and religious organizations helped to fund the organization's activities and to help families who wanted to move. The state of New York also provided office space for the association.

Members of the organization disagreed about what strategies they should use. Should they use confrontational tactics such as burning houses down or be more moderate? By choosing a certain set of tactics over others, they risked losing members. After two years of political pressure, the group was successful in getting the state and federal government to purchase 700 homes in the contaminated area. This success marked a new period for the environmental movement. "The outcome of the Love Canal story would have been very different for the people, and for our society, had there been no citizens' associations at Love Canal."[44]

Love Canal demonstrates that environmentalism is not solely a cause for society's "post-material" thinkers. The physical effects of environmental pollution can and does affect the lives of working-class people. Love Canal is considered one of the starting points for the environmental justice movement,[45] a movement that will be discussed later in this chapter. The environmentalism of the working classes in the United States is similar to environmentalism in less-developed nations. Many environmentalists in the Third World are not moved to act on environmental issues because of post-material values. They become involved in issues because their land, their air, their water, and their health are being degraded. These environmentalists do not have the luxury of moving beyond "material" concerns to "post-material" concerns.

Inglehart's thesis is a close cousin to "new social movement" approaches that partially characterize "new" movements in relation to shifting values. New social movements (NSMs), including the environmental, peace, and feminist movements, are thought to replace "old" movements such as the labor movement—movements based upon "material" values. NSM theorists suggest that social movement activists are responding to a "contradiction in modernity."[46] For example, in materialist thinking, "the good life" is defined by a consumerist culture led by an economic growth ethic in which more is better. However, unfettered economic growth and technological changes have led to environmental degradation. Post-materialists are reinterpreting what the meaning of the good life is. The NSMs are closely related to what Marxist ecologists call "the second contradiction of capitalism"[47]—the situation in which unbridled capitalism threatens to deplete the very resources essential to continue the expansion of the economy.

NSM reasoning emerged in Europe when class-based (Marxist) theoretical explanations ceased to adequately explain a new style of social movements.[48] Fred Buttel comments on the relevance of NSMs to environmentalism. "What is new about NSMs in general, and the environmental movement in particular, is not that qualitatively new issues are the subject of movement mobilization. Rather these movements are manifestly or latently alternative modes of expression and political mobilization to trade unions and labor and working class parties of social-democratic society."[49] New social movement work has been synthesized into three characteristics that new social movements share: (1) values that are anti-modernistic, (2) action forms that are unconventional, and (3) constituents that are middle class, young, and educated.[50] Aspects of new social movement theorizing are subject to the same critiques as Inglehart's thesis (for example, the theoretical focus on middle-class constituents). In addition, NSM theorizing is criticized extensively on other counts,[51] including the fact that there is little empirical work to support its claims,[52] and because much of the work has been more descriptive than explanatory.

New social movement theorists focus on the macro-historical explanation to explain *why* new forms of collective action are emerging at this time in history. Meso-level theoretical explanations of the emergence of the environmental movement focus on the nuts and bolts of *how* the movement actually emerged.

Meso Level

Meso-level explanations focus largely on the topics of organizations and mobilization. A dominant theory at this level is resource mobilization theory. While limited in terms of explaining why movements emerge, the strength of the resource mobilization approach has been the explanation of the dynamics of movements—how social movement organizations mobilize, why tactical choices are made, and how movements adapt and change.

Resource mobilization theorists argue that a social movement emerges when resources become available to a potential group of actors. A "resource" may be money, media attention, or sponsorship from elites. Craig Jenkins and Charles Perrow argue, "What increases, giving rise to insurgency, is the amount of social resources available to unorganized but aggrieved groups, making it possible to launch an organized demand for change."[53] Some argue that the rising wave of environmentalism in the late 1960s and early 1970s was encouraged by an increase in economic support to new and existing organizations. For example, the founding of the Environmental Defense Fund (1967) and the Natural Resources Defense Council (1970) were both assisted through Ford Foundation Grants[54] and the first Earth Day (1970) was sponsored by a federal grant. In their summary of analyses regarding the emergence of environmentalism, Riley Dunlap and Angela Mertig emphasize that "Many of the existing conservation organizations broadened their focus to encompass a wide range of environmental issues and attracted substantial support from foundations, enabling them to mobilize increased support for environmental causes. In the process, they transformed themselves into environmental organizations."[55]

Resource mobilization theory has two sides. The first relates to the development of organizations and mobilization of an aggrieved population through the addition of resources. The second relates to the social context a movement faces, most notably, the political context. A number of empirical assessments of resource mobilization theory have blended with what is called the "political process model."[56] This approach analyzes the relationship between movement organizations, political elites, and political opportunities, what is collectively termed "political opportunity structure." For example, are political elites and media pro-environment? Are there established rules for how to bring an environmental grievance to policymakers? A nation's political opportunity structure is "comprised of specific configurations of resources, institutional arrangements and historical precedents for social mobilization, which facilitate the development of protest movements in some instances and constrain them in others."[57]

Researchers have explained differences in national social movement organizations' outcomes based on variation in the political opportunity structures they face.[58] In relation to the environmental movement, Dieter Rucht argues that the political opportunity structure was a critical variable in explaining why the environmental movement in the former West Germany was more powerful than the movement in France.[59] Favorable factors attributed to the German success included: federalist structure, the possibility of formal participation in political procedures and litigation, openness toward participatory democracy and political reform, a balance of power between political parties, divided political elites, and alliances. A similar type of analysis by Herbert Kitschelt shows that a state's political opportunity structure helps to explain variation in the impacts that anti-nuclear movements in France, Sweden, the United States, and West Germany have had on energy policies.[60] Researchers also suggest that social movement actors strategically choose when to take actions based upon changes in the political opportunity structure that might weaken traditional alliances and be favorable to movement concerns.

The resource mobilization argument contrasts with an older class of theories that suggests that movements emerge due to a "strain" or grievance in society.[61] Resource mobilization adherents contend that grievances always exist in society; however, resources vary. Therefore, grievances cannot be singled out as the cause of peoples' actions, but resources can.[62] The question of the role of grievances in social movement mobilization leads to two critiques of resource mobilization. First, empirical accounts suggest that grievances do have a mobilizing impact in some instances. For example, Edward Walsh documents the importance of grievances in the establishment of movement mobilization after the Three Mile Island incident (near Harrisburg, Pennsylvania) and argues that the resource mobilization perspective is lacking because the perspective treats grievances (discontent) "as a constant rather than a variable in the emergence of social movement organization."[63] Walsh's critique is well taken since environmental mobilizations, in particular, are catalyzed by environmental crises. Since the 1960s, environmentalists have reacted to crises (grievances), such as fires during the Progressive Era, the oil spills in Santa Barbara in 1969 and in Prince Edward Sound in 1989, and toxic contamination such as Love Canal in 1978.

The second critique asks how, when, and why are grievances translated into mobilization? For example, problems at Love Canal existed for decades prior to their interpretation as problems. Similarly, one can argue that environmental injustice has occurred far longer than the environmental justice movement. Why did these movements organize and mobilize when they did? Micro-level theories address these questions.

Micro Level

A renewed area of interest to social movement scholars is the micro level of analysis. This was brought on by a critique of resource mobilization theory and the realization that grievances are not automatically translated into social movement mobilization.[64] "The resource mobilization approach did not take into account mediating processes through which people attribute meaning to events and interpret situations."[65] How, when, and why do actors interpret grievances as worthy for taking action?

David Snow and Robert Benford's work on frame alignment attempts to explain how social movement organizations "frame" their concerns to match with potential movement participants' interests as a means to mobilize participants to action.[66] An effective frame is one that resonates with potential activists. For example, while individuals may not respond to "frames" that abstractly express environmental problems as distant concerns for future generations, they may be more likely to respond to "frames" that draw connections to their own consequences, such as their ability to eat fish from local waters. Environmental sociologists use frame analysis to examine grassroots environmental action, especially how environmental justice movements frame events as "unjust."[67] This type of analysis fits with a micro-interactionist interpretation and emphasizes how actors interpret events and give meaning to circumstances.

While Snow and his collaborators largely focus on face-to-face micromobilization, others evaluate the role of the media in reframing and reinterpreting social phenomenon, suggesting that the media play a powerful role in the mobilization of social movements.[68] A noteworthy study in this genre is Andrew Szasz's study of the toxic waste movement. He argues that local protest coupled with media attention shaped toxic waste as a national social problem. Szasz's analysis, from the perspective of social problems theories, is that of a social constructionist, another theoretical strain arising from the interactionist tradition. Prior to the mobilization at Love Canal, the public's perception of hazardous waste facilities was not negative.[69] Szasz uncovered a U.S. Environmental Protection Agency national survey taken in 1970. The survey indicated, as you may recall from Chapter 1, Figure 1.3, that the majority of Americans would be willing to live within five miles of a waste disposal facility; just over one-third of them would live within one mile of such a facility.

In the period following the Love Canal episode, according to Szasz, the American public literally began to dread the thought of being exposed to the health risks associated with waste treatment facilities. "Practically overnight, hazardous waste went from being a hazy, poorly organized perceptual object in popular imagination to being the most feared of environmental threats."[70] In a second national survey, this one taken in 1980, the majority of Americans would not live within fifty miles of a hazardous chemical waste facility (see Chapter 1, Figure 1.3). The media attention that Love Canal received effec-

Wargo, "DDT Being Applied to Control Flies," in *Our Children's Toxic Legacy*, 1996. Used by permission of Yale University Press.

FIGURE 6.3 Children Frolic in Mists of DDT Sprayed to Control Biting Flies at Jones Beach, New York, 1945

tively reshaped public perception of hazardous waste. In a broad analysis of print media, Szasz finds that stories of Love Canal were widespread. Stories "conveyed the message that the nation was just waking up to a disaster of unknown dimensions, that this could and would happen to others."[71]

Many argue that other forms of media, in particular, a few popular books, had a tremendous impact on the development of the movement in the late 1960s. Rachel Carson's *Silent Spring,* a bestseller about the dangers of pesticides, is an oft-cited causal force in the emergence of the contemporary environmental movement.[72] Other writers, such as Paul Ehrlich and Barry Commoner, also contributed to reframing ecological issues on the basis of new scientific knowledge. These authors gained notoriety by writing about how the growth in human societies causes an imbalance in natural ecosystems. For example, prior to the publication of Carson's book, most Americans viewed the use of pesticides such as DDT as a safe and beneficial method to control insects.[73] The popularity of DDT's use is traced to the chemical's effectiveness against infectious diseases such as malaria.[74] The government produced film footage of trucks spraying DDT in residential neighborhoods and in public pools full of smiling children to convince the public of the safety of the chemical. The photo in Figure 6.3 of the Jones Beach Authority DDT spray truck discharging the chemical to control mosquitoes while children play at the beach is a telling example of official attitudes toward toxic chemicals in the 1940s and 1950s. Today, this seems appalling to us.

The shift in our perception is largely attributed to the popularization of Carson's ecological analysis of the environmental consequences of the use of toxic substances such as DDT and chlordane. Carson argued that while pesticides, especially in American agriculture, can have short-term benefits for society, their long-term effects include irreversible damage to ecosystems. Carson also suggested that toxic substances have health-impairing effects for human beings. Her analysis attracted attention from diverse interests, including preservationists, hunters and fishermen, public health professionals, the chemical industry, and farmers. While her ideas initially met resistance, especially from industry, *Silent Spring* led to the creation of a presidential advisory panel on pesticides. More importantly, perhaps, Carson's work encouraged the U.S. society to reinterpret the cost-benefit balance of pesticides.

Other books in this period also forced Americans to reassess their relationship to the environment. Less than a decade after *Silent Spring,* biologist Paul Ehrlich published *The Population Bomb,* another bestseller suggesting that world population growth was the cause of a decline in environmental quality and other social problems.[75] Barry Commoner, another populizer of environmental knowledge, published *Science and Survival* and *The Closing Circle.*[76] He issued the warning that the energy-intensive technological advances that many Americans enjoy, such as large automobiles and air conditioners, are the cause of unintended and poorly understood degradation of the environment. Carson, Ehrlich, and Commoner all contributed to reframing the way people looked at the environment.

Perceptions about the environment, nature, and ecology have changed over a longer period of time, as well. While issues of "framing" have been discussed in this section on "micro" changes in individual belief systems, certainly, conceptions of such issues are cultural and occur at institutional levels. For instance, David Frank argues that in the period from 1870 to 1990, "There has been a shift away from conceptions of nature as a realm of chaos and savagery, and away from conceptions of nature as a cornucopia of resources, and toward conception of nature as a universal, life-sustaining 'environment' or ecosystem."[77] The idea of "ecosystem" is now the taken-for-granted understanding of the human-nature relationship.

It is unlikely that the mass mobilization that surged on Earth Day 1970, with the participation of 300,000 people, called "the largest environmental demonstration in history,"[78] would have taken place only from the publication of these books. Social movement theorists enumerate a number of conditions that are necessary for a social movement to emerge: social-structural shifts, changes in the political opportunity structure, resource availability, grievances in society, and the redefinition of situations as problematic. No single event caused the emergence of the environmental movement in the late 1960s, rather a confluence of events culminated to bring about the contemporary environmental movement that persists into the twenty-first century.

TRENDS IN THE CONTEMPORARY ENVIRONMENTAL MOVEMENT

In this section, we will examine four trends in the environmental movement: (1) changes in public opinion, (2) changes in organizational styles and ideologies, (3) the increasing diversity of environmental organizations, and (4) globalization of environmentalism. In 1985, fifteen years after the first Earth Day, the environmental movement could be characterized as increasingly professionalized and institutionalized. In the period from 1985 to the present, these trends are being replaced by a trend toward diversification.[79]

Public Opinion

Since the late 1960s, public interest in environmentalism rose, fell, and rose again. Riley Dunlap, the foremost tracker of these trends, identifies three eras in public concern over environmental quality.[80] The first period, from 1965 to 1970, is a period of rapid development of concern that peaked with Earth Day 1970. For instance, the National Environmental Policy Act was passed by Congress in 1969 and signed into law by President Nixon in 1970. This piece of legislation was the result of twelve years of work by Congress. During the 1970s, the second period, concern declined. Why the downturn? Ironically, because the government took steps during this period to acknowledge that environmental problems needed legislative action; the public was appeased and concern declined. During this period, the government institutionalized environmental concern by passing legislation including the Clean Air Act (1970), the Clean Water Act (1972), the Pesticide Control Act (1972), and the Resource Conservation and Recovery Act (1976) and by establishing the Environmental Protection Agency (EPA) (1970).

The perception that the government was attending to the environment, however, changed in the 1980s, the third period, with the entrance of the Reagan administration. President Reagan appointed James Watt as Secretary to the Interior and Ann Gorsuch to head the EPA. Both appointees attempted to limit government involvement in environmental matters through deregulation. Watt was "the darling of the Sagebrush Rebellion, a movement led by western loggers, miners, and ranchers to sell public land to private interests."[81] Gorsuch reduced the funding for EPA, cut staff, and curtailed enforcement. The public and environmental organizations interpreted these moves as anti-environmental. The environmental backlash against Reagan resulted in the resignations of Watt and Gorsuch. Dunlap argues, "the 1980s saw a significant and steady increase in both public awareness of the seriousness of environmental problems and in support for environmental protection, with the result that by the twentieth anniversary of Earth Day in 1990, public concern for environmental quality reached unprecedented levels."[82] The environmentalism of the 1980s culminated in Earth Day 1990. The event was not sponsored by the government, but by a group of celebrities including Carol Burnett,

Table 6.2 Environmental Organization Memberships, 1960–2000

		MEMBERSHIP (THOUSANDS)					
Organization	Year Founded	1960	1969	1972	1979	1990	2000
Sierra Club	1892	15	83	136	181	560	>600
National Audubon Society	1905	32	120	232	300	600	>500
National Parks & Conservation Assn.	1919	15	43	50	31	100	>400
Izaak Walton League	1922	51	52	56	52	50	>50
The Wilderness Society	1935	(10)	(44)	(51)	48	370	200
National Wildlife Federation	1936	—	(465)	(525)	(784)	975	4,500*
Defenders of Wildlife	1947	—	(12)	(15)	(48)	80	>380
Environmental Defense Fund	1967	—	—	30	45	150	>300
Friends of the Earth	1969	—	—	8	23	30	40†
Natural Resources Defense Council	1970	—	—	6	42	168	>400
Total		123	819	1,109	1,554	3,083	>7,370

*The National Wildlife Federation has 4.5 million "members and supporters."

†Friends of the Earth has "15,000 active members in the United States and over 40,000 supporters worldwide."

NOTE: Data for 1960, 1969, 1972, 1979, and 1990 are taken from Mitchell (1992). Data in parentheses are estimates. Data for 2000 were obtained from organizations' websites and membership departments.

SOURCE: Adapted from Mitchell, 1992. Used by permission of Taylor and Francis.

Chevy Chase, Bette Midler, Burt Reynolds, and others. During the 1980s, membership in environmental organizations grew dramatically. Table 6.2 indicates that the Sierra Club grew from 181,000 members in the late 1970s to 560,000 in 1990; the Wilderness Society grew from 48,000 in 1979 to 370,000 thousand in 1990. Media also helped to fuel the surge in environmentalism. For example, in 1988, rather than name the traditional "Man of the Year," *Time* magazine named the Endangered Earth the "Planet of the Year." Internationally, the United Nations Conference on Environment and Development (popularly known as the "Earth Summit") held in Rio de Janeiro in 1992, marked the attention being given to the environment, not just in the United States, but around the world.

Was the increase in public concern for the environment due to a shift in values toward post-materialism? Research demonstrates that the rise in environmental concern during the period from 1980 to 1990 can be attributed to two types of effects: (1) a "cohort effect," that is, a new generation of young adults with post-materialist values, including more concern for environmental issues (which supports Inglehart's notions discussed earlier), and (2) "period effects," that is, specific events occurring during the decade; for examples the James Watt scandal (1983); the scientific evidence of ozone depletion over Antarctica (1985); the Chernobyl nuclear accident (1986); and the backlash against Reagan's anti-environmentalism.[83] The "period effects" have an important role in sustaining environmental concern. Crisis events at various points over the last quarter century, such as hospital wastes washing up on New Jersey beaches and railcars with toxic chemicals overturning into rivers, keep the environmental movement alive. This interpretation supports Walsh's critique of resource mobilization theory—that suddenly imposed grievances are important mobilizing events. This type of explanation also suggests that it is not simply the construction of environmental issues by the media and by politicians that kept environmentalism alive during this period, but that, in fact, environmental problems are real and they are increasing in "frequency, scale and seriousness."[84]

Students of the environmental movement foresee a fourth wave of environmentalism continuing into the early twenty-first century. The fourth wave builds on the momentum of the third wave in the post–Reagan-Bush era. Fourth wave environmentalism is a double-sided movement. On one side, the wave involves declining or flattening membership in the most popular national environmental organizations of the 1980s, called the Group of Ten, though the absolute number of members in these organizations will still be very substantial. Dowie reports that membership in the Sierra Club declined by 20 percent between 1990 and 1994; the Wilderness Society declined by 30 percent; and international membership in Greenpeace declined by 30 percent to 800,000 members.[85] It is not clear how widespread this has been, however. Between 1990 to 2000 a flattening of memberships occurred for some groups including the Sierra Club, National Audubon Society, the Izaak Walton League, the Wilderness Society, and Friends of the Earth. A few groups have grown during that decade: the National Parks and Conservation Association, Defenders of Wildlife, Environmental Defense Council, and the Natural Resources Defense Council. Though from Table 6.2 it appears that the National Wildlife Federation has grown dramatically to over 4.5 million members, note that the figures for 2000 include members and supporters. If NWF had one million members in 2000, the total number of members for the groups listed on Table 6.2 for 2000 would be 3,870, an increase from 1990 by roughly 800,000 members. The slow down in joining may reflect disillusionment with the national organizations' willingness to take contributions from organizations whose activities in waste management or the financing of polluting industries ironically create the very problems that caused these environmental organizations to soar into the public sphere in the first place.[86]

While there may be a flattening out of membership in national environmental organizations, environmental concern among Americans is still high. In a comparison of the beliefs and values of five groups: sawmill workers, dry cleaners, lay public, members of the Sierra Club, and members of Earth First!, researchers found a high degree of cultural consensus of environmental issues, even among these diverse groups. The difference among the groups was willingness to act. For example, "It is not that Earth First! members have unusual values but that they are willing to make greater personal sacrifices for their environmental values."[87]

People may be taking their environmental concern to other organizations. In the fourth wave, people are joining more specialized groups, groups with different ideologies and tactical choices than the national organization.[88] Today, many organizations are more locally based. There are literally hundreds of local protests over proposed waste treatment facilities in working-class and minority neighborhoods and reservations, and local efforts to protect forests. Such groups include Headwaters Charitable Trust in Clearfield, Pennsylvania, working to protect northern hardwoods and Hoopla Tribal Forestry dedicated to protecting 88,000 acres of forests on the Hoopla's California reservation. There are also thousands of groups around the country working to set up land trusts, reflecting the older causes of wilderness protection. The Land Trust Alliance reports that over 1,200 such groups exist.[89] Some of these new groups also reflect a shift in ideology. For example, working-class, Native-American, and African-American groups working to integrate social concerns with environmental concerns have presented what some call a new paradigm in environmental thought—the Environmental Justice Paradigm, which will be discussed later.[90] Many groups have also formed on college campuses. See Box 6.1 about student environmentalism.

Organizational Professionalization and Managerialist Ideology

While membership in the national organizations may be stagnating, the nationals still represent a formidable social force. The number, scope, and power of environmental organizations grew dramatically over the last quarter century. In the early 1990s, the environmental movement in the United States included 12,000 grassroots groups, 150 formal organizations, and an annual budget of $600 million.[91] Some of the more popular organizations include the World Wildlife Fund, the Environmental Defense Fund, and Greenpeace. A number of influential organizations were established in the progressive era, such as the Sierra Club and the National Audubon Society, while many others emerged in the late 1960s and early 1970s (for examples, Environmental Defense Council in 1967 and Natural Resources Defense Council in 1970).

Like the conservation organizations of the progressive era, contemporary organizations differ in terms of their organizational form, composition, tactics, and ideologies. In this section we will review these differences moving across a spectrum with "professional" organizations on one end and "grassroots" groups on the other end, as shown in Table 6.3.

As the environmental movement evolved since the 1960s, organizations became professionalized. Social movement theorists have a precise structure in mind when they speak of "professional" or "formal" social movement organizations. A distinction is made between "professional movement organizations" and "indigent movement organizations." McCarthy and Zald define

BOX 6.1 Focus on U.S. Environmentalism on Campus

Focus on the United States: Students have played an important role in the development and maintenance of the environmental movement. Student efforts include environmental "teach-ins," work with administrators to make campuses "greener" through recycling and purchasing of recycled products, rallies in support of pro-environmental legislation, and campaigning for corporate accountability, to name a few.

The Student Environmental Action Coalition (SEAC, pronounced "seek") is a network of student organizations founded in 1988. Its goal is: "to uproot environmental injustices through action and education . . . By challenging the power structure which threatens [the environment], . . . SEAC works to create progressive social change on both the local and global levels." SEAC links environmental issues to social issues.

SEAC has fourteen organizational principles:

- Fight environmental degradation.
- Recognize the impact of the environment on human individuals and communities.
- Support human rights.
- Support animal rights.
- Demand corporate responsibility.
- Fight class inequalities.
- Fight racism.
- Fight sexism.
- Fight homophobia and heterosexism.
- Fight imperialism and militarism.
- Have a diverse membership.
- Develop an activist rather than a volunteer approach.
- Link our issues to local, community concerns.
- SEAC National exists to empower the grassroots through training and education. We view national campaigns as one of the tools to accomplish these goals.

In 2000, SEAC was involved in a number of national campaigns. One is a health-related campaign to get dioxin out of tampons. According to SEAC's website, "Feminine hygiene products contain two things that are potentially harmful: dioxin (a by-product of chlorine bleaching) and rayon (used for absorbency)." It recommends a number of actions: changing products, writing letters to newspapers and tampon manufacturers, learning more about the issues, and taking direct action by posting orange stickers warning of the danger of dioxin in tampons (more information is available on SEAC's website).

SEAC is also working on an anti-World Trade Organization (WTO) campaign. SEAC believes that WTO threatens the environment as well as democracy and workers rights. SEAC is part of a Student Coalition for Fair Trade that includes other student activist groups, such as United Students Against Sweatshops. Students in these organizations protested against the WTO in Seattle. In addition to protesting, they advocate organizing locally and education about the WTO. One form of education they promote is "guerilla theatre"—acting out skits in public spaces that portray the fate of various nations in a world governed by the WTO. Their website contains the text for one skit.

SOURCE: *www.seac.org,* 11/6/00.

Table 6.3 Professional Movement Organizations versus Grassroots Movement Organizations

	Professional	Grassroots
Leaders and members	White, educated professional leaders; "paper" members	Local volunteers; includes working class and racial minorities
Ideology	Typically managerial	Varies, some radical
Tactics	Primarily institutionalized tactics including political lobbying, litigation	Multiple tactics including direct action
Key issues	National and international policy	Local health, economic livelihood, local sustainability
Example	Environmental Defense Fund, World Wildlife Fund, Colorado Environmental Coalition	Southside Community Action Association, Chipko movement, Earth Liberation Front
Understanding of causes of environmental problems	Lack of proper policies and lack of effective enforcement	Owners of the means of production and their state partners do not take into account the negative impacts of production on local people

the professional organizations as having full-time leaders, paid staff, paper memberships, and professional representation.[92] These organizations are more centralized, hierarchical, and bureaucratic than indigent organizations. For example, the Environmental Defense Fund (EDF), a professional movement organization, has a full-time and paid staff that does not rely on volunteer time and efforts to accomplish their goals. The EDF is organized like a corporation with presidents, vice presidents, regional specialists, and support staff. The paper members of EDF do not directly engage in environmental action; instead, they send money to the organization that allows EDF to represent their concerns with professionals, including lawyers in court and lobbyists in Congress.

EDF was not always this type of organization.

> The EDF began as a local environmental discussion group, the Brookhaven Town Natural Resources Coalition (BTNRC). . . . Similar to a number of other locally based antipollution or antidevelopment groups that sprang up in middle-class residential communities during the 1960s, the Long Island group addressed a wide range of issues. . . . The DDT issue was of particular importance for the group, given the widespread use of the pesticide throughout Long Island's Suffolk County.[93]

After winning a court case that blocked DDT spraying in Suffolk County, Long Island, the local group asked the National Audubon Society to establish a legal defense fund to continue the type of work they had been engaged in. Audubon's conservative board refused, so the BTNRC formed the Environmental Defense Fund. The Ford Foundation offered funding to the organization. As part of the agreement, the EDF had to clear its actions with Ford.

"Though the review committees were designed to partially shield Ford from EDF actions that might be deemed too radical, it was another step to further professionalize the organization."[94]

Other groups that fit the "professionalized" characterization are the Colorado Environmental Coalition, and the "Group of Ten" that include the Sierra Club, the Wilderness Society, the Audubon Society, Natural Resources Defense Council, National Wildlife Federation, Izaak Walton League, Defenders of Wildlife, Environmental Defense Fund, National Parks and Conservation, and the Environmental Policy Institute. The Group of Ten formed in 1981 as a collaborative strategy to address the threats of the new Reagan administration and eventually moved on to attempt to improve environmental policies. These groups have narrowed their tactical repertoire by institutionalizing themselves within a political sphere that operates using the respected and accepted tactics of social change such as lobbying political leaders (as did Muir and Pinchot).[95] These groups are reformist. "Reformists accept the general framework of an institution or social arrangement, but consider it capable of improvement."[96] The Group of Ten helped shift the movement's battleground to the field of institutional politics.[97] "Social protest action now takes place in Congressional hearings, regulatory commission hearings and reports, environmental impact meetings, and in legislative lobbying and formal litigation."[98] This type of framework is analogous to the "managerial" paradigm. However, not all organizations within the movement are reformist. Groups like Earth First! and the Sea Shepherd Society do not accept the social arrangements related to the environment. Their goals are "revolutionary" as evidenced by Earth First!'s slogan, "Subvert the dominant paradigm!" Whereas reformists wish to improve the system that is in place, "revolutionaries [radicals in our terminology] insist that the [system] must be fundamentally transformed, or replaced."[99]

From a managerial standpoint, professionalization brought the environmental movement closer to meeting its goals. Professionalization of organizations legitimized environmentalism and kept environmental issues on the policy agenda. The Group of Ten tried to make changes from within the system and could only do so by being accepted by those who controlled the policy process.

Despite their success in policy issues, these professional environmental movement organizations have generated a number of criticisms. As part of the environmental movement's quest for legitimacy from the late 1960s to the present, professional organizations entered the political realm. A major change occurred from "participation strategies," which urged people to make changes voluntarily in their consumption habits (a conservative approach) to a power orientation of active pressure to influence public policy (a state-based managerial approach). Here are some of the negative consequences of this professionalization and institutionalization of environmental organizations:

> The necessity of managing large budgets and complicated, multi-faceted programs made possible by the recruitment of large numbers of new members has resulted in managerial skills becoming more highly valued

Table 6.4 Environmental Organizations and Funding Sources

National Wildlife Federation

$63 million; Members 22%; Magazine Subscriptions 15%; Corporate donors: Amoco, ARCO, CocaCola, Dow, Duke Power, DuPont, Exxon, GE, GM, IBM, Mobil, Monsanto, Tenneco, USX (formerly U.S. Steel), Waste Management, Westinghouse, Weyerhauser. Matching grants from Boeing, Chemical Bank, Citibank, Pepsi, the Rockefeller Group, United Technologies.

Audubon Society

$38 million; Members $10 million. Corporate donors include the Rockefeller Brothers Fund, Waste Management Inc., General Electric, GTE, Amoco, Chevron, Dupont, Morgan Guaranty Trust. Donations under $5000 from Dow Chemical, Exxon, Ford, IBM, Coca Cola.

Sierra Club

$19 million, Members 64%. Corporate donors matching gifts programs through which companies match employee contributions include funds from ARCO, British Petroleum, Chemical Bank, Morgan Guaranty Trust, Pepsi, Transamerica, United Technologies, Wells Fargo.

Natural Resources Defense Council

$11 million; Members 40%

Wilderness Society

$9 million; Members 50%; Corporate donors include Morgan Guaranty Trust and Waste Management.

SOURCE: Tokar, Brian, 1997.

> than are the visionary goals of charismatic leaders. . . . Increased professionalization also carries with it the dangers of routinization in advocacy, careerism on the part of staff members, and passivity on the part of volunteers, all of which have been detected in national organizations. . . . An inevitable result has been that the national organizations are increasingly criticized as being bureaucratic and unresponsive to members, too eager to engage in political compromise, and more concerned about their budgets than the state of the environment.[100]

A major criticism is that environmental organizations have compromised or "sold out." This critique has been made of other professional movement organizations in other movements, such as the women's movement. The professional environmental movement organizations have become managerial by choosing to work within the system. For instance, these groups receive much of their funding from the corporations who radicals would argue are the cause of environmental problems, as evident in Table 6.4. This makes it difficult, if not impossible, for the professional groups to oppose the practices of corporate funders. In the process of their own growth, environmental organizations are charged with loosing sight of the grand mission in favor of their mission to simply grow. A second leading criticism is that the large organizations are now unresponsive to local needs and that they privilege national concerns over local ones. For example, "From the perspectives of many grassroots activists, the national organizations still seem to be more interested in protecting threatened animal species from extinction than in protecting children from toxic pollutants in their own backyards."[101] A third charge against these professional organizations is that they are elitist in terms of their composition, ideologies, and the impacts of their reforms.[102] Together, these three critiques characterized the leading environmental movement organizations of the 1980s, which appear to be narrowly focused and elite.

In response to these charges against professional environmental movement organizations and in conjunction with local environmental crises, groups with radical ideologies and tactics, local membership, and diverse race and class membership emerged in the 1980s and 1990s. Examples of these groups will be discussed next.

Increasing Diversity of the Environmental Movement: Radical Actors

One of the better known "radical" environmental "nonorganizations" is Earth First! Earth First! differs in organization, tactics, and ideologies from the managerialist Group of Ten. Earth First! calls itself a movement, not an organization. Earth First! does not attempt to gain members through mass mailings, nor are staff members paid to organize. Earth First!ers are volunteers who engage in direct action in "defense of Mother Earth" including Ghandian civil disobedience (such as chaining oneself to a tree and physically blocking bulldozers) and monkeywrenching. Monkeywrenching, a term popularized by Edward Abbey in his novel *The Monkeywrench Gang,* is "sabotage in the name of the environment, also called 'ecotage.' "[103] *Ecodefense: A Field Guide to Monkeywrenching* describes techniques to defend wilderness including tree spiking, disabling heavy equipment, and burning billboards.[104] (See Figure 6.4.) While monkeywrenching is a noninstitutionalized tactic, Earth First! also uses institutional tactics such as education and letter writing.

Earth First!'s actions are often described as "radical" because they are extreme; however, their ideology is not necessarily radical as we have defined the term in Chapter 1. In popular use, radical denotes noninstitutionalized tactics. In our use, radical refers more to a belief about the causes of environmental degradation. Our definition states, "The radical paradigm rejects the notion that capitalism can be reformed to provide meaningful solutions to environmental problems. Indeed, users of the radical paradigm argue that the very institutions of capitalism, including a captured state apparatus, are the ultimate cause of environmental degradation; an environmentally benign capitalism is thus a contradiction in terms."

Earth First!'s actions are built upon the biocentric ideology of deep ecology. Deep ecology, as discussed in Chapter 2, is a philosophy built upon two core ideas. First is the concept of "ecocentric identification": "Out of identification with forests, rivers, deserts, or mountains comes a kind of solidarity: 'I am the rainforest' or 'I am speaking for this mountain because it is a part of

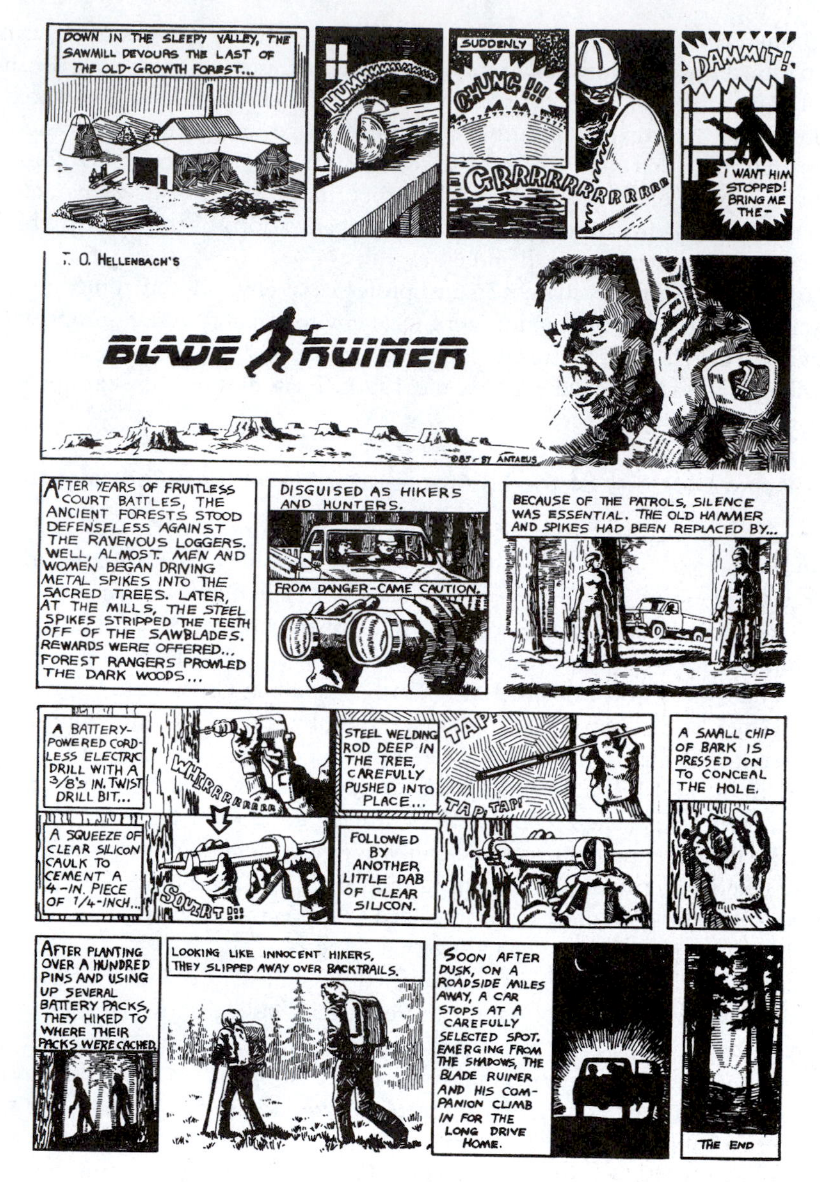

FIGURE 6.4 Blade Ruiner: A Guide to Earth First! Monkeywrenching

SOURCE: Foreman and Haywood, 1987. Used by permission of David Foreman.

me.' "[105] Second is the concept that all life is interconnected and has intrinsic value. "Biocentrism, or ecocentrism . . . suggests that humans are part of the 'web of life'—not at the top of creation but equal with the many other aspects of creation."[106] This ideology can be interpreted as having aspects of conservative, managerial, and radical perspectives, depending upon how individuals' interpret the demands of such a belief system.

In a conservative interpretation, deep ecology requires a reorientation from an anthropocentric worldview to a biocentric worldview based on individual actions, such as those espoused by the proponents of green consumerism and by the voluntary simplicity movement:

> Practicing an ecologically-centered lifestyle, for example, means simplifying one's life, living with minimal impact on the environment. In practical terms, this may mean growing a garden, using public transportation rather than a personal car, being aware of how our choices as consumers affect the environment, and ultimately consuming as little as possible. In essence, deep ecology tells us that we should limit what we take from the environment to that which is necessary to fulfill our 'vital needs' . . . Deep ecologists also call for reductions in the Earth's human population to lessen the strain on non-human nature.[107]

Other aspects of this ideology have managerial aspirations: "The deep, long-range ecology movement is a philosophical movement with implications for personal lifestyles *and public policy.*"[108] Another direction this philosophy can lead is to environmental solutions at a social structural level that include decentralization, small-scale communities, and participatory democracy. At a structural level, "Deep ecologists fault both capitalist and communist societies, based as they are on huge industries producing food for an ever growing population."[109]

Some radical actions may have radical roots. For instance, ELF's denouncement of Vail Associates as greedy [capitalists] and their destruction of private property in the name of the lynx suggests that these deep ecologists are also against the logic of capitalism that structures human greed. The actions of groups like ELF and Earth First! are not able to be placed neatly into one of the three paradigms we consider. They draw upon aspects of the radical, managerial, and conservative understandings of environmental issues. Environmental organizations with similar organizational structures, direct tactics and ideologies include Greenpeace, which recently dumped four tons of Monsanto's genetically engineered soybeans in front of British Prime Minister Tony Blair's home, and The Sea Shepherd Society, an organization that uses its ships to "ram" boats that are fishing illegally.

As we illustrate in the case of the Vail action, many mainstream groups denounce organizations that use radical tactics. Radical groups, nonetheless, serve to legitimize mainstream ones in what is called the "radical flank effect." Organizations with radical demands can aid other organizations within a movement whose reform goals, in comparison, seem mild. The radical goals and tactics of groups like Earth First!, Sea Shepherds, and Greenpeace, for instance, serve to legitimize the goals and tactics of mainstream groups like the Sierra Club, Nature Conservancy, and the Audubon Society. While Earth First! members are spiking trees and chaining themselves to bulldozers shouting, "No compromise for Mother Earth!" Sierra Club's letter-writing campaigns and lobbying appears reasonable, timid, and nonthreatening. The history of organizational development in the environmental movement is that groups factionalize on the bases of these concerns. For example, the Sea Shepherds broke off from Greenpeace because they were not radical enough. The most recent example of groups with radical tactics begetting groups with more radical tactics is ELF, the group responsible for the Vail fires, that broke off from Earth First!

While the ideology of most of the professional movement organizations is managerial, Earth First! illustrates that the environmental movement is not uni-ideological. Other ideologies that are associated with the growing ideological diversity of the environmental movement include ecofeminism, bioregionalism, and social ecology. Bruelle classifies six environmental discourses that include: preservationism, conservationism, ecocentrism, political ecology, deep ecology, and ecofeminism.[110, 111]

Increasing Diversity of the Environmental Movement: Local Actors, Race, and Class Diversity

In the 1980s, grassroots environmental organizations emerged in response to local crises and to the perception that the national, professional environmental organizations were not assisting people at the local level. Many of these groups emerged to resist locally unwanted land uses (LULUs) and took on what has been called a "not in my back yard" (NIMBY) orientation. These volunteer groups tend to be more racially and class diverse than the mainstream environmental organizations.[112] These groups believe in the value of citizen participation in decision making, are concerned about human life, grow to distrust scientific and governmental expertise, and question the benefits of economic growth.[113] This sector of environmentalism is called the environmental justice movement. It combines two major influences: "tough women against toxics" (i.e., Lois Gibbs of Love Canal) and the civil rights movement.[114] The urban public health movement in the earlier part of this century could have formed an earlier environmental justice movement, but the health movement was not framed in this way.[115]

The Southside Community Action Association (SCAA), based in Columbus, Ohio, is another example of an environmental justice organization. SCAA formed in 1984 in response to a crisis. Georgia-Pacific, a forest products and chemical company that manufactures resins, is located adjacent to the community. In 1984, Georgia-Pacific had an explosion at the Columbus plant that blanketed the community in a chemical downpour. Roberta Booth, leader of the SCAA, recalls, "When we drove home that day, there was a detour, but no one was telling us what was wrong. On the six o'clock news, they advised everyone to stay inside. There were people in space suits walking around. We were angry. We have a Health Department and an EPA that wouldn't tell us what we were exposed to."[116] What has ensued is a sixteen year long attempt by this group to force Georgia Pacific to either "get out and clean up" or relocate the 1,600 families in the community. Since 1984, fifteen additional explosions have taken place at the plant. One of the latest occurred in 1997 when an 8,500 gallon tank exploded and damaged the plant and houses in the area, and killed an employee. Residents are also concerned about the high cancer rate in their area, the highest in Franklin County. Booth feels that Georgia-Pacific chose to locate on the Southside because the community is low income and primarily black. Racial minorities make up 77 percent of the neighbors surrounding the plant. Georgia-Pacific moved to the area in 1970, well after the residential area was established. SCAA has protested, written letters, and pursued legal channels to get Georgia-Pacific out of their neighborhood, but sixteen years later, they are still fighting.

Local groups like SCAA emerge to combat what is called "environmental racism." "Environmental racism refers to any policy, practice, or directive that differentially affects or disadvantages (whether intended or unintended) individuals, groups, or communities based on race or color."[117] One of the first instances of mobilization against environmental racism occurred in 1982 when the mostly African-American community of Afton in Warren County, North Carolina, protested against the disposal of PCB-contaminated waste in their community.[118] Since then, numerous campaigns have taken place.[119] For example, in Kettleman City, California, a community made up of 95 percent Mexican-Americans, residents are fighting against the citing of a hazardous waste incinerator; in North Carolina, poor blacks and whites are working with the NAACP to limit corporate hog farms adjacent to their homes.

Some argue that these new groups emerged onto the environmental scene because racial minorities are at a greater risk of living in areas with environmental problems that threaten their health and quality of life. A government study, for instance, showed that minorities and the poor are more likely to live near hazardous waste landfills than other social groups.[120] Some argue that this is not coincidental, but, in fact, the siting of LULUs follows the path of least resistance, which leads to a place-in-blacks'-backyard (PIBBY) principle.[121]

Other research, however, suggests that there is no correlation between negative environmental sites, such as hazardous waste facilities, and areas with high proportions of minorities.[122] The economic geography of dumping involves a number of critical factors and understanding the history of sites is important. For instance, unlike the case of SCAA, in some cases, hazardous environmental sites existed prior to residential areas. In some areas, toxic dumps were located on the periphery of cities. Then, suburbanization occurred. The most inexpensive housing was found next to dumps. Minorities and poor people moved into these areas. Early studies of geographies of environmental injustices focused on points in time. More recent work has become increasingly sophisticated. David Pellow argues that it is critical to look at "environmental inequality formation."[123] He argues that research in environmental justice must take into account the process and history of events. Szasz and Meuser conducted a historical case study of Santa Clara County, California (which includes Silicon valley), to understand how environmental inequalities were created. They found that there was not an intentional racist decision-making process that occurred. Instead, the inequalities "were the result of the combination of several 'normal' processes: economic boosterism, unregulated development, and racial and ethnic differences in education, occupation, and income."[124]

Environmental activists who emerge because of local environmental justice issues usually begin their quest for justice through narrow single-issue campaigns and then shift to a broader social critique.[125] For example, many NIMBY movements lead to "not in *anybody's* backyard" movements. The Love Canal Homeowners Association began as a single issue organization—getting Homeowners relocated. From that group, Lois Gibbs formed the Citizens' Clearinghouse for Hazardous Waste, an organization that assisted grassroots groups around the country. Their newsletters is called Everyone's Backyard. Recently, the Citizens' Clearinghouse has been renamed the Center for Health, Environment, and Justice (CHEJ). It does not take corporate or government donations. According to its website:

> CHEJ believes in environmental justice, the principle that people have the right to a clean and healthy environment regardless of their race or economic standing. Our experience has shown that the most effective way to win environmental justice is from the bottom up through community organizing and empowerment. When local citizens come together and take an organized, unified stand, they can hold industry and government accountable and work towards a healthy, environmentally sustainable future.

Dorecta Taylor argues that environmental justice thought represents a paradigm shift away from the New Environmental Paradigm (NEP) that has been followed by mainstream, national environmental organizations toward an Environmental Justice Paradigm (EJP). There is overlap between these two paradigms. However, Taylor argues that despite NEP rhetoric about supporting a view that all are interconnected in the web of life, that NEP groups still hold social problems as separate from environment problems. The environmental justice paradigm integrates social concerns of racism, classism, and sexism with environmental concerns (see Table 6.5). Proponents of EJP emerged from existing church, community, and social justice groups. In contrast to middle-class whites, Taylor argues that people of color took a different pathway: "They linked social justice concerns like self-determination, sovereignty, human rights, social inequality, access to natural resources, and disproportionate impacts of environmental hazards with traditional working-class environmental concerns like worker rights and worker health and safety to develop an environmental justice agenda."[126]

Aspects of the environmental justice movement are radical. A number of researchers argue that this wing of the environmental movement may be the basis for a progressive, radical critique of society. Szasz, for example, argues that the toxic waste movement and the movement for environmental justice is creating a new kind of environmentalism—"radical environmental populism."[127] He explains how the radicalizing process takes place:

> When the problem was conceived of as "our contaminated community," the cause was "that landfill" or "that careless chemical firm." When the problem was conceived of as hazardous industrial waste, generally, the

Table 6.5 Comparisons of the Exploitative Capitalist Paradigm (ECP), the New Environmental Paradigm (NEP) and the Environmental Justice Paradigm (EJP)

Characteristics	ECP	NEP	EJP
Valuation of nature—nature has intrinsic value, environmental protection over economic growth	No	Yes	Yes
Generalized compassion to other species, other people, other generations	No	Yes	Yes
Environmental planning and risk avoidance—support government regulations to protect humans/nature, emphasize the development of safe technology	No	Yes	Yes
Limits to growth—limited resources, conserve resources, limit consumption	No	Yes	Yes
Completely new society—openness and participation, emphasis on public goods, cooperation, postmaterial values, simple lifestyles, job satisfaction	No	Yes	Yes
New politics—citizen participation, discussions about human relationship with nature and about the management of the economy	No	Yes	Yes
Spirituality—religion and religious institutions incorporated	No	No	Yes
Biocentrism—emphasize animal rights, animal liberation, vegetarianism and veganism	No	Yes, limited	Yes, limited
Environmental Justice—emphasis on limiting racism, sexism, classism	No	No	Yes
Environmental Rights—right to clean air, land, water, food, safe and healthy work environment	No	No	Yes
Autonomy, self-determination—recognize native people's treaties and all people's rights to self-determination	No	No	Yes
Cultural diversity—respect and celebrate other's culture and language, honor cultural integrity of all communities, respect other's belief system in the natural world	No	Limited	Yes
Corporate-community relations—producers held accountable, compensation for those harmed by toxins, consumer protection, waste reduction and elimination	No	Yes	Yes
Oppose military occupation, repression, and exploitation of land	No	No position	Yes
Oppose experimental reproductive and medical procedures on humans	No	No position	Yes

SOURCE: Adapted from Taylor, 2000.

cause of the problem was Waste Management, Inc., Browning-Ferris, the disposal industry, polluting firms, do-nothing state and federal officials. As the movement grew and addressed an ever larger set of problems, the cause came to be defined very broadly, in terms of a whole system of technology and chemical production, driven by profit, unchecked by a government that serves private wealth rather than public interest.[128]

Daniel Faber sees the goals of the environmental justice movement being a form of "ecological democracy." He characterizes the movement in this way:

> Linking struggles for social/economic justice and environmental quality and invoking the direct-action tactics typical of local initiatives, these advocates of environmental justice are addressing issues of ecological racism, the economic and environmental exploitation of working people, and unequal capitalist political-economic power and control over community planning and natural resource development.[129]

As discussed in Chapter 2, the logic behind the concept of the growth coalition or the treadmill of production (a coalition between the state, corporations, and labor for the goal of economic growth) makes it difficult for environmentalists and the working class to form alliances. The growth coalition expresses environmental concern in a "jobs versus the environment" frame. Given this structural configuration and the movement's prior race and class elitism, this new alliance must overcome a number of barriers. However, it can be done, as was the case in Gary, Indiana, where a coalition was formed across race and class boundaries in the late 1960s and early 1970s to successfully force U.S. Steel to limit air emissions, a few decades before the concept of environmental justice or environmental racism emerged.[130] Local concern over health concerns, as in Love Canal and Three Mile Island, can generate fear in communities, and start a new period of reframing environmental concerns.[131]

Since the 1980s, new environmental organizations have attracted individuals from groups that have not formerly been attracted to an environmental movement, such as the working class and African Americans. The environmental justice movement has the potential to forge links between environmentalists, civil rights activists, and other groups concerned with other social issues. These are also some of the concerns of outside the United States, particularly in less-developed countries.

Globalization of Environmentalism

North Americans are not alone in their environmental concern. The environmental movement has grown to a global level. "Whether rich or poor, industrial or agrarian, authoritarian or democratic, socialist or capitalist, almost every society [feels] compelled to reassess its attitudes toward resource management and the condition of the human environment."[132] We see this in the Green Parties of Western Europe and Scandinavia; the environmental movement in Japan; the resistance to un-ecological development in Central America, Asia, and Africa;[133] and broad participation in the United Nations' 1992 Earth Summit.

This discussion of the globalization of environmentalism will focus on the "Third World," also called the "South" or less-developed countries (LDCs). A number of authors suggest that Third World environmentalism is a different brand of environmentalism than that expressed in the wealthy First World.[134] For example, Juan Martínez-Alier suggests that Inglehart's "post-material" thesis for explaining environmental concern does not explain environmentalism in the South.[135] The main difference between environmentalism of the North and environmentalism of the South is that Northern environmentalists are more concerned with preserving recreation areas and Southern environmentalists are engaged in "livelihood struggles."[136] "Most commonly, [Southern] groups have their genesis in the survival efforts of persons and communities living at the margins of existence, especially peasants and indigenous peoples in rural areas."[137] There are a number of examples of "poor" environmentalism in the South that arise from threats to economic survival. Some of the most popular of these movements are the rubber tappers movement in Brazil, discussed in Chapter 1, the Chipko movement in India, discussed in Chapter 2, the Green Belt movement in Kenya, and movements by indigenous groups in Borneo to halt logging.[138] Like the "new" environmentalists in the United States that consist of working-class and racial minorities, at the global level, environmental concern is not limited to elites.

An example of Southern environmentalism is the case of the indigenous U'wa people of Colombia struggling against California-based Occidental Petroleum and their partner, Royal Dutch Shell, ongoing since 1997.[139] The government of Colombia granted these corporations the right to drill for oil in the U'wa's migratory territory. Five thousand U'wa threatened to commit mass suicide if oil exploration continued. The threat stopped the companies, at least temporarily.

One of the reasons that the U'wa achieved partial success is that they have organized internationally. For social movements, this is a tactical innovation that is made possible through communication technologies such as faxes and the Internet. Local environmental actors are able to mobilize transnational environmental support. The U'wa Defense Project brings together a number of organizations around the world including: Amazon Watch, Center for Justice and International Law, FIAN Germany, National Indigenous Organization of Colombia, and Rainforest Action Network.[140] One observer argues,

> Before its debut on the internet, the U'wa struggle had already seen the formation of a solidarity committee of environmentalists and indigenous rights activists in Colombia, failed negotiations with the government and the oil company, debates on the militarization and inevitable degree of violence that the proposed oil exploration would entail, and some degree of mobilization by the U'wa themselves. As a result of the internet postings, the U'wa struggle branched out in many directions—from lengthy newspaper articles in the world press that highlighted the U'wa's alleged traditional nonviolence and ecological knowledge, to the establishment of international support groups. Adopted by several international NGOs [nongovernmental organizations], the U'wa's struggle spread spatially and socially in unexpected directions. This included international travel by U'wa leaders themselves . . . to disseminate knowledge of their struggle and gather support for it. They arrived with their concerns even at the door of Occidental's headquarters in Los Angeles, with the support of a transnational U'wa defense project.[141]

Other indigenous groups threatened by the effects of the search for resources (often for use in the North) have also drawn upon North American environmentalists' support. For example, Ecuador's indigenous Huaorani allied with the National Resources Defense Council in their ongoing battles against U.S.-based Texaco.[142] Chico Mendes and his group allied with Environmental Defense Fund to create extractive reserves. In India, local groups fighting against dams were assisted by Northern nongovernmental organizations.[143]

The U'wa's biocentric conception of nature contrasts with the anthropocentric view of the government and the petroleum companies. One of the U'wa's leaders, Berito Kuwar U'wa, explains:

> We look at the way that life works and the way life is interrelated . . . The money that the U'wa have is the Earth. Everything that we make, that we sow, that we grow, we also consume in our community. We don't have to sell it, and we don't have to buy things. As we say, "The sun is the money." The Earth is also money—it's our gold. The water is also our gold. That's what we value . . . There are many laws in the world, but no one thinks to protect Mother Earth. But I think that if the petroleum companies continue to exploit the petroleum, they will take all of the strength and spirit out of Mother Earth. If they do this, if they take it all, then we're all going to die. That's why I said to one of these petroleum men, "Take all of that money you make and stuff it into the Earth, and see if it sustains life. That money won't sustain anyone."[144]

The resistance of the U'wa to transnational corporations and their government has come at a price. U'wa leaders have been threatened by masked gunmen to sign agreements with the petroleum companies, and three human rights activists from the United States were murdered in the region. This is not the first time such actions have taken place. For example, in Nigeria, Ken Saro-Wiwa fought against Dutch Shell's petroleum extraction on the lands of the indigenous Ogoni. Saro-Wiwa and eight other activists were tried for murders against their own people (crimes the international community believes they did not commit) and executed by the government.[145]

Environmental resistance to development tends to be radical, as we use the term throughout the book, in that it questions conversion of communal land to private ownership and elite desires for industrial development.[146] In addition, "The perception (usually accurate) within popular environmental movements [in the South] is that the land is being exploited for and by outsiders—either multinational commercial interests, or more commonly, commercial elites within the nation in question—interested in quick profits and not the ecologically sustainable use of land."[147]

Recent scholarship has linked movements such as these to the environmental justice framework and to human rights. For instance, Francis Adeola argues that the Ogoni people of Nigeria suffered both environmental injustice and human rights abuses.[148] If flows of hazardous materials follow the path of least resistance, and move from countries in the North to countries in the South, this is a global problem of environmental injustice.

As noted earlier, globalization of movements such as the U'wa's can bring in actors outside of the conflict, such as Northern nongovernmental organizations, that can use their power to alter the decision-making process. Thus, alliances between North and South can be fruitful. However, analysts of globalization ask to what degree the globalization of the environmental movement contributes to the homogenization of environmental concerns. Some evidence suggests that Northern environmentalists would like their brand of environmentalism to be the dominant one; however, Northern interpretations of environmentalism have also had negative effects in the South. Two examples illustrate these negative effects.

First, in the 1960s and 1970s, Southern countries developed national parks and protected areas following the United States' model and with the help from funding through agencies like the United States Agency for International Development and the United Nations Food and Agriculture Organization. By definition, parks excluded human inhabitants. In 1962 the Ugandan government established Kidepo National Park in an area where the nomadic Ik tribe dwelled. Ugandan officials relocated the Ik, banned hunting, and essentially destroyed the Ik.[149] In the U.S. understanding of land protection, "culture" was understood as separate from "nature." This idea manifested itself in uninhabited parks. This is not how the Ik had constructed their understanding of nature. More recent national park projects, often labeled "integrated conservation and development projects," attempt to avoid the nature-culture dichotomy by permitting people to live on the edges of parks and extract resources "sustainably."[150] The theory behind these projects is that local people are best suited to protect biodiversity when they are also permitted to use the fruits of biodiversity to survive simultaneously. The motivation for this reconception of national parks is partially built upon alternative local conceptions of the nature-culture relationships. In this way, Northern prescriptions for environmentalism such as the creation of parks without people has not homogenized Southern environmental actions. Instead, in the South, Northern ideas have been transformed to meet local needs.[151]

The second example shows how groups in the South resist the globalization of Northern conceptions of environmentalism.[152] In 1995, researchers at the Charles Darwin Research Station in Ecuador's Galapagos Islands were rounded up by fishermen armed with machetes and held hostage for four days. Why? Fishermen were angry that the government, at the recommendation from Northern environmentalists and scientists working at the research station, was trying to keep the fishermen from continuing to harvest sea cucumbers from the Galapagos Marine Reserve. Sea cucumbers are a delicacy in Japan and have a high market value. The allowable catch set by the government is half a million per season. At the time of the takeover, at least 7 million sea cucumbers had been harvested. "Gringo" [white, Northern] professionals convinced the government to cut the fishing season short, and the fishermen reacted against what they believed was unnecessary protection that limited their ability to economically survive. The costs of environmental protection were falling on the fishermen. They did not feel they had input in the process. Recommendations for environmental protection were perceived to come from the top down, from North to South. A common critique of the Northern brand of environmentalism is that it does not take local populations into account.[153]

The globalization of environmentalism has not been generic. Around the world, actors respond to environmental crises and organize. The increasing diversity that is becoming evident in U.S.-based environmental groups is a subset of what is occurring around the globe. Class divisions between environmentalists in the United States are also apparent in interactions between environmentalists in the North and environmentalists in the South. For effective coalitions to form, the differing interests of environmentalists (North versus South and rich versus poor) need to be recognized.

Finally, it would be inaccurate to speak only of this globalization of environmentalism in the last quarter century. International environmental organizations have existed since the late 1800s and the numbers of such organizations has grown tremendously over the last century.[154] These include organizations such as the International Friends of Nature, established in 1895, and more recently, the International Society for Environmental Ethics, founded in 1989.[155] The number of these international groups working for social change, "transnational environmental movement organizations," has increased dramatically in the latter part of the century. In 1973 there were ten such organizations; in 1993, there were ninety.[156]

SUMMARY: THE OUTCOMES OF THE ENVIRONMENTAL MOVEMENT

Has the contemporary environmental movement been a success? Yes and no. The answer to this question depends upon how we define and evaluate success. The success of social movement organizations has been measured in a number of ways: (1) by membership growth, (2) organizational survival and longevity, (3) attainment of goals, (4) acceptance into mainstream life, and (5) acceptance into the political system.[157] Which of these are the most useful measurement of success has been debated.[158] To evaluate the environmental movement, one might add a criterion that the environment be "improved" or that, at least, degradation be slowed.

According to the first two criteria, the movement has been successful: environmental organizations, large and small, mobilize large numbers of people to join in their efforts. While there has been a decline in the membership of some of the major national organizations, their total membership still remains far greater than enrollment was in the initial period of the contemporary environmental movement. The number of members in environmental organizations grew (though not steadily) from the 1970s to the present.[159] Growth has been documented in national level organizations, local and grassroots organizations, and organizations in other nations. In addition, the overall number of environmental organizations grew, and older organizations (such as the Sierra Club) survive and continue to be healthy organizations in terms of members and resources.

The third criterion, attainment of goals, is trickier to assess given the diversity of expressed goals. Goals of environmentalists range from radical attempts to change the economic structure of capitalist society, to reformist/managerialists work to create better environmental regulations, to conservative aims to convince individuals to reduce personal consumption. Reformist goals are, by definition, narrower than radical goals. Because of this, organizations with radical goals are less likely to succeed than reformist organizations. If we consider that reformists can single out individual issues to campaign for, such as individual laws to enact, they are able to achieve partial goals within a reformist framework. However, radical goals of transforming the capitalist structure of American society, obviously, is an immense challenge, to say the least. Since the movement as a whole does not have one unified goal, scholars have difficulty assessing this criterion. Analyzing individual organization's successes, such as whether or not they stop oil development or whether or not they prevent the building of a new waste incinerator, might be more practical.[160]

In reference to the fourth and fifth criteria, aspects of the environmental movement have achieved acceptance in mainstream life and political spheres. For instance, opinion polls suggest that people's attitudes across race, class, gender, and nationality are now more sensitive to the environment and more favorable to environmental protection.[161] In everyday life, evidence of environmentalism is present. For example, many communities have curbside recycling programs and environmental education in their schools. At the political level, the movement has also been successful in establishing policies and regulations (for example, the National Environmental Policy Act and Clean Air Act). Politicians now often take stands (at least rhetorical ones) in favor of environmental protection.

Some social movement theorists suggest that the environmental movement is a consensus movement rather than a conflict movement.[162]

> Conflict movements—such as the labor movement, poor people's movements, the feminist movements, and the civil rights movement—are typically supported by minorities or slim majorities of populations and confront fundamental, organized opposition in attempting to bring about social change. Consensus movements, on the other hand, are those organized movements for change that find widespread support for their goals and little or no organized opposition. . . .[163]

The concept of a consensus movement is useful in understanding the broad trend of the environmental movement and the idea that relatively few people would argue against environmental protections. Dunlap points out that "In the terminology of public opinion analysts . . . environmental protection had become a consensual issue by 1970, as surveys found a majority of the public expressing pro-environment opinions and typically only a small minority expressing anti-environment opinions."[164] One journalist has quipped, "In the 1990s, everyone wants to be an environmentalist. Concern for nature is 'in' these days."[165]

Despite this evidence for success, the movement still has a long way to go. There are four areas that the environmental movement needs to address. First, the movement is not entirely a consensus movement, but in fact environmentalism generates overt (as well as covert) opposition. The counter-environmental movement, or "green backlash," has a history that is traceable to the turn of the century. The two most discussed flanks of the contemporary movement are the "wise use" movement and the property rights movement.[166]

> Although difficult to neatly characterize either [of the movements] in terms of their membership or goals, Wise Use and related ideologies generally attack environmental protection and conservation efforts as harmful to the economy and job creation, insensitive to the needs and desires of local communities, and inconsistent with certain traditional American values, including constitutionally protected property rights.[167]

Those who are opposed to the goals of the environmental movement have diverse interests "including extractive resource industries like timber and mining, property rights activists (developers and individual landowners), recreationists, western ranchers and corporate farmers, businesspeople, and militia members and conspiracy theorists."[168] Despite the organizing by grassroots groups or by industries with economic interests in preventing regulation, there is little evidence that anti-environmental campaigns have lessened public support for protection. However, outright attacks against environmentalism, as in the case of attacks and killings of activists working for the U'wa and Ogoni, and Chico Mendes, may be increasing. Cases of this are not limited to the Third World. Journalist Rowell cites numerous cases from the United States.[169] For example, Judy Bari, who conducted actions for Earth First!, received death threats and her car was bombed by those who opposed her work.

A second limitation of the movement is that although it has made its way into mainstream life, much of what is "green" is diluted. Business and government coopt the environment and "greenwash" their actions. Some argue that corporations wave the environmental flag as a way to reap profits when in fact their actions are anything but "green." Greenpeace has published a guide to "Anti-Environmental Organizations" that attempts to knock down the green facade of organizations who claim to be environmentalist. For example, Greenpeace discusses the Environmental Conservation Organization (ECO) based in Illinois:

> The acronym ECO suggests an environmentally friendly agenda. In truth, the Environmental Conservation Organization was started in 1990 as a front group for real estate developers and other businesses opposed to wetlands regulations. It shares a suburban Chicago office with the Land Improvement Contractors of America, a national trade association for real estate developers.[170]

A third concern is voiced by critics of managerial environmental organizations who point out that the reformist, single-issue approach to environmentalism has not been successful in actually improving the quality of the environment. While environmentalists have been extremely successful at getting their concerns on the political agenda, the policies that have been enacted have not necessarily improved environmental quality. In some instances, analysts argue, legislation has had the unintended consequence of worsening the environment.[171] During the last twenty-five years, "we have increases in species extinction world-wide, acid rainfall and forest destruction in industrial nations, deforestation in large parts of the developing world, and desertification in much of the developing world, arising from agricultural and extractive investments."[172] In addition, environmental policies have unintended consequences including the export of environmentally destructive industries to the Third World.[173]

This ties into the fourth limitation of the consequences of the environmental movement: environmental protection for whom? The movement needs to continue to look at the regressive impacts of environmental protection within nations and internationally. For example, Ken Gould, Adam Weinberg, and Allan Schnaiberg argue that, in the seeming environmental successes of recycling, lake cleanup and environmental legislation, there have also been significant losses. For example, recycling leads "to socially regressive financial outcomes for lower-income urban residents. Local property and sales taxes, which are primary means of supporting local curbside pickups, tend to be regressive in nature."[174] This means that low-income households are paying a higher proportion of their take-home pay for recycling than higher income residents. Another dilemma exists when we look at environmental policies in a global framework. For instance, stronger legislation for toxic wastes in North America may simply move toxic waste problems to Central and South America. While North Americans may be protected by some legislation, the effect of the legislation may be intensified contamination for less-empowered communities.

Faber and O'Connor present a radical critique of the movement. They argue that the environmental movement should continue to look for ways to unite with the labor movement to

> democratize the state and the workplace; to fight against ecological racism and incorporate oppressed racial minorities and broader segments of the working class; and to develop environmental solidarity with those movements and governments in the Third World which know that capitalist economic development, ecological degradation, and human poverty are different sides of the same general problem.[175]

These are some of the key challenges that the environmental movement will need to address in the future. Given the diversity of the movement, it is unlikely that a unified movement will emerge to focus on one or all of these challenges, though alliances between groups is likely to be a key ingredient for success. The diversity of the movement reflects differences in defining problems, communicating ideologies, and taking actions. Are environmental problems based on individual beliefs, values, and actions? Lack of political will and adequate regulation? Racism? Classism? Are they the result of our capitalist productive system? Of globalization? These are issues environmentalist will continue to evaluate and act upon.

CITATIONS AND NOTES

1. Defenders of Wildlife, 1999.
2. White, 1998.
3. Eddy, 1998.
4. Switzer, 1997, p. 294.
5. As noted in Chapter 1, public attention to the environment and the emergence of the environmental movement played a critical role in the development of environmental sociology. In summarizing the literature, Dunlap and Catton (1994) note that, "It is generally agreed that the field of environmental sociology developed largely in response to the emergence of widespread societal attention to environmental problems in the early seventies."
6. McAdam and Snow, 1997, p. xviii.
7. Zald and McCarthy, 1987, p. 20.
8. Zald and McCarthy, 1987, p. 161.
9. Zald and McCarthy, 1987, p. 21.
10. One scholar dates the roots of Western conservation to scientific and economic concerns (especially in colonized tropical land) emerging over 200 years ago (Grove, 1992). An environmental historian suggests, "The environmental movement had no clear beginning. There was no single event that sparked a mass movement, no great orator or prophet who arose to fire the masses, few great battles lost or won, and few dramatic landmarks. The movement did not begin in one country and then spread to another; it emerged in different places at different times, and usually for different reasons" (McCormick, 1989, p. 1).
11. Petulla, 1977; Hays, 1959.
12. McCormick, 1989, p. 13.
13. Hays, 1959, p. 47.
14. The terms biocentric and ecocentric are used interchangeably.
15. Yosemite did not become a national park until 1890.
16. McCormick, 1989, p. 11.
17. Hays, 1959, p. 36.
18. Merchant, 1984.
19. Merchant, 1984, p. 57.
20. Merchant, 1984.
21. Hays, 1959, p. 47.
22. Schnaiberg, 1980.
23. Hays, 1959, p. 2.
24. Hays, 1959.
25. Faber and O'Connor, 1989, p. 26.
26. Faber and O'Connor, 1989, pp. 26–27.
27. Faber and O'Connor, 1989.
28. Commoner, 1990.
29. Potter, 1996, 1997.
30. Hays, 1987.
31. Potter, 1999.
32. Buttel, 1992; Inglehart, 1977; McAdam, 1982.
33. Zald and McCarthy, 1987; Jenkins and Perrow, 1977.
34. Steinhart and Steinhart, 1972; Walsh, 1986.
35. Snow and Benford, 1992; Szasz, 1994.
36. Inglehart, 1977, p. 42.
37. Schmid, 1987, p. 35.
38. Inglehart, 1977, pp. 13–14.
39. Kidd and Lee, 1997, p. 1.
40. Brechin and Kempton, 1994.
41. Dunlap, Gallup, and Gallup, 1993.
42. Levine, 1982.
43. Inglehart, 1995.
44. Levine, 1982, p. 209.
45. Edwards, 1995.
46. Olofsson, 1988.
47. Faber and O'Connor, 1989.

48. Laraña, Johnston, and Gusfield, 1994.
49. Buttel, 1992, p. 11.
50. Klandermans and Tarrow, 1988.
51. See Pichardo, 1997.
52. An exception is Kriesi, 1989.
53. Jenkins and Perrow, 1977, p. 25.
54. Mitchell, Mertig, and Dunlap, 1992, p. 14.
55. Dunlap and Mertig, 1992, p. 3.
56. McAdam, 1982.
57. Kitschelt, 1986, p. 58.
58. Jenkins and Perrow, 1977; McAdam, 1982; Rucht, 1989; Kitschelt, 1986.
59. Rucht, 1989.
60. Kitschelt, 1986.
61. For example, Smelser, 1962.
62. Zald and McCarthy, 1987.
63. Walsh, 1981.
64. Morris and McClurg Mueller, 1992.
65. Klandermans, 1992.
66. Snow and Benford, 1992.
67. Capek, 1993; Edwards, 1995; Taylor, 2000.
68. For example, Gamson and Modigliani, 1989.
69. Szasz, 1994, p. 14.
70. Szasz, 1994, p. 38.
71. Szasz, 1994, p. 50.
72. Carson, 1962; McCormick, 1989; Milbraith, 1984; Petulla, 1980.
73. However, scientists were amassing evidence of the problems of DDT. For example, Commoner (1963) provides evidence from the 1940s.
74. Wargo, 1996.
75. Ehrlich, 1968.
76. Commoner, 1963, 1971.
77. Frank, 1997, p. 411.
78. McCormick, 1989, p. 47; Gottlieb, 1993.
79. Dowie, 1995; Dunlap and Mertig, 1992, p. 5.
80. Dunlap, 1992.
81. Kline, 1997, p. 105.
82. Dunlap, 1992, p. 106.
83. Kanagy, Humphrey, and Firebaugh, 1994.
84. Dunlap and Catton, 1994, p. 13.
85. Dowie, 1995, p. 175.
86. Dowie (1995) believes this was the reason for the decrease in membership in the early 1990s.
87. Kempton, Boster, and Hartley, 1995, p. 209.
88. In the late 1970s, a number of groups split up over tactical disagreements. For instance, the founders of the Sea Shepherd Society, a group that advocates destroying ships that illegally fish, were once members of Greenpeace International, a group that promotes direct action strategies, but does not advocate Sea Shepherd's approach.
89. *www.lta.org.*
90. Taylor, 2000.
91. Sale, 1993.
92. Zald and McCarthy, 1987.
93. Gottlieb, 1993, p. 136.
94. Gottlieb, 1993, p. 139.
95. Social movement scholars have differentiated between institutionalized tactics and noninstitutionalized tactics, which vary over time and across cultures (Tilly, 1978). Currently, in North American culture, institutionalized tactics include demonstrating, lobbying, petitioning, etc. Noninstitutionalized tactics fall outside the realm of accepted behavior for a given time and place.
96. Hobsbawm, 1959, p. 11.
97. Rucht, 1989.
98. Schnaiberg, 1980, p. 389.
99. Hobsbawm, 1959, p. 10.
100. Mitchell, Mertig, and Dunlap, 1992, pp. 23–24.
101. Freudenberg and Steinsapir, 1992, p. 32.
102. See Buttel, 1987, for a summary of elitism charges.
103. Scarce, 1990, p. 72.
104. Foreman and Haywood, 1985.
105. Devall, 1992, p. 52.
106. Devall, 1992, p. 52.

107. Scarce, 1990, p. 37.

108. Devall, 1992, p. 55, emphasis added.

109. Scarce, 1990, p. 38.

110. Bruelle, 1995.

111. For example, adherents of ecofeminism believe that humans have placed themselves above nature in a hierarchy that places culture above nature and men above women. Like deep ecologists, cultural ecofeminists identify core cultural beliefs as the source of unjust inequality. As we discussed in Chapter 3, cultural ecofeminists draw parallels between patriarchy, a tradition that creates and legitimizes gender inequality, and male-dominated ecological changes such as species extinction and the buildup of nuclear waste. As with deep ecology, cultural feminists also advocate change in societal values toward feminine traditions such as nurturing and caregiving to others, including the natural world.

112. Freudenberg and Steinsapir, 1992.

113. Freudenberg and Steinsapir, 1992, pp. 31–32.

114. Edwards, 1995.

115. Gottlieb, 1993.

116. Booth, 1998.

117. Bullard, 1994, p. 98.

118. Bullard, 1994.

119. Bullard and Wright, 1992; Bullard, 1993; Bryant, 1995.

120. USGAO, 1983.

121. Bullard, 1994.

122. For example, see Anderton, Anderson, Oakes, and Fraser, 1994. See Szasz and Meuser, 1997, for an excellent review of the environmental justice literature and a special issue of the *American Behavioral Scientist* (Vol. 43, No. 4, January 2000) dedicated to the topic.

123. Pellow, 2000.

124. Szasz and Meuser, 2000, p. 602.

125. Szasz, 1994.

126. Taylor, 2000, p. 525.

127. Szasz, 1994, p. 6.

128. Szasz, 1994, pp. 80–81.

129. Faber, 1998, p. 7.

130. Hurley, 1995, Chapter 6.

131. Gould, Schnaiberg, and Weinberg (1996) call these emerging actors local citizen-workers. Even with the emergence of worker-environmentalists, however, the control capacity of industries affects whether or not workers mobilize over local environmental concerns (Gould, 1991).

132. McCormick, 1989, p. 171.

133. Kamieniecki, 1993; Broadbent, 1998; Taylor, 1995; respectively.

134. Guha and Martinez-Alier, 1997.

135. Martinez-Alier, 1997.

136. Redclift, 1987, p. 35.

137. Taylor, Hadsell, Lorentzen, and Scarce, 1993, p. 69.

138. Hecht and Cockburn, 1990; Guha, 1989; Michaelson, 1994; Brosius, 1997; respectively.

139. This case was constructed using the following websites: U'wa Defense Project (www.solcommunications.com/uwa/udpreport.html.); Native Americas Journal (nativeamericas.aip.cornell.edu): and Yes!A Journal of Positive Futures (*www.futurenet.org*).

140. *www.solcommunications.com/uwa.*

141. Escobar, forthcoming, pp. 12–13.

142. Kimerling et al., 1991.

143. Rothman and Oliver, 1999.

144. U'wa, 1999.

145. Rowell, 1996.

146. Taylor, Hadsell, Lorentzen, and Scarce, 1993, pp. 70–71.

147. Taylor et al., 1993, p. 69. For other examples of "Green Guerillas" who are working to solve local environmental conflicts, see Collinson (1996).

148. Adeola, 2000.

149. Harmon, 1987, p. 152.

150. Gezon, 1997.

151. This may be shifting again. "A sea change has occurred in the literature on international biodiversity conservation. This new wave of writing from well-known conservationists argues that current approaches, which promote the integration of sustainable development and habitat protection with greater local control,

have failed to maintain biological diversity. . . . Some . . . members of the conservation community are advocating a renewed emphasis on strict protection through strong environment practices" (Brechin, Wilshusen, Fortwangler, and West, 2000).

152. See also Grueso, Rosero, and Escobar, 1998.

153. Lewis, 1996.

154. Frank, Hironaka, Meyer, Schofer, and Tuma, 1999.

155. Frank, Hironaka, Meyer, Schofer, and Tuma, 1999, p. 83.

156. Smith, 1997, p. 48.

157. Nelkin and Poulsen, 1990, p. 1.

158. Gamson, 1975; Goldstone, 1980.

159. Mitchell, Mertig, and Dunlap, 1992.

160. Walsh, Warland, and Smith, 1997.

161. Brechin and Kempton, 1994; Dunlap, 1992; Dunlap, Gallup, and Gallup, 1993.

162. Michaelson, 1994.

163. McCarthy and Wolfson, 1992, pp. 273–274.

164. Dunlap, 1992, p. 92.

165. Tokar, 1990, p. 15.

166. Gifford Pinchot coined the term wise use (Switzer, 1997).

167. Echeverria and Eby, 1995, p. xi.

168. Switzer, 1997, p. 13.

169. Rowell, 1996.

170. Deal, 1993, p. 51.

171. Commoner, 1990.

172. Gould, Weinberg, and Schnaiberg, 1993, p. 210.

173. Faber and O'Connor, 1989, pp. 26–27.

174. Gould, Weinberg, and Schnaiberg, 1993, p. 214.

175. Faber and O'Connor, 1989, p. 36.

7

The Sociology of Sustainable Development

> While it's obvious the Earth Summit in Rio de Janeiro did not solve the world's environmental problems, it did lay a foundation for continued progress. The emphasis of the event was to integrate economic and environmental issues into the philosophy of sustainable development. . . . Within the chemical industry, we're beginning to see more and more examples of how specific companies are making great strides toward sustainability. The chemical industry has the means and the desire—not to mention the technological expertise—to become part of the solution. I believe that by working with governments and the environmental community in a productive and cooperative manner, the chemical industry can help to make sustainable development a reality.
>
> (DOW CHEMICAL CEO, FRANK POPOFF IN *CHEMICAL WEEK,* 24 JUNE 1992, P. 18).

HISTORICAL CONTEXT OF SUSTAINABLE DEVELOPMENT

The positive spin on environmental protection that Mr. Popoff promotes in his commentary in *Chemical Week* deviates sharply from the responses the chemical industry has typically taken to plans by government and environmentalists to curb the negative effects of chemical production. In the 1960s, the industry belittled Rachel Carson and her claim that DDT was harmful to both environmental and human health. The industry regularly lobbies Congress to limit the amount of legislation placed on chemical production. What happened for the CEO of Dow Chemical to write a call-to-arms to fellow industry leaders to work with the government and environmentalists, traditional foes, in a "productive and cooperative manner . . . [to] help make sustainable development a reality"? What does sustainable development promise that earlier attempts at environmental protection did not?

The most commonly used definition of sustainable development (SD) comes from the 1987 report prepared by the World Commission on Environment and Development (WCED, also known as the Brundtland Commission) titled, *Our Common Future.* Sustainable development is "Development that meets the needs of the present without compromising the ability of future generations to meet their own needs."[1] This term became a buzzword at the 1992 United Nations Conference on Environment and Development (the "Earth Summit"). The 178 heads of state that gathered at this forum sought to address both the "environment problem" and the "development problem." The concept of sustainable development presented a paradigm in which officials viewed environment and development as partners rather than adversaries. The WCED's sustainable development presumed that economic growth and environmental protection could be reconciled. The idea was not new, it harked back to Pinchot's utilitarian view of nature as a resource; as providing the "greatest good for the greatest number over the longest time."

The idea of sustainable development contrasts with development that focuses on economic gain often at the expense of the environment. Some natural resource extractive industries, such as mining and fishing, have depleted resources in the name of promoting social and economic concerns. However, unsustainable development can have devastating effects for the environment and humans. For example, in 1992 the northern cod collapsed in Newfoundland due to overfishing. In light of this, the government called for a two-year moratorium on cod fishing so that the stocks could recover. This action affected "40,000 workers and hundreds of communities."[2] In this case and others like it, the tension between biological/ecological concerns and human social/economic concerns highlights the importance of finding a balance between these systems.

While WCED's definition has the greatest recognition, a range of definitions are associated with SD. For example, David Pearce and colleagues present a thirteen-page annex of definitions of the term.[3] What WCED's brief and vague definition has in common with other treatises on SD is that the WCED identifies three main, but not equal, goals of sustainable development: (1) economic growth, (2) environmental protection, and (3) social equity. Different interest groups highlight different aspects of the three part sustainable development definition. The economic concerns of industrialists, such as Mr. Popoff, are incorporated into the definition, as are the environmental concerns of environmentalists and the social concerns of nongovernmental organizations and some governments wishing to alleviate poverty and injustice.

While the WCED popularized the concept, the term SD has been around for at least ten years prior to the report. The International Union for the Conservation of Nature, for instance, used the term in its 1980 publication, *World Conservation Strategy.* The *Strategy,* however, emphasizes ecological sustainability, not the integration of ecological, economic, and social sustainability.[4] The ideas embodied in the term sustainable development were, likewise, not new in 1987. Sustainable development draws upon "limits to growth," "appropriate and intermediate technology," "soft energy paths," and "ecodevelopment" discourses from the 1970s and 1980s.[5]

The limits to growth debate centers around the much-publicized *The Limits to Growth* study produced by the Club of Rome.[6] In a nutshell, the book presents evidence that severe biophysical constraints would impinge upon the growth and development of societies. *The Limits* predicts ecological collapse if current growth trends continued in population, industry, and resource use. The study generated tremendous debate, attention, and critique. The leading criticisms of the study are threefold: (1) it assumes that there were fixed amounts of exploitable resources, (2) it does not account for technological innovation and substitution, and (3) no resource limits have been reached or documented.[7] In addition to these problems, the limits to growth idea became politically unpopular in the less-developed countries (LCDs, or, the South) "on the grounds that it was unjust and unrealistic to expect countries of the South to abandon their aspirations for economic growth to stabilize the world environment for the benefit of the industrial world."[8]

While the limits to growth debate asks whether environmental protection and continued economic growth are compatible, the mainstream sustainable development rhetoric assumes that the two are complimentary and instead focuses on *how* sustainable development can be achieved.[9] The SD discourse does not assume there are fixed limits; it is pro-technology, pro-growth, and compromise oriented. The WCED report clearly states, "The concept of sustainable development does imply limits—not absolute limits but limitations imposed by the present state of technology and social organizations on environmental resources and by the ability of the biosphere to absorb the effects of human activities. But technology and social organization can be both managed and improved to make way for a new era of economic growth."[10]

The emphasis of sustainable development on meeting the needs and desires of multiple constituencies also relates back to a key theme of "appropriate," "intermediate," and "soft" technologies. This theme, discussed earlier in Chapter 5, is that technology should be suited to cultural contexts. While consulting with the LDCs to assist them in expanding production and decreasing unemployment, E. F. Schumacher developed a critique of the transfer of energy- and capital-intensive technologies from the more-developed countries (MDCs, or, the North) to the LDCs.[11] The technologies are directed to maximizing output per worker. For Schumacher, this transfer is not appropriate for nations with high rates of unemployment. Schumacher suggests intermediate technologies (using human labor and efficient methods) could increase production and employment, thus addressing economic and social concerns of the LDCs. Unlike some of the political disagreements generated by the notion of limits to growth, Allan Schnaiberg argues that there was support for Schumacher's ideas from both the MDCs and the LDCs. He says, "What makes [appropriate technology] or its institutionalized form of intermediate technology . . . so valuable as a comparison to sustainable development is that it drew favourable attention in both North and South among citizens, politicians and even some private-sector agents of the treadmill of production. Interestingly, like sustainable development, appropriate technology also generated little overt political resistance." Unfortunately, the political acceptability of appropriate/intermediate technologies did not result in real changes in the system of production.[12]

In 1987, the discourse of sustainable development presented a shift in thinking about development. SD presented a solution to the problems of economic development and environmental degradation. International aid agencies, such as the United States Agency for International Development (USAID) and the World Bank, adopted the SD framework for the design of their development programs. The emergence of the concept came at the same time that environmental policymakers began framing environmental problems, such as biodiversity loss, the greenhouse effect, and the thinning of the ozone layer, as "global problems." No longer was it enough to "think globally, act locally." In an era of globalization, the new interpretation of environmental problems suggested that we must "think globally, act globally."

Sustainable Development's Definitional Problems

While critics of sustainable development from the radical, managerial, and conservative viewpoints are concerned with a range of problems related to the concept of SD, a criticism that unifies their thoughts is the lack of clarity in the meaning of the term. What should be "sustained" in sustainable development? the economy? the environment? human welfare? What should be "developed"? Is "development" the same as growth? Whose "needs" and whose "development" should be promoted?[13]

As an example of the definitional problem, Paul Ekins considers the issue of "needs" and argues that the term is "an imprecise formulation which makes no distinction between the vastly different 'needs' in the First and the Third Worlds nor between human needs and the consumer wants towards the satisfaction of which most of the First World consumption, at least, is directed."[14] Similarly, development has a number of possible connotations. Does development refer to production growth, as is typically indicated by growth of gross national product or gross domestic product; does it refer to environmental growth, such as an improvement of environmental resources; or does development refer to growth in human welfare, including health, working conditions, and income distribution?[15] " 'Development' is conceptually an empty shell which may cover anything from the rate of capital accumulation to the number of latrines, it becomes eternally unclear and contestable just what exactly should be kept sustainable."[16]

Sustainable development and "sustainability" are not synonymous. SD analysts argue that sustainable development is not a neutral term; it is a political concept that represents a political agenda.[17] Sustainable development fits into a global conversation about the best way for nations to "develop," often thought of as poverty alleviation. John Dryzek argues that sustainable development is a discourse. "And it is not just any discourse. Since the publication of the report of the Brundtland Commission . . . it is arguably *the* dominant global discourse of environmental concern."[18] Sustainable development presents a strategy for development, an agenda for a style of development. The term sustainability, at least as related to ecological sustainability, is more neutral or "scientific" in that whether or not an ecological process can be said to be "sustainable" can be related to objective criteria. Ecological and social sustainability could also be constructed along more "objective" criteria; nonetheless, cataloguing these two types of sustainability is more problematic and more prone to debates as to what is/is not sustainable.

In part due to the lack of consensus of meaning, critics argue that being in favor of sustainable development comes relatively commitment-free.[19] For example, "Sustainable development is a mother-and-apple-pie formulation that everyone can agree on; there are no reports of any politician or international bureaucrat proclaiming his or her support for unsustainable development."[20] Akin to this criticism is that the term "sustainable" is used to describe so many desirable institutions that the word has lost meaning. Who could argue against sustainable society, sustainable economics, sustainable democracy, sustainable cities, or sustainable tourism, to name a few? The "sustainable" tag is integrated into many aspects of life. For example, the following definition of sustainable agriculture (from the 1990 Farm Bill) touches on all three of the aspects of SD—economic, environmental, and social.[21] Sustainable agriculture is:

> An integrated system of plant and animal production practices having a site-specific application that will, over the long term: satisfy human and fiber needs; enhance environmental quality and the natural resource base upon which the agricultural economy depends; make the most efficient use of non-renewable resources and integrate, where appropriate, natural biological cycles and controls; sustain the economic viability of farm operations; and enhance the quality of life for farmers and society as a whole.[22]

Individuals and institutions in powerful positions embrace the idea of SD as it is popularly interpreted. This rendition of sustainable development fits squarely into a managerial interpretation of social life[23] in that SD only requires slight modifications to existing modes of production, existing political structures, and existing values.[24] Radical interpretations, such as that put forward by Sharachandra Lélé, point out that the concept "Does not contradict the deep-rooted normative notion of development as economic growth. In other words, SD is an attempt to have one's cake and eat it too."[25]

Fred Buttel, nonetheless, points out some of the advantages of a "vague" notion of sustainable development:

> SD still does focus our attention on the two great contradictions of the world today: the long-term compromising of the integrity of ecosystems (local as well as global ones) and the tendency toward reinforcement of

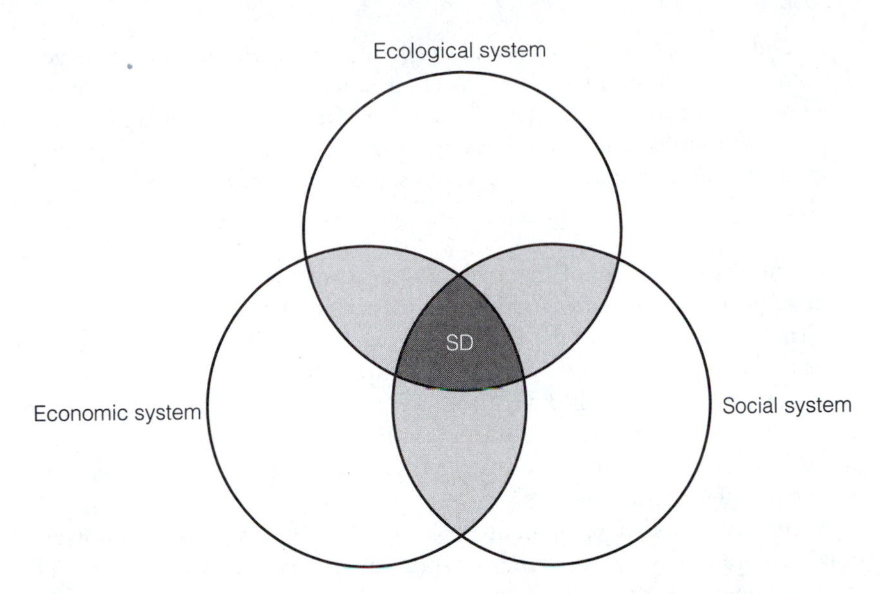

FIGURE 7.1 The Goal of Sustainable Development Is to Maximize Three Systems

SOURCE: From Edward B. Barbier, "The Concept of Sustainable Economic Development." *Environmental Conservation,* vol. 14, no. 2 (1987): 101–110. Reprinted with permission of Cambridge University Press.

> the socioeconomic processes of social exclusion of billions of the world's people. Because of its relevance to spotlighting attention on these two great institutional failures of our epoch, SD allows a range of groups to contest structures and policies and to develop alternative visions of the future.[26]

Our treatment of sustainable development works from the premise that there are three systems involved in sustainable development that must be resolved: ecological, economic, and social. Barbier asserts that the objective of SD is "to maximize the goals across all these systems through an adaptive process of trade-off."[27] (See Figure 7.1.) In sum, for development to be sustainable, the environment should be protected, people's economic situation improved, and social equity achieved.

Overview of Conservative, Managerial, and Radical Interpretations of Sustainable Development

Conservative interpretations can be either anti-sustainable development or pro-sustainable development. On the one hand, conservatives condemn sustainable development because the concept suggests tampering with the free-market economy. William Sunderlin summarizes this position:

> "Free market environmentalism" (FME) is the theoretical spearhead of pluralist opposition to sustainable development. . . . According to FME, environmental problems are caused by government interference of the free operation of the marketplace. Government ownership and control of

> natural resources, it is held, are the principal obstacles to sound management of natural resources. . . . Well-specified private property rights to all resources and an unrestrained market, FME argues, are the essential preconditions for wise custodianship of the environment. . . . The individual is seen as the key unit of analysis and as the critical agent of social change.[28]

At the international level of free trade, the anti-sustainable development conservative position favors free-trade agreements, such as the North American Free Trade Agreement and the General Agreement on Tariffs and Trade (now under the World Trade Organization), as long as they do not impose restrictions on corporations. This differs from a managerial approach that favors the agreements as long as they would "include environmental and social safeguards" so there would be an even playing ground in which trade would be less environmentally damaging.[29]

A pro-sustainable development flank of conservatives places the power of social change largely in the hands of individuals' voluntary behaviors. This group is hopeful that individuals, through green consumerism and boycotting harmful products (voting with one's dollar), can pressure producers to change environmentally harmful processes, thus changing corporate behavior. Proponents of this perspective also favor lifestyle changes. "Voluntary simplicity," for example, is the ideology of many such individuals and groups who do not believe that governments or corporations can be relied upon to enact significant changes. The proponents believe, instead, that everyday citizens transform culture by putting their beliefs into action through commitments to ideals such as "sustainable consumerism." These groups have numerous strategies to aid individuals to consume less, among other things, and to develop a way of life that is "outwardly simple, inwardly rich."[30]

The managerial account of sustainable development is also the mainstream approach to this goal, at least in the United States.[31] Sunderlin argues that "sustainable development is essentially a managerial and reformist concept."[32] Managerial accounts of SD do not question existing political or economic structures. As we indicated earlier in Chapter 2, managerialists advocate incremental changes. In this way, managerial approaches to SD enforce the existing power structure and reinforce an economy built on the ideology of growth. Actors currently in control of economic development processes, such as the World Bank and the USAID at a transnational level, and national governments and corporations at the state level, tend to take a managerial course to sustainable development. An assumption of the managerial approach is that poverty is linked to environmental degradation; thus, ending poverty through economic development (aka economic growth) will also curb environmental damage. Managerial actors are concerned with how the "theory" of SD can be put into action, especially through existing "development" programs. Rather than reconstruct their entire way of doing things, these actors instead try to adapt the themes of SD into their existing development programs.

Much of the work of Lester R. Brown and the Worldwatch Institute focuses on the managerial "nuts and bolts" of sustainable development. Brown and colleagues' recommendations for a "sustainable society" include a discussion of using more efficient technologies, decentralizing energy production, and reusing and recycling.[33] At an international level, sociologist Michael Cernea works with the World Bank to promote "putting people first" in development projects if one wants "lasting social sustainability for development programs, and better environmental management."[34] Other organizations, such as the World Health Organization, the Food and Agriculture Organization, and the Nigerian Environmental Action Team, produce managerial-style "what to do texts," reviewed in Diana Mitlin's guide of the literature on SD.[35]

The radical critique of sustainable development examines the degree to which the mechanisms of sustainable development serve to reproduce global inequality. In particular, the critique focuses on three MDC-LDC relationships—trade, aid, and debt. Critics of sustainable development argue that unequal relationships between the MDCs and the LDCs, entrenched in the post–World War II "developmentalist" period, are reproduced in the sustainable development paradigm; thus, the term SD is just a new guise for continued imperialism.[36] Michael Redclift argues that sustainable development serves to support those with power in the international world order.[37] This world order is one that was defined during the post–World War II development period in which ". . . the industrialized nations of North America and Europe were supposed to be the indubitable models for the societies of Asia, Africa, and Latin America, the so-called Third World."[38] Radicals do not believe that sustainable development offers a real alternative to old practices that serve those in power at the expense of the "have nots."

The radical interpretation does not present an agenda for sustainable development like the conservative and managerial perspectives do. Instead, radicals attempt to unpack the assumptions of each of the other approaches. The next section will present a variety of strategies that are being used to promote sustainable development. The radical position will be more fully fleshed out in terms of what it opposes of these strategies, since, from the radical perspective, there are serious problems with being "for" sustainable development.

Finally, another model for thinking about approaches to sustainable development that is similar, but lacks a one-to-one correspondence to the conservative-managerial-radical model we present, is used by Susan Baker and colleagues.[39] They evaluate approaches ranging from least to most social restructuring, and from most anthropocentric to most biocentric (Table 7.1). Roughly, their "treadmill" SD corresponds with our conservative model; in the range between their "weak sustainable development" and their "strong sustainable development" fits the managerial model; and their "ideal model" for SD has aspects of the radical model calling for profound structural changes in economic and political systems, and aspects of the conservative model calling for value changes that would align society more with deep ecological values.

Table 7.1 Another Conception of Sustainable Development

Approach in Sustainable Development	Role of Economy and Nature of Growth	Geographical Focus	Nature	Policies and Sectoral Integration	Technology	Institutions	Policy Instruments and Tools	Redistribution	Civil Society	Philosophy
Ideal Model of sustainable development	Right livelihood, meeting needs not wants; changes in patterns and levels of production and consumption	Bioregionalism; extensive local self-sufficiency	Promoting and protecting biodiversity	Holistic intersectoral integration	Labour-intensive appropriate technology	Decentralization of political, legal, social and economic institutions	Full range of policy tools; sophisticated use of indicators extending to social dimensions	Inter- and intra-generational equity	Bottom-up community structures and control. New approach to valuing work	Ecocentric/biocentric
Strong sustainable development	Environmentally regulated market; changes in patterns of production and consumption	Heightened local economic self-sufficiency, promoted in the context of global markets	Environmental management and protection	Environmental policy integration across sectors	Clean technology; product life-cycle management; mixed labour- and capital-intensive technology	Some restructuring of institutions	Advanced use of sustainability indicators; wide range of policy tools	Strengthened redistribution policy	Open-ended dialogue and envisioning	↑
Weak sustainable development	Market reliant environmental policy; changes in patterns of consumption	Initial moves to local economic self-sufficiency; minor initiatives to alleviate the power of global markets	Replacing finite resources with capital; exploitation of renewable resources	Sector-driven approach	End-of-pipe technical solutions; mixed labour- and capital-intensive technology	Minimal amendments to institutions	Token use of environmental indicators; limited range of market-led policy tools	Equity a marginal issue	Top-down initiatives; limited state-environmental movements dialogue	↑
Treadmill	Exponential growth	Global markets and global economy	Resource exploitation	No change	Capital-intensive production technologies; progressive automation	No change	Conventional accounting	Equity not an issue	Very limited dialogue between the state and environmental movements	Anthropocentric

SOURCE: Baker et al., 1997. Used by permission of International Thompson Publishing Services, Ltd.

CONSERVATIVE STRATEGIES FOR SUSTAINABLE DEVELOPMENT

This section looks at concrete examples of sustainable development projects. The purpose is twofold: (1) to highlight different types of sustainable development strategies, and (2) to present critiques of the strategies.

Corporate Sustainable Development

The Business Council for Sustainable Development prepared an influential publication on business and the environment, *Changing Course: A Global Business Perspective on Development and the Environment,* in 1992.[40] *Changing Course* represents a conservative strategy for sustainable development—one that operates within the boundaries of free-market capitalism. The book opens with the council's declaration for sustainable development. "Business will play a vital role in the future health of this planet. As business leaders, we are committed to sustainable development, to meeting the needs of the present without compromising the welfare of future generations."[41] Over fifty corporations from around the world, including Germany's Volkswagen, Japan's Mitsubishi, Kenya's First Chartered Securities Ltd., and Brazil's Aracruz Celulose, endorse this declaration. These organizations that comprise the Business Council assembled prior to the Earth Summit to make recommendations to the national leaders meeting in Rio. Critics of the Business Council argue that a more apt name for the group would be the Sustainable Council for Business Development because they present themselves "as part of the solution to the global environmental crisis rather than as part of the problem."[42] *Changing Course* offers examples and recommendations for managing "cleaner" production and improving corporations' environmental records. Interest in such corporate sustainability has grown since 1992 as evidenced by a plethora of publications on the subject of business and sustainable development, including a new publishing company (Greenleaf Publishing) dedicated to helping businesses "make profit while performing sustainably," as well as a new journal dedicated to such endeavors, the *Journal of Industrial Ecology.*

An example presented by the Business Council for Sustainable Development of a corporation taking "successful steps toward sustainable development" is the case of the U.S.-based transnational corporation, Minnesota Mining and Manufacturing (3M). 3M produces a number of consumer products, including tape. 3M was an early initiator of voluntary environmental actions through its Pollution Prevention Policy ("3P"), that the corporation implemented in 1975. According to 3M's own literature, "3P was established because it is more environmentally effective, technically sound and economical than conventional pollution controls."[43] The company tries to prevent pollution at the source rather than by managing its outputs, the company's wastes. 3M has four strategies to reducing pollution: "product reformulation, process modification, equipment redesign, and recycling and reuse of waste materials." An example of such a strategy was the redesign of a resin spray booth that cost $45,000 to implement but saves $125,000 a year in resin incineration disposal. 3M has won awards from the U.S. Council on Sustainable Development and the National Wildlife Federation. In addition, in the twenty-five year life of the 3P project, 3M reduced corporate pollutants by 771,000 tons and saved $810 million. 3M's goals are to move toward zero emissions to the environment. This has also been the goal of "industrial ecology" and "ecological modernization" policies in general.

As introduced in Chapter 5, the premise of the ecological modernization approach to environmental protection is that there is a material environmental problem that can be improved through industrial production that is cleaner, more efficient, and more profitable.[44] The academic proponents of ecological modernization speak of it in much the same way that the corporate proponents speak of sustainable development. For example, Arthur Mol and Gert Spaargaren state, "More production and consumption in economic terms (GNP, purchase power, employment) does not have to imply more environmental devastation (pollution, energy use, loss of biodiversity)."[45] David Sonnenfeld states, "In simple form, it [ecological modernization] might be thought of as industrial restructuring with a green twist."[46] EM believes a green capitalism is possible. Mol and Spaargaren explain:

> It is not that Capitalism is considered to be essential for environmentally sound production and consumption (as neo-liberal scholars want us to believe), nor that Capitalism is believed to play no role in environmental deterioration. But rather that (i) Capitalism is changing constantly and one of the main triggers is related to environmental concerns, (ii) environmentally sound production and consumption is possible under different "relations of production" and each mode of production requires its own environmental reform program, and (iii) all major, fundamental alternatives for the present economic order have proved unfeasible according to various (economic, environmental and social) criteria.[47]

Theorists writing in the radical tradition, such as Schnaiberg and Gould on the "treadmill of production"[48] and O'Connor on the "second contradiction of capitalism"[49] would disagree with such arguments by arguing that it is within the capitalist logic to maximize profit. Thus, any action, voluntary or not, that would limit profit making, would not appeal to capitalists. However, Buttel argues (and 3M's example supports this) that the corporate capitalist logic can work in favor of efficiency and conservation.[50] The "win-win" notion that sustainable development and industrial ecology/ecological modernization theory touts, in other words, may work well within the capitalist logic. Buttel qualifies this argument:

> Although we can, of course, exaggerate the extent to which capital can be expected to embrace industrial ecology and related forms of capitalist environmentalism, it must be kept in mind that one feature of capitalist competition is that efficiency in the use of resources and even in the minimization of the waste stream can be means for capitalists to reduce their

costs. This logic may be particularly strong if state policies are structured so as to penalize privately caused pollution or resources destruction.[51]

A case study of pulp and paper manufacturing based on Sonnenfeld's work in Southeast Asia shows ecological modernization in practice as well as the social processes by which corporate change takes place.[52] Sonnenfeld summarizes objectives that can be used to gauge materially whether production is proceeding according to ecological modernization principles: "in the short-term, waste reduction and elimination, resource recovery and reuse, and dematerialisation; in the long-term, resource conservation and clean production."[53] Pulp and paper manufacturing is criticized from an environmental standpoint because the industry uses chlorine in its processes, which releases dioxin, a toxin. In Indonesia, Malaysia, and Thailand, manufacturing has been cleaned up. One of the main ways that improvements occurred was through the adoption of "green" technological innovations. Though these improvements came about in a "voluntary" fashion, a number of social actors pressured the manufacturers for change. Sonnenfeld summarizes the process by which environmental technologies were adopted:

> The core dynamics of adoption of environmental technologies . . . include [an] original "landmark" conflict [in one case he examines, it is a chemical spill]; the establishment of new standards/levels of expectations for industry environmental performance; the encouragement of both firm and supplier innovations; and implementation/adoption of the new, cleaner production technologies. Key participants in these processes are local community groups, domestic and international business interests, non-governmental organizations, regulatory agencies, bi- and multi-lateral aid agencies, and "green consumers."[54]

In Southeast Asia, pulp firms are presently "among the most efficient in the world."[55] Wastes have been significantly lowered and two resources, water and chemicals, are being reduced and recovered. However, another key resource, fibrous raw materials, which had historically been reused, are no longer. "As the scale of production has increased . . . the industry has moved away from recycled inputs to greater reliance on virgin raw materials from native forests and tree plantations. . . . Perhaps the biggest Achilles heel of Southeast Asian pulp producers with regard to ecological modernisation is the criterion of dematerialisation."[56] Thus, despite significant gains in ecoefficiency brought about by technological improvements and social pressures, "resource conservation, one of the long-term objectives of ecological modernisation, thus remains in the distant future of Southeast Asian pulp industries."[57]

This case identifies another issue that has not been adequately addressed by the practitioners of ecological modernization. "A further concern is the applicability of ecological modernisation theory to small- and medium-sized enterprises [SMEs], some of them government-owned. . . . In Southeast Asia's pulp and paper industries, many SMEs are older, use poorer technology, and are more polluting. While it may make environmental sense to phase out some or many of such firms, doing so would have high social costs."[58] Radicals would push this critique. If ecological modernization is only possible for large corporations, what does this mean for the accumulation of capital? Managerial critiques would, likewise, question the equity of a system that favors the largest producing groups over small-scale operators. Concerns about production of paper products and the limits of ecological modernization for small companies leads to the next examples of corporate sustainability: small companies attempting to use "clean" processes.

A number of relatively small companies have also taken steps to use raw materials in their products that promote environmental sustainability. Gardening supply company Smith and Hawken, for example, notes of its garden furniture, "All Smith and Hawken's teak is ecologically grown and responsibly harvested on the island of Java, where replanting programs are strictly enforced."[59] Other companies, such as Ben and Jerry's (ice cream), built a reputation on being environmentally, economically, and socially sustainable. Ben and Jerry's mission statement is built around: (1) producing high-quality ice cream products made from Vermont ingredients, (2) profitable growth, and (3) socially "initiating innovative ways to improve the quality of life of a broad community—local, national, and international."[60] The annual report for 1998 notes, "Financial returns to shareholders continued to improve in 1998, reflecting the impressive growth in sales and earnings. Long-term investors in Ben and Jerry's can now take satisfaction knowing that the Company is capable of accomplishing both social and financial objectives."

While not explicit in its mission, Ben and Jerry's is considered a "proenvironmental" corporation. The milk used to produce their ice cream, for example, is free of bovine growth hormone, a hormone that many environmental and social groups consider to be negative both for the health of cows and for small family farms that attempt to farm sustainably. Ben and Jerry's pay local farmers extra to produce their milk and cream hormone-free. Recently, the corporation switched their packaging material from white bleached paper to unbleached paper. Their website boasts, "This is a bigger deal than you might think. Bleaching paper with chlorine to make it whiter is one of the largest causes of toxic water pollution in the United States."[61] The company has also successfully lowered its solid and dairy wastes and continues to find strategies to do this. For example, they are attempting to use what they call "totes," reusable containers, to receive shipments from their suppliers.

Economically, the company provides very good worker benefits and attempts to limit the income disparity between its highest paid and lowest paid workers. However, this has been difficult to do. For example, in their 1998 annual report, the social auditor notes that female senior nonexecutive managers earn 12 percent less than their male counterparts and "the income disparity between the highest and lowest paid employees is near its historical high at 16–1." The company also promotes their social concerns by allocating more than 7 percent of annual profits to the Ben & Jerry Foundation that supports grassroots organizations. The recipients of foundation grants range from environmental justice groups (the Community Coalition for Environmental Justice in Seattle), to groups fighting for affordable housing (Mutual Housing Association in New York), to those lobbying against sweatshops in China (National Mobilization Against Sweatshops).

One of the reasons that Ben and Jerry's, along with companies such as the Body Shop and Seventh Generation, are considered to be "pro-environmental" is because they were the first to sign the CERES principles (see Table 7.2). CERES stands for Coalition for Environmentally Responsible Economies. The CERES principles, originally drafted in 1989 under the title the "Valdez Principles" (after the Exxon Valdez's oil spill in Prince Edward Sound), are based upon the ideas that corporate environmental responsibility, in addition to legislation, is necessary for "environmental progress." "Success . . . depends on the willingness of corporations to lead, rather than be led, in the transition to a more ecologically sound economy."[62]

Corporate signers pledge to participate in environmental reporting and ongoing improvement. These are voluntary actions that are also driven by the fact that in today's culture, a green image sells. At first, only small corporations like Ben and Jerry's and Aveda signed the principles. However, since 1993, over fifty corporations have signed including American Airlines, Bethlehem Steel, Coca-Cola, General Motors, and Sunoco. Other voluntary corporate measures such as CERES exist. For example, in response to the Earth Summit's call for sustainable development, the International Organization for Standardization (ISO) has developed a framework (ISO 14000) for industries to use to measure and evaluate their environmental program intended to promote "sustainable business development." According to the U.S.'s representative to ISO, the American National Standards Institute, "These international standards are voluntary standards for establishment of a common worldwide approach to management systems that will lead to the protection of the earth's environment while spurring international trade and commerce. They will serve as tools to manage corporate environmental programs and provide an internationally recognized framework to measure, evaluate, and audit these programs."[63]

The Voluntary Simplicity Movement

In addition to corporate strategies for sustainable development, there are conservative strategies that are more value-based and individually directed. Proponents of the voluntary simplicity movement, which is connected to the philosophy of deep ecology, promote behaviors that could be considered conservative strategies for sustainable development. In the United States, some of the key national organizations that lead the voluntary simplicity movement are the Northwest Earth Institute in Portland, Oregon; the New Road Map Foundation in Seattle, Washington; and the Center for a New American Dream in Takoma Park, Maryland. Founded in 1993, the Northwest Earth Institute describes itself as, "Motivating individuals to examine and transform personal values and habits, to accept responsibility for the earth and act on that commitment."[64]

Table 7.2 The CERES Principles

Protection of the Biosphere

We will reduce and make continual progress toward eliminating the release of any substance that may cause environmental damage to the air, water, or the earth or its inhabitants. We will safeguard all habitats affected by our operations and will protect open spaces and wilderness, while preserving biodiversity.

Sustainable Use of Natural Resources

We will make sustainable use of renewable natural resources, such as water, soils and forests. We will conserve non-renewable natural resources through efficient use and careful planning.

Reduction and Disposal of Wastes

We will reduce and where possible eliminate waste through source reduction and recycling. All waste will be handled and disposed of through safe and responsible methods.

Energy Conservation

We will conserve energy and improve the energy efficiency of our internal operations and of the goods and services we sell. We will make every effort to use environmentally safe and sustainable energy sources.

Risk Reduction

We will strive to minimize the environmental, health and safety risks to our employees and the communities in which we operate through safe technologies, facilities and operating procedures, and by being prepared for emergencies.

Safe Products and Services

We will reduce and where possible eliminate the use, manufacture or sale of products and services that cause environmental damage or health or safety hazards. We will inform our customers of the environmental impacts of our products or services and try to correct unsafe use.

Environmental Restoration

We will promptly and responsibly correct conditions we have caused that endanger health, safety or the environment. To the extent feasible, we will redress injuries we have caused to persons or damage we have caused to the environment and will restore the environment.

Informing the Public

We will inform in a timely manner everyone who may be affected by conditions caused by our company that might endanger health, safety or the environment. We will regularly seek advice and counsel through dialogue with persons in communities near our facilities. We will not take any action against employees for reporting dangerous incidents or conditions to management or to appropriate authorities.

Management Commitment

We will implement these Principles and sustain a process that ensures that the Board of Directors and Chief Executive Officer are fully informed about pertinent environmental issues and are fully responsible for environmental policy. In selecting our Board of Directors, we will consider demonstrated environmental commitment as a factor.

Audits and Reports

We will conduct an annual self-evaluation of our progress in implementing these Principles. We will support the timely creation of generally accepted environmental audit procedures. We will annually complete the CERES Report, which will be made available to the public.

Disclaimer

These Principles establish an environmental ethic with criteria by which investors and others can assess the environmental performance of companies. Companies that endorse these Principles pledge to go voluntarily beyond the requirements of the law. The terms "may" and "might" in Principles one and eight are not meant to encompass every imaginable consequence, no matter how remote. Rather, these Principles obligate endorsers to behave as prudent persons who are not governed by conflicting interests and who possess a strong commitment to environmental excellence and to human health and safety. These Principles are not intended to create new legal liabilities, expand existing rights or obligations, waive legal defenses, or otherwise affect the legal position of any endorsing company, and are not intended to be used against an endorser in any legal proceeding for any purpose.

SOURCE: *www.ceres.org.* Used by permission of CERES.

There are a number of such groups in Europe, as well. For example, the Northern Alliance for Sustainability consists of organizations in six European nations. Their goal is "to make consumption and production . . . patterns in the North more sustainable."[65] They provide information to consumers so that through consumer pressure, producers will change environmentally unsound products. Their main "sustainable product campaign" focuses on food. They argue that they cannot rely on government to assist with organic agriculture, for example, because of the close connection between government and the agriculture industry. The Alliance argues that consumer awareness and pressure works.

> The appearance of more and more organic food in Western European supermarkets is a direct result of increased consumer awareness of health hazards of eating industrially produced food. In the UK, food scares concerning BSE (Mad Cow Disease), *E. coli* and salmonella in eggs have increased consumer pressure on retailers to stock organic produce. This has forced retailers to respond by demanding that their suppliers switch to organic agriculture or by importing organic produce from abroad.[66]

Voluntary simplicity groups and safe food groups in the United States, such as the Organic Consumers Association, also believe that consumer and pluralist democratic actions work to change corporations and government standards. The U.S. Department of Agriculture, for example, has currently revised its national organic food standards, making them more stringent, after receiving an unprecedented number of comments from consumers on what were considered to be lax standards. USDA's revised guidelines (that are not yet final regulations) take into account the leading concerns of consumers.

Radical critiques of the conservative position focus on the assumptions of conservatives' version of SD. Radicals would argue that while one may be attracted to sustainable development for its vision of compromise, SD must be critically assessed to understand its inherent bias toward concepts of "economic progress" and "growth" and the underlying assumption that growth benefits all sectors of society. Radicals come at this problem from an international scale, with much of the critique arising from the LDCs. A leading critique is that, by adopting sustainable development as "the" development

BOX 7.1 Focus on the United States: Sustainable Communities

Communities are looking for alternatives to unsustainable development. "Intentional communities" are one way people are attempting to proceed along sustainable paths. Intentional communities are the 1990s' term for what were called "communes" and "back-to-the-land" movements in the 1960s and 1970s. "An `intentional community' is a group of people who have chosen to live together with a common purpose, working cooperatively to create a lifestyle that reflects their shared core values" (*www.ic.org*). Many intentional communities are focused on ecological values. According to data from 1990, over 8,000 people in North America live in intentional communities. The popularity of intentional communities has grown since then.

A long-standing intentional community, the Farm, founded in 1971 in Tennessee, has worked toward creating a sustainable lifestyle with ecological building, permaculture, and sustainable forest management. One way its 200 residents earn a living is through their Ecovillage Training Center, which offers instruction in sustainable living, including courses on mushroom cultivation, composting, solar water heating, cob construction, hybrid vehicles, organic gardening, and social justice.

Intentional communities don't only exist in rural areas. In an inner-city neighborhood in Los Angeles, 500 neighbors have created the Los Angeles Eco-Village. Residents describe it: "We are a neighborhood in the built-out mid-city area working toward becoming a demonstration of healthy urban community. Our whole-systems approach to community development integrates the social, economic and physical aspects of neighborhood life to be sustainable over the long term. Eco-villagers intend to achieve and demonstrate high-fulfillment, low-impact living patterns, to reduce the burden on government, and to increase neighborhood self-reliance in a variety of areas such as livelihood, food production, energy and water use, affordable housing, transit, recreation, waste reduction and education. We also plan to convert the housing in the neighborhood from rental to permanently affordable cooperative ownership" (*www.ic.org*). The village gardens, composts, reuses materials, has an environmental education program for children, and has weekly community potlucks to develop a sense of community.

Efforts to create sustainable communities can also be found on college campuses. For example, at Denison University, students, faculty, and administrators created "The Homestead," an alternative living option for twelve students. Founded in 1977, the purpose of the Homestead is for students to live in a cooperative manner utilizing an agriculture-based, low technology lifestyle. A primary objective is to reduce dependence on fossil fuels and mass production. Students live "off the grid" in solar-powered cabins, grow their own organic food, raise chickens, ride their bikes to campus, and engage in participatory democracy. Residents and members of the Denison community are currently building a community center on the Homestead land out of strawbales and cob.

Other efforts are being made outside the intentional communities label. A good resource is the Sustainable Communities Network (*www.sustainable.org*).

SOURCE: Intentional Communities website (*www.ic.org*).

paradigm, developers secure the place of economic growth and progress as the international development strategy without deeply questioning what development means or whom development should benefit. In this way, sustainable development is not a significant shift away from traditional development schemes based in modernization theory. SD works within the same paradigm of market-oriented growth and sustains the "treadmill of production." Cynics suggest that those whose ultimate goal is really economic growth have coopted green thinking, shaped themselves into "eco-" growthists, and called for sustainable development. Thus, sustainable development is traditional development disguised by a new name.[67] The focus on growth comes at the expense of environmental and social aspects of sustainable development. In relation to this concern is that conservative SD shifts attention from the problems created by the "haves" to the problems created by the "have nots."

> On the surface, Rio was a considerable success, united North and South through the concepts of free-market environmentalism and growth based on the position and policies advocated by the major multinational corporations (MNCs) and the Business Council for Sustainable Development. But in ecological or biocentric terms Rio was a failure, doing nothing to reverse the historic process whereby trade-led growth has led to ecological degradation through the overexploitation of natural resources. Thus there was a convention on biodiversity, but none on free trade; a convention on forests, but none on logging; a convention on climate, but none on cars. . . . In other words, the reality of UNCED was that it was concerned with defending the power, interests and living standards of the "haves" of the industrialized North at the expense not only of the "have-nots" of the industrializing South but also of Gaia.[68]

Conservatives do not address the power and economic differentials between the "haves" and the "have nots." Critics argue that this makes sense given that the proponents of the conservative, free-market approach are currently those who are at the top of the stratification system and who have an interest in maintaining the status quo. Chatterjee and Finger argue that it was by no mistake that business groups contributed a significant percentage of the total cost to pay for the Earth Summit.

> Business and industry are not to be blamed for having sponsored UNCED and taking advantage of it. They were basically profiting from an opportunity offered on a golden plate. However, they must be criticized for double-speak, and for using the Earth Summit as a strategic event without being willing even to consider the profound changes that would be necessary in order to take significant steps toward a sustainable society. Indeed, many of the corporations that paid for the Earth Summit had appalling environmental management records. Perhaps more insidious still, many of these corporations funded anti-environmental lobbying groups in the United States and probably elsewhere. In short, while promoting themselves through the Earth Summit as the solution to the environmental and

> developmental problems, they simultaneously opposed environmental protection standards and legislation at the national and the local levels. . . . This is what turned their sponsorship of UNCED into a greenwashing farce.[69]

In sum, radicals criticize the conservative approach to sustainable development for focusing primarily on the growth element of SD while glossing over concerns about environmental sustainability and social equity.

MANAGERIAL STRATEGIES FOR SUSTAINABLE DEVELOPMENT

Managerial approaches to sustainable development can be studied by looking at the programs and actions of national governments and international development agencies. The first part of this section provides an overview of their actions followed by a radical critique of such actions. Following this, we examine an issue that is the target of state and development agencies' sustainable development efforts: biodiversity conservation. The final part of this section looks to a less controversial project of states and development agencies: finding new ways to measure sustainable development.

States' and International Development Agencies' Sustainable Development

One of the outcomes of UNCED was encouraging countries to develop what are called "National Environmental Action Plans (NEAPs)" that focus on sustainable development. Many nations have established and are currently trying to implement these plans. Their purpose is summarized by a statement about Kenya's plan: "The NEAP will identify the major environmental problems, lay out an overall strategy to deal with the problems and provide a very specific plan for action to be taken by government and the private sector, including NGOs."[70] As an incentive to receive long-term, no interest loans, the World Bank encourages countries to create NEAPs. International agencies, such as USAID, provide funding for developing countries, including Haiti, Madagascar, and Ukraine, to develop and implement these plans.[71]

National governments have also taken independent actions to incorporate sustainable development into their actions. For example, in 1992, President Clinton established the President's Council on Sustainable Development. This council has a broad constituency: corporate leaders (CEOs of Ciba-Geigy, Georgia Pacific, and Chevron), environmentalists (leaders of National Resources Defense Council, National Wildlife Federation, and the Nature Conservancy), and government officials from agencies including the Department of the Interior, the Environmental Protection Agency, and the Department of Agriculture. According to two key documents produced by the Council (*Sustainable America: A New Consensus for Prosperity, Opportunity, and a Healthy Envi-*

ronment for the Future [1992] and *Towards A Sustainable America* [1999]), the vision of the council:

> . . . is of a life-sustaining Earth. We are committed to the achievement of a dignified, peaceful, and equitable existence. A sustainable United States will have a growing economy that provides equitable opportunities for satisfying livelihoods and a safe, healthy, high quality of life for current and future generations. Our nation will protect its environment, its natural resource base, and the functions and viability of natural systems on which all life depends.[72]

For MDCs like the United States, incorporating sustainable development also means refocusing the activities of its bilateral aid agency, the United States Agency for International Development (USAID). In Chapter 3, we discussed the important role that USAID plays in promoting family planning programs. USAID also supports numerous "environmental" projects in developing nations, including pollution prevention in India and Chile, biodiversity protection in Madagascar and Peru, and the training of energy professionals in Nigeria and Ecuador.[73] One of the two strategic environmental goals is "Promoting sustainable economic growth locally, nationally, and regionally by addressing environmental, economic, and developmental practices that impede development and are unsustainable."[74] Among others, Canadian and German aid agencies also incorporate environmental emphases in their aid programs.

USAID's underlying assumptions are very similar to those presented in the Bruntland Commission's report. They believe that poverty can be alleviated by economic growth and that if poverty is eliminated, environmental quality will improve. This account does not consider how growth is distributed or the degree of inequality between rich and poor. A summary of USAID's premise follows:

> Environmental problems are caused by the way people use resources. . . . Environmental damage often is driven by poverty and food insecurity . . . [which] force individuals and communities to choose short-term exploitation over long-term management. . . . Economic growth cannot be sustained if the natural resources that fuel growth are irresponsibly depleted. Conversely, protection of the environment and careful stewardship of natural resources will not be possible where poverty is pervasive. This is the conundrum and the opportunity of sustainable development.[75]

The focus on the degrading activities of the poor shifts attention away from the degrading activities of the MDCs' consumers and capitalists. The focus also directs attention to population issues. Here, too, USAID assumes that the cause of environmental problems is the sheer number of people rather than the way the people produce and consume. Managerial agents operate within the assumption that continued economic growth is desirable and that only slight modifications and incremental changes are necessary to achieve continued growth, and thus, sustainable development.

Along with aid agencies, international development agents embrace the concept of sustainable development. Over the last few decades there has been a gradual "greening" of both bilateral and multilateral development agencies, including the World Bank.[76] In 1992, for example, the World Bank's annual report was subtitled "Development and Environment."[77] In the report, the Bank "strongly endorses"[78] the work of the Brundtland Commission for two reasons: (1) it argues that a degraded environment is antithetical to development; (2) it notes that environmental problems undermine future productivity.

In line with the World Bank's mission of alleviating global poverty, most of the programs proposed by the World Bank are framed around what they consider to be the environmental problems of the poor—sanitation, air pollution, soil erosion, and loss of tropical forests.[79] The World Bank also acknowledges that nations with different income levels produce different types of environmental problems (Figure 7.2). The Bank's proposed solutions to environmental problems focus on tactics that they have used to address purely "development" (without the environment) problems in the past: new technologies, increased investment, selective debt relief, and reduced population growth, to name a few. The Bank's main lines of action to incorporate the environment into its work include assisting nations in developing environmental policies, incorporating environmental conditions in its lending process, and assisting members "to build on the complementarity between poverty reduction and the environment."[80] Examples of environmental loans include improving environmental information systems in Uganda, promoting pollution control efforts in India, and assessing the environmental impacts of energy projects in Colombia.[81] While much of the work of the World Bank is focused on "global" problems such as biodiversity, "local" issues are also a central concern of projects. A visit to the World Bank's website will demonstrate how the Bank is paying increasing attention to the environment with information ranging from the issues of biodiversity conservation, pollution management, and green accounting to explanations of its many initiatives and partnerships with environmental organizations such as the World Wildlife Fund.[82]

One of the World Bank's initiative/partnerships is with the Global Environment Facility (GEF). GEF was established in 1990 through a collaboration of the Bank with the United Nations Environmental Program, and the United Nations Development Program to facilitate environmental aid transfers from the MDCs to the LDCs.[83] Examples of programs include assistance to national governments to comply with international treaties such as the Convention on Biological Diversity (including Senegal and South Africa), to projects to lower ozone depleting substances (as in Slovenia), to projects for reducing carbon dioxide emissions through the promotion of renewable energy (as in China). GEF funds are often coupled with funds from other development agents. For example, in 1998, GEF coupled US$5 million with US$62.5 million from the International Development Association to fund a national park project in Zimbabwe. The park is intended to promote sustainable development by protecting biodiversity, boosting tourism, and improving opportunities for local communities.[84]

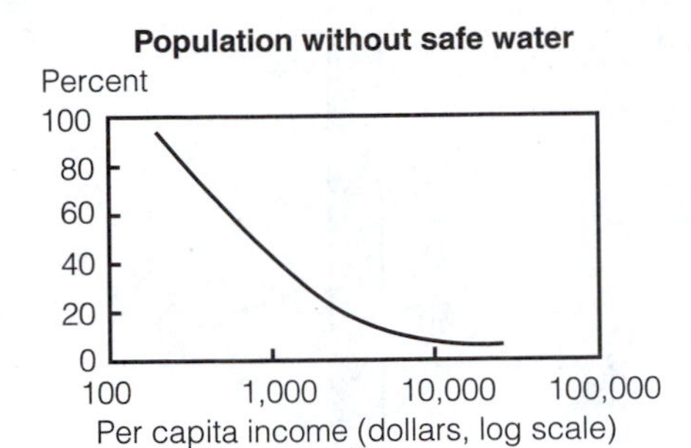

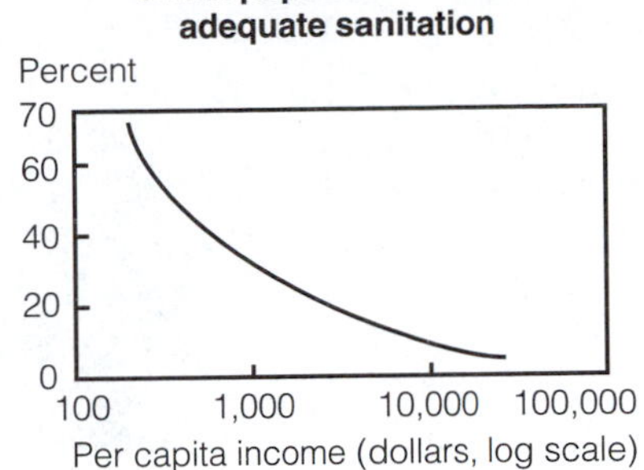

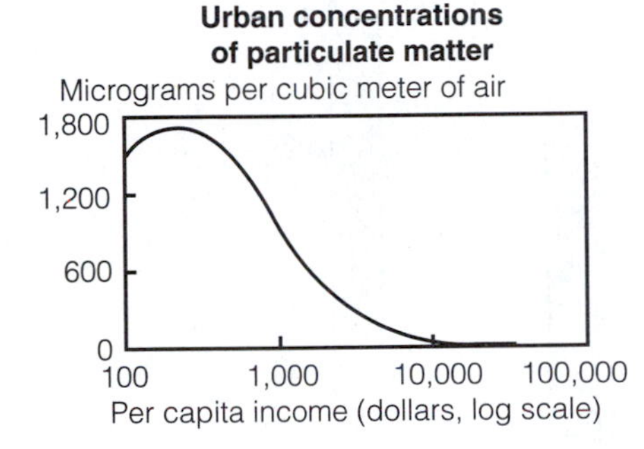

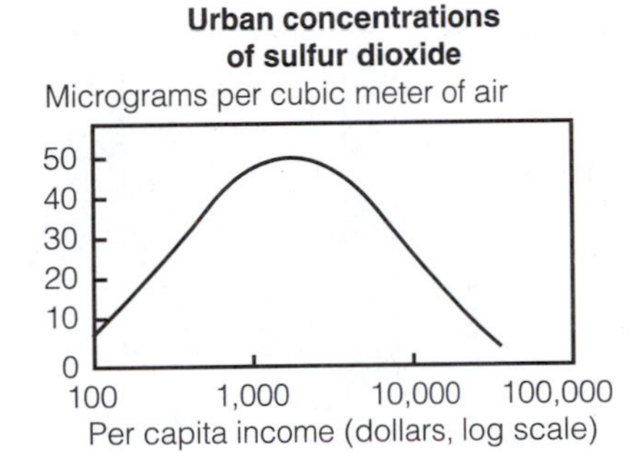

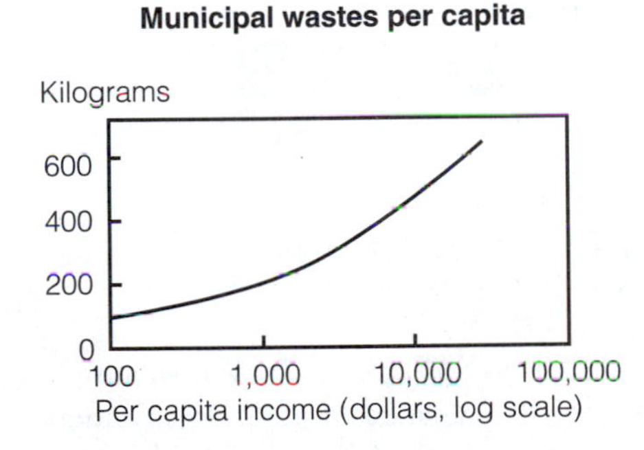

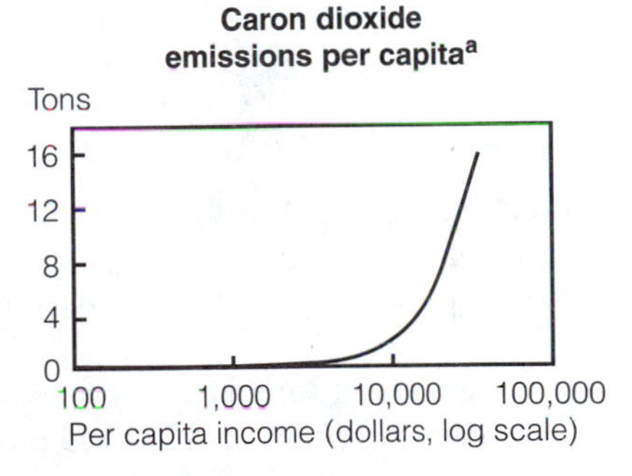

FIGURE 7.2 Various Indicators for Environment and Development

[a] Emissions are from fossil fuels.

NOTE: Estimates are based on cross-country regression analysis of data from the 1980s.

SOURCE: World Development Report, 1992, p. 11. Used by permission of Oxford University Press.

Despite the promising sound of these projects, many critics attack SD projects on the grounds of the projects' records of achievement. These critics measure the success of international development agencies on their own terms, in other words, in relation to agencies' goals. The evidence suggests they are not meeting their goals. Critics argue that both bilateral and multilateral development assistance transferred to the LDCs in the form of loans creates more, not

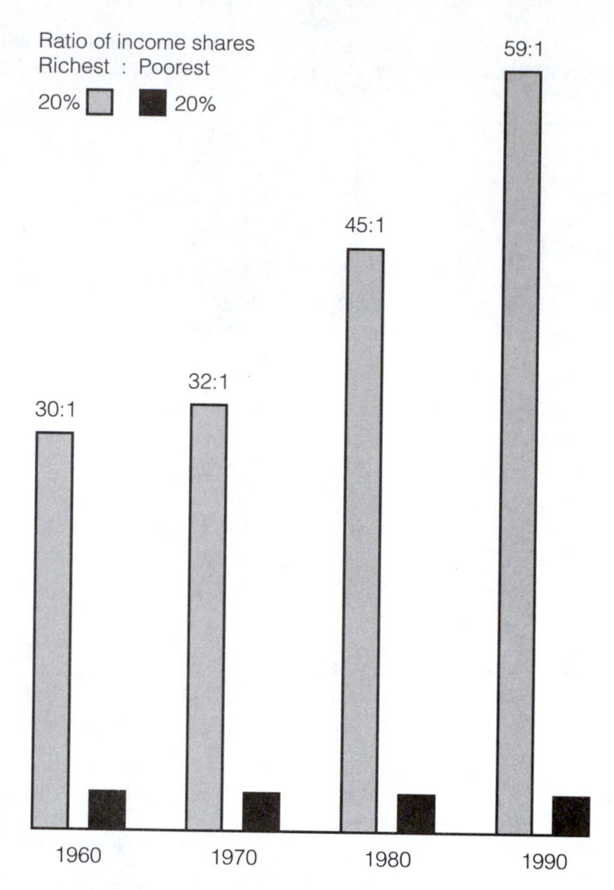

FIGURE 7.3 Income Disparity

SOURCE: From *World Development Report 1992* by World Bank, copyright © 1992 by The International Bank for Reconstruction and Development/The World Bank. Used by permission of Oxford University Press, Inc.

less, inequality in the LDCs and between nations[85] (Figure 7.3). Vandana Shiva summarizes this position. "The old order does not change through environmental discussions, rather it becomes more deeply entrenched."[86] Critics call this the "myth of development."[87] In addition, a growing literature denounces the negative effects of bilateral and multilateral development projects on the environment[88] and the negative effects of aid-produced debt on the environment.[89] For example, development assistance to increase the carrying capacity in Kenya backfired; the assistance actually reduced the capacity of range lands to support people, thus intensifying famines.[90] By admission of the World Bank's own Director of the Environmental Department, the Bank notes that the effect of its policy changes toward the environment "has been less than was hoped for at the time of Rio, and the achievements of various programs have been mixed."[91]

A radical critique of these managerial sustainable development strategies examines the degree to which the mechanisms of development agents reproduces global inequality. Radicals argue that the debt relationship between the

MDCs and the LDCs is a source of continued inequality and domination of the LDCs by the MDCs. The debt of the South has increased dramatically since the 1980s. For example, for those nations that the World Bank categorizes as "low income," in 1980s their total debt was US$102 billion; in 1997 the debt increased to US$387 billion. The debt of "middle income" nations increased over the same period from US$580 billion to over US$2 trillion[92] (Figure 7.4). Since the 1980s, the amount that the developing countries paid back on their loans exceeded the amount they received in loans, thus resulting in a net gain for the MDCs. The debt is so high in some nations that countries cannot even pay off the interest amounts, let alone the principal.

In addition to the problem of paying back loans, critics point out that the projects that the loans are intended for are ill-conceived ones that produce environmental problems rather than improving the quality of the environment. A number of critics point out the devastating environmental and social effects of many World Bank projects.[93] For example, in Brazil and Indonesia, World Bank loans encourage clear-cutting tropical forests to create new cropland—a short-term view resulting in unarable land in only a few seasons.[94] The World Bank is aware of these problems and is trying to make adjustments. For example, in fiscal year 1991, the Bank approved ninety-four projects with environmental components. Of these, thirteen of the programs had over 50 percent of the total costs or benefits of the project related to environmental protection benefits. The objective of a forest development project in Kenya is to "Conserve and protect indigenous forest resources, soil, and water on the forest, farm, and range land; provide technical assistance in forestry extension and agroforestry; prepare a forestry development master plan; strengthen planning and implementation capacities of forest agencies."[95]

Another significant problem with development agents' strategies for SD is the indirect relationship between debt and social/environmental degradation. The debt load carried by the LDCs is a significant factor for explaining the decline in environmental quality. Debt affects SD in two ways. First, countries often attempt to meet debt repayments by intensifying economic practices, turning to new investments, and increasing exports. These actions can result in environmentally risky development since they include resource-exploiting activities—mining, use of dangerous agricultural chemicals, and increased planting of cash crops often on deforested land.[96] The increased need for export earning to pay off the debt can excelerate natural resource extractions. Buttel and Taylor point out that

> Third World countries that are most "debt-stressed," and thus that are most in need of hard-currency export revenues, are most likely to see little alternative but to aggressively "develop" their tropical rainforests and other sensitive habitats in order to maintain their balance of payments and service their debts.[97]

National debt also often forces governments to limit social and environmental services, thus decreasing funds for environmental protection.[98] If nations respond to debt by reducing government expenditures, the poor and the

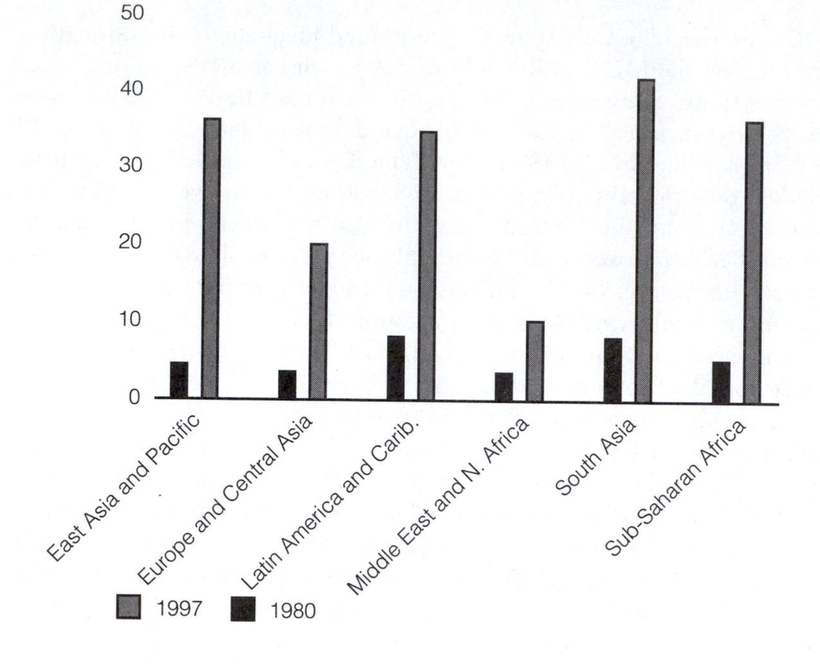

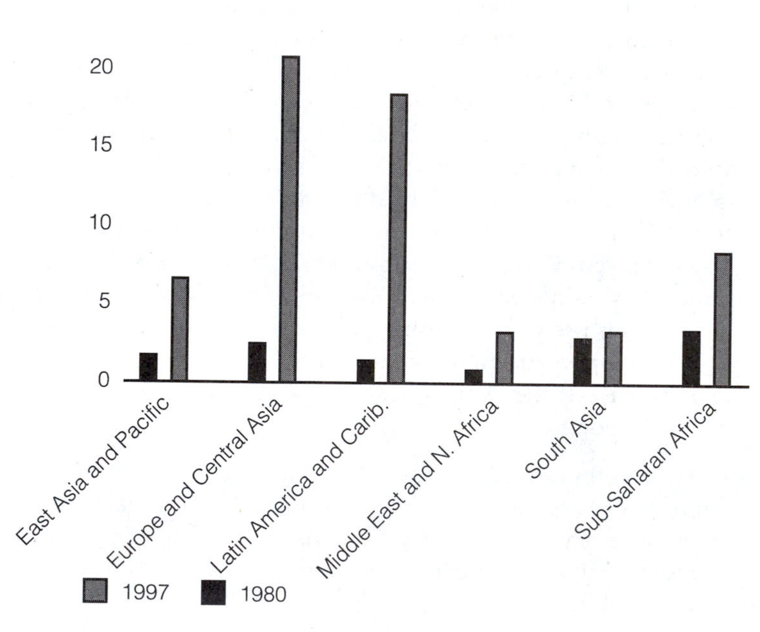

FIGURE 7.4 Growth in World Bank Loans: World Bank and IMF Lending Expands in the Regions Most at Risk of Financial Crisis

SOURCE: Used by permission of the Copyright Clearance Center for the World Bank.

environment are often the losers since less funds go to social and environmental agencies and services.

Debt, resource exploitation, and cutbacks in public services have been the pattern in a number of cases, such as Ecuador. Like many other Latin American nations, Ecuador is severely indebted. The country's political-economic history is similar to that of other LDCs that followed the traditional development trajectory. Ecuador borrowed large amounts of money in the 1970s, which led to a crisis in the early 1980s, because the country was unable to meet debt payments. In 1970, Ecuador's total foreign debt was US$242 million; by 1982 debt had increased to US$12.5 billion, more than a fifty-fold increase. Debt problems reached their height in 1979 when debt payments demanded 45 percent of export earnings.

In the late 1980s and early 1990s, Ecuador renegotiated loans with the International Monetary Fund in exchange for agreeing to make "structural adjustments." These included steps to liberalize their economy, increase exports, and reduce social spending. Environmental protection and other pro-sustainable development actions have been squeezed out of Ecuador's economic equation. The funding for Ecuador's protected areas (including national parks), for example, has been limited. The options of Ecuador's government, to preserve or exploit, in the face of immediate problems to service the debt, led them down the path of exploitation. Ecuador's three biggest foreign exchange earners—oil, bananas, and shrimp—are all clearly linked to land degradation and resource depletion. Oil extraction has been the most obviously troublesome. Petroleum's negative environmental effects on the Amazon have been well documented by both the state's own environmental agency and the World Bank. The photo in Figure 7.5, from Ecuador's capital, Quito, illustrates the dismay of many Ecuadorians in regard to oil exploitation. Debt exacerbated Ecuador's environmental problems of polluted land, air, and fish kills, and indigenous people have suffered from negative health effects.[99]

Similar processes and results of the debt cycle—high debt, structural adjustment, and environmental and social degradation—occur in other regions, such as Sub-Saharan Africa.[100] Structural adjustment policies, in particular, receive much criticism. Ted Lewellen summarizes critics' concerns. "In essence, the debt crisis has given the United States—through the [International Monetary Fund]—the power to impose its particular philosophy of growth on much of the Third World. . . . The focus of conditionality is on the economic policies of individual countries, with little recognition of the need for structural adjustments at the international level."[101] In sum, the development system does not serve the poor, the system serves nations at the top of the economic hierarchy.

Some interventions are being established to slow the growth of debt and to reduce the total debt loads of the poorest nations. Many governments in the LDC and nongovernmental organizations are calling for debts to be cancelled. In 1996, the World Bank responded by creating a program— "Debt Relief for Sustainable Development"—that aims to reduce the debt of Heavily Indebted Poor Countries (HIPCs) (Table 7.3). The use of the term sustainable development here is not clearly linked to the Bruntland

Photo by T. Lewis.

FIGURE 7.5 Graffiti in Quito. "The gasoline prices rise and the jungle cries."

Table 7.3 Debt of Selected Heavily Indebted Poor Countries (HIPCs) in US$million

	TOTAL DEBT STOCK			TOTAL DEBT/GNP (%)		
	1980	**1992**	**1997**	**1980**	**1992**	**1997**
Bolivia	2,702	4,235	5,248	101	80	68
Congo, Rep	1,526	4,770	5,071	98	187	278
Ethiopia	824	9,341	10,079	19	169	159
Guyana	—	1,897	1,611	—	711	236
Honduras	1,472	3,614	4,698	62	111	103
Kenya	3,383	6,907	6,486	48	91	65
Lao, PDR	—	1,917	2,320	—	170	132
Mali	727	2,898	2,945	45	103	119
Mauritania	840	2,088	2,453	125	186	235
Nicaragua	2,190	11,178	5,677	112	828	306
Niger	863	1,517	1,579	35	66	86
Vietnam	—	24,332	21,629	—	247	89

SOURCES: Data compiled from Global Development Finance, 1999; World Bank, 1992; and World Bank, 1999. Used by permission of the Copyright Clearance Center for the World Bank.

Commission's definition. Instead, nations that cannot meet their debt payments are classified as unsustainable. Nations are deemed eligible for this program by an economic calculation (debt-to-export ratio), not in terms of environmental or social issues.

Despite the intentions of the programs, the World Bank received criticism for not moving ahead fast enough, not providing enough relief, and not linking debt to social issues. Critics believe there is a lack of commitment to truly helping the poor and highlight hypocrisies. For example, by contrast, "In 1997 the Group of Seven countries responded to East Asia's crisis with extraordinary resolve, mobilizing in a few months more than $100 billion of loans. Equal resolve is now needed for finding the mere $7 billion needed to implement the HIPC initiative in more than 20 African Countries."[102] Despite the Bank's attempts to ameliorate their failures, critics contend that the international model of sustainable development is not working by anyone's standards.

Debt-for-nature swaps are another form of debt reduction, specifically related to environmental concerns. In a swap, a transnational organization, such as the Nature Conservancy or the World Wildlife Fund, buys a portion of a developing country's debt in exchange for a commitment to environmental projects and establishing a "Conservation Trust Fund." This reduces the developing country's foreign debt and provides funding for the conservation and management of protected areas, usually channeled through a nongovernmental organization in the LDC. In the period from 1987 to 1994, thirteen nations participated in thirty-one transnational debt-for-nature swaps. Over US$128 million in conservation funds have been generated at a cost of US$46 million. The face value of the debt that has been reduced is US$187 million. While this is a miniscule amount in relation to the total debt of the involved countries, the swaps generate previously nonexistent funds for conservation activities.

In addition to the problems of debt, another criticism against development agencies' sustainable development practices questions the assumptions of agencies' logic. This critique, which focuses on power relationships between the MDCs and the LDCs, reflects the criticisms against modernization theory waged by dependency and world-systems theorists.[103] The logic of the argument, which arises from the LDCs, follows: If the North blames the poor for environmental degradation, this justifies their intervention in the South. The MDCs frame themselves as heroes of the environment and bring their agents, knowledge, and technologies to the LDCs to "solve" their problems. This top-down approach, despite the rhetoric of "participation" and "democracy," demobilizes local, Southern actors. The approach also shifts attention away from the North's destructive activities and from structural problems with the global economic system.

Development agencies, as noted, do not place the blame for environmental degradation on the desires of the affluent; rather the poor are blamed for seeking their basic needs. This is seen as a hypocritic flaw in the managerial position. Anthropologist Arturo Escobar comments:

> Over the years, ecosystems analysts have discovered the "degrading" activities of the poor but seldom recognized that the problems are rooted in development processes that displaced indigenous communities, disrupted peoples' habitats and occupations, and forced many rural societies to increase pressure on the environment. Although in the seventies ecologists saw that the problem was economic growth and uncontrolled industrialization, in the eighties many of them came to perceive poverty as a problem of great ecological significance. The poor are now admonished for their "irrationality" and their lack of environmental consciousness. Popular and scholarly texts alike are populated with representations of dark and poor peasant masses destroying forests and mountainsides with axes and machetes, thus shifting visibility and blame away from the large industrial polluters in the North and South and from the predatory way of life fostered by capitalism and development to poor peasant and "backward" practices such as swidden agriculture.[104]

Many of the sustainable development strategies proposed by the MDCs, such as land conservation, focus on environmental problems in the LDCs, and the solutions stress what the LDCs should do. This shifts the blame for environmental destruction away from the unsustainable economic development that took place in the MDCs during the nineteenth and twentieth centuries toward the LDC's strategies for present and future economic growth. This shift of blame masks the related issue of equity, which is at the center of the debates surrounding international efforts to attain sustainable development. The MDC's understanding of sustainable development stresses intergenerational equity (for future generations), while the LDC's understanding emphasizes current intragenerational equity (between countries).

A disagreement occurring during preparations for the Earth Summit illustrates these tensions. During negotiations over how to address the problem of greenhouse gases, the MDCs stressed the environmental side of sustainable development. They focused on the importance of slowing the clearing of tropical forests (most of which are in the South), since rain forests are important "sinks" that absorb greenhouse gases. The LDCs, stressing the development side of sustainable development, responded by pointing out that the greenhouse gas problem arose largely from the fossil fuel habits of the MDCs. Other effective sinks, such as nontropical forests found in the North, have already been deforested. The LDCs resisted writing legislation that would limit their ability to use their resources for economic development.

Shiva, a scientist and activist from India, points to how the MDCs shift attention away from their own harmful activities to the degrading activities of the LDCs through the example of the globalization of the problem of ozone depletion.[105]

> CFCs, which are a primary cause of ozone depletion, are manufactured by a handful of transnationals, such as Dupont, with specific locally iden-

> tifiable manufacturing plants. The rational mechanism to control CFC production and use was to control these plants. That such substances as CFCs are produced by particular companies in particular plants is totally ignored when ozone depletion becomes transformed into a "global" environmental problem. The producers of CFCs are apparently blameless and the blame laid instead on the potential use of refrigerators and air-conditioners by millions of people in India and China. Through a shift from present to future, the North gains a new political space in which to control the South. "Global" concerns thus create the moral base for green imperialism.[106]

Blaming the LDCs for global ozone depletion justifies the MDC's intervention in the South through the North's knowledge systems and technologies.[107] The imperative question here is sustainable development by whom? There is a presumed "expertise" in the MDCs that critics would argue is unwarranted since the MDCs are the cause of much damage. One of the main ways that development agencies propose to help the LDCs, nonetheless, is through technical expertise, education, and technology transfer.[108] This discounts the value of knowledge in the LDCs despite evidence that a number of indigenous groups have lived more sustainably than we have, that groups have adapted to changing environments without depleting resources, and that ecological systems must be geographically, culturally, and ecologically specific.[109] Managerial solutions to environmental problems are uncritical of the "global" and universal constructions of such problems. This managerial approach allows "the factors that lead to global constructions of ecological knowledge to be privileged over 'sub-global' frameworks."[110] Technological transfers from the LDCs to MDCs rarely occur. "Few Northerners are proposing that Senegalese peasants be allowed to have a say in American energy consumption, or that Ecuadorian tribal peoples form groups to help protect German forests"[111] and "there are no Latin American networks advising how to deal with, say, Canadian and U.S. Pacific forests."[112]

Presumed solutions come from the top down rather than bottom up despite development agencies' rhetoric regarding the importance of grassroots organizations, women, and NGOs. Feminist critiques of managerial projects argue that women are used by development agencies. For example,

> The imagery of women as "valuable resources" and "assets" has now prompted development planners to seriously consider women's roles in environmental projects and in virtually all environment-related project documents there is at least rhetoric about women. . . . [But, for example,] while they [women] invest their valuable time planting and weeding tree plantations, they have no legal control over the resources created. Women rarely benefit from tree planting scheme . . . when the trees are sold men reap the benefits and get the money. Hence, the imperative for women's involvement in environmental projects clashes with the market orientation propagated in most development projects.[113]

The radical critique of managerial sustainable development strategies is summarized by Ekins:

> The Northern establishment must recognize its countries' primary responsibility for the present environmental crisis and determine to take radical action to address it. . . . The North must further recognize that current structures of interdependence, of trade, aid and debt, make Southern sustainable development impossible. They must, therefore, embark on wholesale reform of such institutions as [General Agreement on Tariffs and Trade], the World Bank and [International Monetary Fund].[114]

Biodiversity Conservation

An issue prompting collaboration between national governments, bilateral and multilateral agencies, and international and local nongovernmental organizations is biodiversity conservation. Conservation is identified as an important sustainable development strategy. Protecting land preserves biological diversity and can provide long-term social and economic benefits through sustained resources use and tourism. The World Conservation Union and other agencies frame land protection as a form of sustainable development in the World Conservation Strategy aforementioned. Since the Strategy, other international actions have linked the conservation of protected areas to sustainable development and strengthened the World Conservation Strategy. The Convention on Biological Diversity signed at the Earth Summit, for example, is designed to prevent the "destruction of biological species, habitats and ecosystems."[115] USAID, the World Bank, GEF, and nongovernmental organizations have all promoted land conservation. USAID funds a program called Parks in Peril (PiP), for example, which is executed through the Nature Conservancy. The program is designed to enforce park protection. From 1990 to 1997 the program received $14 million from USAID and $5.5 million in matching funds from NGOs and developing nations. The program also promotes ecotourism, such as that in the Ecuadorian "selva" (jungle) in Figure 7.6.[116]

As noted in Chapter 6, national parks and protected areas have a long history in the United States. The United States currently has over 10 percent of its land under protection. With increased concern over biodiversity loss in the tropics, most of which exist in developing countries, LDCs have established parks at a rapid rate over the last twenty years as a strategy of sustainable development. In addition to protecting biodiversity, officials expect that parks have the economic potential of earning foreign exchange from tourism.[117] In 1985, Kenya earned $300 million from wildlife associated tourism.[118] According to the World Tourism Organization, tourism is the fastest growing industry in the world.[119] Tourism development is a controversial economic strategy;[120] nonetheless it is supported by the World Bank and other developers.[121] Safaris and other nature specific tourism that are dependent upon protected areas are called "ecotourism." These tourism programs are meant to be ecologically sound and many believe ecotourism has potential as a sustainable development strategy, though it, too, is criticized.[122] The main criticism is that the estab-

Photo by T. Lewis.

FIGURE 7.6 Sacha Lodge in Ecuador's Amazon Region

lishment of parks in the LDCs, while often ecologically and economically sound, has not always been socially sound; ecotourism has not benefited local people. For example, in 1962 the Ugandan government established Kidepo National Park in an area where the nomadic Ik tribe dwelled. Since, by definition, people cannot live in designated national park areas, the Ik were relocated, forbidden to hunt, and were essentially destroyed.[123]

The costs of biodiversity protection are often social, and disproportionately paid by those living closest to biodiversity sites.[124] Especially in LDCs, the local residents are the ones forbidden to cut down trees, grow food, or raise animals in protected areas. In the United States, since much of the land under protection has long been protected, there are less dislocations than there are in LDCs where new protected areas are currently being established. However, those who are dependent on natural resource extraction in the North—such as the fisher folk in Newfoundland who were forbidden to fish and the loggers in the Pacific Northwest banned from logging old growth forests—are often displaced by biodiversity protection, as well.

Despite these negative examples, humans are becoming more important in national park planning. Conservation organizations are acutely aware of the problems associated with limiting human access to lands. Groups like the World Wildlife Fund and Conservation International have sought ways to integrate the social and ecological systems. The "pure" preservation ideas of parks has shifted to a vision that includes human development. Newer projects, associated specifically with sustainable development efforts, are called "integrated conservation and development projects." The theory behind these projects is that local people are best suited to protect biodiversity when they are also permitted to use the fruits of biodiversity to survive economically. The extractive reserve associated with Chico Mendes' work with Brazilian rubber tappers, described in Chapter 1, is a good example of an integrated conservation and development project.[125] Other efforts have been made by international conservation organizations to demonstrate that indigenous people, in particular, have long coexisted with nature and through their traditional knowledge of local ecologies and their land and resource management, they can provide a path toward sustainability.[126]

An example from Ecuador, the Tagua Initiative, is considered a successful approach to sustainable development in conservation. In a coastal region of Ecuador, Esmeraldas, the United States-based environmental organization, Conservation International, supports the initiative, which "links rural harvesters of the ivory-like nut of the tagua palm—which grows in coastal rain forests from Panama to Ecuador—with manufacturers of buttons, jewelry, and arts and crafts made from nuts. Key members of the Tagua Initiative include Esprit, L.L. Bean, Smith & Hawken, and more than 45 other U.S. and international clothing manufacturers."[127] The Initiative takes place in an area that is a top conservation priority, a biodiverse "hot spot." Information from Conservation International suggests that this program is a great success; the program employs over 1,800 people, protects the land in the Cotacachi-Cayapas Ecological Reserve, and the initiative has generated over $1.5 million in Tagua button sales. The initiative fits all three criteria of SD: the social, economic, and environmental. The program is currently being expanded to include more than twenty similar products in eight biodiversity-rich nations.

In its best cases, biodiversity conservation can bring together the social, economic, and ecological spheres of SD. Many conservation conflicts, however, play out as contests between economics and ecology, industrialists and environmentalists. These conflicts do not only occur in developing nations. In just the last few years in the United States, conflicts in and around protected areas of Yellowstone (mining versus wilderness protection) and the Pacific Northwest (jobs versus the spotted owl) are framed in this way. For example, industrialists and environmentalists are debating whether or not the Arctic National Wildlife Refuge should be opened for oil exploration. Industrialists argue that, if the area were opened, employment opportunities would increase, that the state would benefit from these taxable incomes, and that, if oil were actually discovered, then there would be even more jobs and more taxable incomes. An editorial in the *Oil and Gas Journal* (1995) states, "By not leasing [the Arctic National Wildlife Refuge], the U.S. government deprives itself and its citizens of an economic opportunity because an environmentalist cause forecloses discussion of what few real environmental questions apply. . . . Their [the government's] refusal is a triumph of obstructionist environmentalism."[128]

Environmentalists call Don Young, the Alaskan Congressman who is leading the effort to explore, "an attack dog for development interests."[129] Citing concerns over species preservation and the problems of dependence on nonrenewable resources, environmental organizations, including the Wilderness Society, have urged the government to change the area's protection status to that of a national monument, which would legally prohibit oil exploration. Federal agencies are split on the issue due to competing missions. For example, the Interior Department's Mineral Management Service promotes oil development, while the Fish and Wildlife Service is charged with protecting the environment.[130] Despite the rhetoric of SD, the two sides have different, not necessarily complementary values. One values economic benefits; the other values the benefits of wildlife preservation. The mainstream view of SD ignores these critical differences.

From a global perspective, radicals are concerned that biodiversity is being commodified and only valued for its economic benefits. Shiva argues that pharmaceutical companies earn billions of dollars from the preservation of rain forests from which they extract chemicals from tropical plants, often with no benefit to the LDCs that protect the land. The companies re-create chemical compounds, patent the compounds, and sell the drugs back to the LDCs. Between 1990 and 2000, according to Shiva, the value of the LDC's germplasm grew from US$4.7 billion to US$47 billion;[131] others project even higher future values.[132]

Biodiversity conservation highlights the tensions between ecological, economic, and social systems and the trade-offs that need to be addressed for sustainable development. Despite hopeful efforts to simultaneously alleviate problems of poverty and biodiversity loss through land conservation, conflicts arise. For sustainable development to work, tough value decisions need to be made.

Measuring Sustainable Development: New Social Indicators

Finally, we examine a managerial strategy that verges on conservative and radical approaches to social change that is less controversial than those just examined. The managerial approach takes a step toward examining the value structure that we have built our ideas of "success" upon and seeks to redefine our measures of success. Arguments of this type suggest that we need to change our system of accounting for such things as "sustainability" and "development." Finding indicators for SD is part of this project. At the time of the Earth Summit, national reports made to the International Commission on Sustainable Development were poor. The New Economics Foundation is working to refine national sustainability indicators, especially as they relate to UNCED's Agenda 21.[133] This strategy has the potential to move beyond simply supporting the status quo since to some degree it forces a reexamination of the meaning of "development" and the value of measuring it in terms of economic growth by critically examining the drawbacks of using indicators of development that focus on gross national product or gross domestic product.

Brown argues, "as the transition to a more environmentally benign economy progresses, sustainability will gradually eclipse growth as the focus of economic policy making."[134]

A number of measures that can be used to compare nations have been proposed. These include measures like the Physical Quality of Life Index, the Human Development Index, the International Indicator of Social Progress, the Sustainable National Income, and the Genuine Progress Indicator.

The Human Development Index (HDI) is intended as an alternative to GNP to measure human development. It is reported yearly in the United Nations' *Human Development Report.* The HDI does not equate development with economic growth. The *Report* states, "The concept of human development provides an alternative to the view of development equated exclusively with economic growth. Human development focuses on people."[135] HDI takes into account "three basic dimensions of human development—longevity, knowledge and a decent standard of living. It is measured by life expectancy, educational attainment . . . and adjusted income."[136] Figure 7.7 illustrates that there is not a perfect correlation between GNP and HDI. In fact, these two measures differ significantly in a number of cases. Another measure, the Human Progress Indicator for the LDCs, is similar to the HDI, but also adds an element of equity by looking at the percentage of people without access to water and health services and the percentage of underweight children.[137] Other variants on these themes take into account other dimensions of human life, such as gender equality (for example, the Gender-Related Development Index and the Gender Empowerment Measure). Unfortunately, the HDI does not include an environmental element or an equity element.

The Sustainable National Income (SNI) is a measure used to compare actual levels of economic activity with "sustainable" levels of activity.[138] By taking additional costs into account, such as costs of environmental restoration and of developing alternatives to natural resources, the measure adjusts national income statistics. While higher consumption levels lead to traditionally "better" statistics, the SNI accounts for the environmentally degrading effects of some consumption. For example, "Consumption patterns in the West . . . includ[ing] consuming large amounts of meat, heating the whole house, extensive use of vehicles, and consuming summer vegetables in winter . . . overburden the environment."[139] These activities raise the GNP, but lower the SNI. Despite the SNI's inclusion of an environmental component, this measure also has shortcomings. For example, it cannot be used for cross-national comparison and it is not a direct measure of national sustainability—it's an environmental correction for GNP.[140] Other measures that account for environmental costs and benefits are the Adjusted National Product,[141] the UN's System for Integrated Environmental and Economic Accounting, and the Index of Sustainable Economic Welfare.[142]

Redefining Progress, a public policy organization, created the Genuine Progress Indicator (GPI). The group's mission is to "ensure a more sustainable and socially equitable world for our children and our children's children.

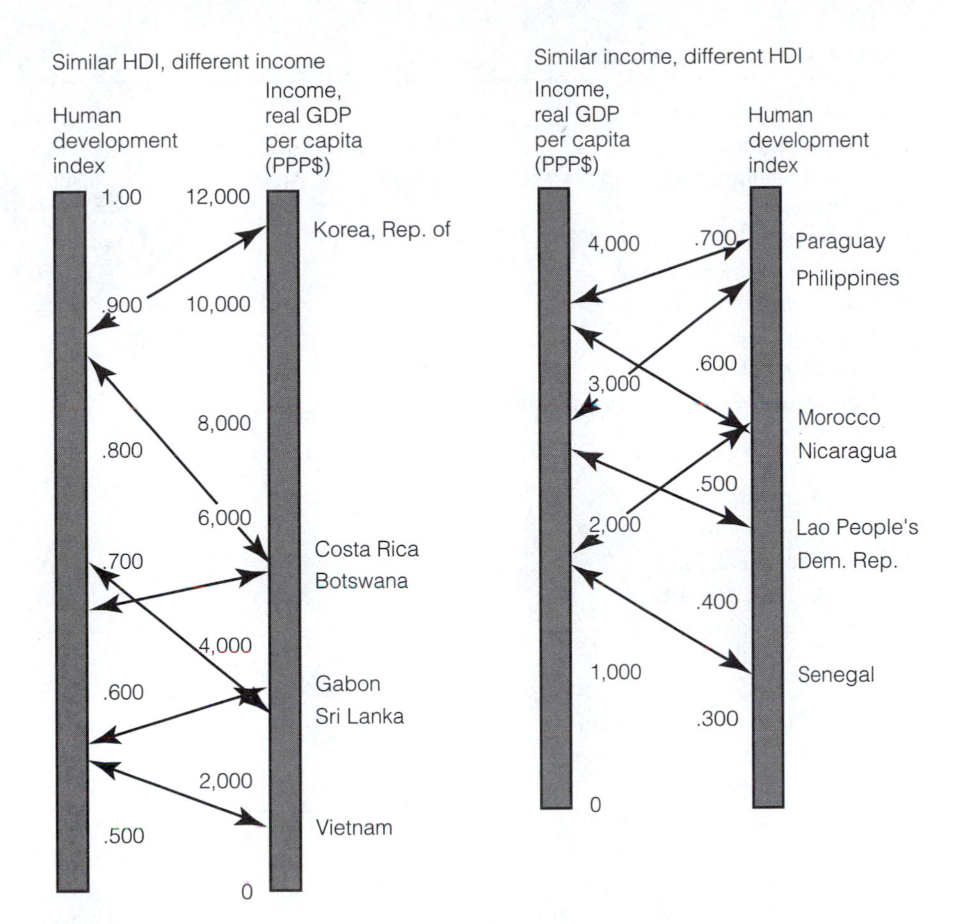

FIGURE 7.7 HDI versus GNP

SOURCE: From *Human Development Report 1998,* by United Nations Development Program, copyright © 1998 by The United Nations Development Program. Used by permission of Oxford University Press, Inc.

Working both within and beyond the traditional economic framework, Redefining Progress generates and refines innovative policies and ideas that balance economic well-being, the environment, and social equity so that those living today and those living in the future can have a better quality of life."[143] The organization has constructed the GPI in contrast to the gross domestic product, which they argue is not a good measure of progress. In particular, they argue:

> The GDP fails to distinguish between monetary transactions that genuinely add to well-being and those that diminish, try to maintain the status quo, or make up for degraded conditions. . . . For example, the GDP treats crime, divorce, legal fees, and other signs of social breakdown as economic gains. Car wrecks, medical costs, locks and security systems,

Table 7.4 The 1998 GPI Account

Personal consumption	5,153
Income distribution	118
Personal consumption adjusted for income inequality	4,385
Adjustments	
Value of housework and parenting	+1,911
Services of consumer durables	+592
Services of Highways and Streets	+95
Value of volunteer work	+88
Net capital investment	+45
Cost of household pollution abatement	–12
Cost of noise pollution	–16
Cost of crime	–28
Cost of air pollution	–38
Cost of water pollution	–50
Cost of family breakdown	–59
Loss of old-growth forests	–83
Cost of underemployment	–112
Cost of automobile accidents	–126
Loss of farmland	–130
Net foreign lending or borrowing	–238
Loss of leisure time	–276
Cost of ozone depletion	–306
Loss of wetlands	–363
Cost of commuting	–386
Cost of consumer durables	–737
Cost of long-term environmental damage	–1,054
Depletion of nonrenewable resources	–1,333
Net genuine progress	**1,770**

SOURCE: Cobb, Clifford, Gary Sue Goodman, and Mathias Wackernagel, 1999. *Why Bigger Isn't Better: The Environmental Progress Indicator,* 1999 update. San Francisco, CA: Redefining Progress. *www.rprogress.org.*

and insurance are also pluses to the GDP. Further, the GDP ignores the environmental costs of economic activities. . . . The GDP counts pollution as a double gain to the economy: The production of oil that creates pollution adds to the GDP; then the clean up of toxic waste sites or the Exxon Valdez oil spill ups the GDP even more.

Like the SNI, the GPI corrects for these by adjusting the GDP in terms of consumer spending that increases or decreases well-being but does not allow for cross-national comparisons. Table 7.4 lays out the factors that are considered. The GPI, like other "alternative" measures, paints a very different pic-

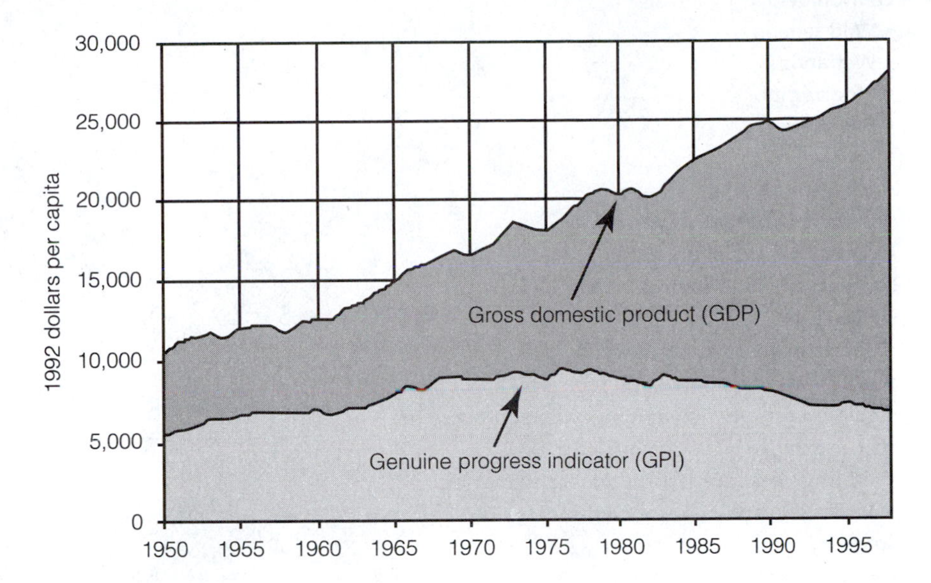

FIGURE 7.8 GPI versus GDP, 1998

SOURCE: Used by permission of Redefining Progress. *www.reprogress.org.*

ture than the GDP (Figure 7.8). While GDP per capita has been steadily increasing since the 1950s, the GPI per capita peaked in the 1970s and has declined since.

National governments have also attempted to adjust their national accounting to include environmental resources. The National Research Council, for example, has recommendations for how the U.S.'s National Income and Product Accounts should integrate environmental components.[144]

A number of cities have attempted to become "sustainable cities" and a great deal has been written about urban sustainability in the United States and abroad.[145] The Sustainable Seattle project shows how national and international level indicators can be applied at a local level. Seattle's example differs from the indicators above in that those indicators are "top down" whereas Seattle's project has been "grassroots."[146] Hundreds of volunteers from Seattle worked over a three-year period to create "Indicators of Sustainable Community."[147] The group defined sustainability as, "long-term health and vitality—cultural, economic, environmental and social." The community identified forty indicators of sustainability fitting into five categories: (1) environment, (2) population and resources, (3) economy, (4) youth and education, and (5) health and community (see Table 7.5). These indicators can be assessed annually to ascertain whether Seattle is moving in a sustainable or unsustainable direction and in which areas of sustainability the city is doing best. Other cities have engaged in similar community projects, for example, Sustainable San Francisco, Sustainable Manhattan, and Sustainable Boston; and there are similar statewide programs: Sustainable Maine and New Jersey's Sustainable State project.[148]

Table 7.5 Sustainable Seattle Indicators, 1995

Environment
- Wild salmon
- Wetlands
- Biodiversity
- Soil erosion
- Air quality
- Pedestrian friendly streets
- Open space in urban villages
- Impervious surfaces

Population and Resources
- Population
- Residential water consumption
- Solid waste generated and recycled
- Pollution prevention and renewable resource use
- Farm acreage
- Vehicle miles traveled and fuel consumption
- Renewable and nonrenewable energy use

Economy
- Employment concentration
- Real unemployment
- Distribution of personal income
- Health care expenditures
- Work required for basic needs
- Housing affordability ratio
- Children living in poverty
- Emergency room use for non-ER purposes
- Community capital

Youth and Education
- Adult literacy
- High school graduation
- Ethnic diversity of teachers
- Art instruction
- Volunteer involvement in schools
- Juvenile crime
- Youth involvement in community service

Health and Community
- Equity in justice
- Low birthweight infants
- Asthma hospitalization rate for children
- Voter participation
- Library and community center usage
- Public participation in the arts
- Gardening activity
- Neighborliness
- Perceived quality of life

SOURCE: Used by permission of Sustainable Seattle. *www.scn.org/sustainable/indicators.*

ALTERNATIVE PATHS TO SUSTAINABLE DEVELOPMENT

There is no single "radical" strategy that addresses the problems identified in the mainstream sustainable development strategies presented in this chapter. Instead, activists and academics discuss principles that should be embodied in alternatives. These place attention on redistribution of power and control in a manner that is also consistent with the managerial focus on power. Three interlinked themes are contained in the "alternatives" literature. Alternatives should be: (1) nonhegemonic, (2) grassroots and participatory, and (3) locally and ecologically based.

That sustainable development strategies should be nonhegemonic is a reaction to the "one size fits all" development that has been applied throughout the LDCs. This argument is connected to anti-development and anti-colonial movements. For example, Ramachandra Guha's explanation of the social forestry movement in India's Himalaya (the Chipko movement) shows how the history of peasant movements against colonial powers in this region is directly related to what is now thought of as an ecological movement.[149] Responses to colonial powers are similar to the contemporary responses to mainstream sustainable development practices. Arturo Escobar, for example, argues against the globalization of a dominant development ideology:

> There are no grand alternatives that can be applied to all places or all situations. To think about alternatives in the manner of sustainable development, for instance, is to remain within the same model of thought that produced development and kept it in place. One must then resist the desire to formulate alternatives at an abstract, macro level.[150]

Escobar is optimistic that alternatives to the dominant sustainable development discourse will arise from multiple locales that reflect the collective concerns of local people. He believes that grassroots social movements will be the driving force behind such new articulations of "development" and SD.[151]

Proponents of these new articulations believe that the best strategies will be developed by the people who will be most affected by them "outside the control of foreign governments, international institutions, and domestic elites. Popular mobilization and nongovernmental organizations (NGOs) are seen as a key part of the process of change in the defense of the environment of the South."[152] Wolfgang Sachs believes that the MDC's centralizing development strategies have been and are disempowering.[153] Others, such as Steve Barkin, also stress the importance of autonomous, self-sufficient, sustainable development through a democratic process.[154]

Not only will these solutions be more politically feasible, according to bioregionalists, they will be more ecologically feasible. Bioregionalists promote decentralized decision making, and production and consumption based on local resources. Sachs calls this type of development that focuses on "local livelihoods" the "home perspective."[155] Those who promote cultural pluralism assume that one of the reasons cultures have developed differently is because societies have had to develop and adapt to local ecosystems. Thus, a precursor to cultural pluralism is the freedom to interpret, adapt, and develop in response to unique ecosystems.[156]

Successful alternative strategies of achieving sustainable development are often small, locally based grassroots efforts, not top-down development attempts.[157] Rosi Braidotti and colleagues argue "What is becoming increasingly clear is that people marginalized by the development process are carving out their own paths in solving their problems. . . . [They are] reviving their old methods of farming, recovering their subjugated knowledges and forms of local organization. They again grow their indigenous crops to become independent of expensive Western seeds and fertilizers and claim control over their local forests."[158] A collection of cases from both the MDCs and the LDCs edited by Bron Taylor points out how local struggles against hegemonic and environmentally destructive forces are producing new forms of development. Taylor calls these movements popular ecological resistance movements and demonstrates how the basis of such movements is the need for sustainable livelihoods based on local ecologies.[159]

Thomas Rudel reports on a case from the tropical rain forests in Esmeraldas, Ecuador that has achieved some success in moving toward sustainable development.[160] Esmeraldas is an area of great concern because the tropical forests of this region are being rapidly deforested. Efforts to reach sustainable development here work at two levels. First, new sustainable forestry techniques are being implemented. External assistance was brought in from a number of groups, including USAID, which has worked with Ecuadorian ecologists "who have designed a plan for the sustainable harvesting of wood . . . in the Cotacachi-Cayapas Ecological Reserve."[161] Second, is the creation of "civic arenas"—"encompassing organizations whose members include all of the stakeholder groups in the region of interests."[162] The encompassing organization in this case includes all of the groups that have an interest in the rain forest: "members from fifty Afro-Ecuadorian comunas, the lumber companies, [the government agency in charge of protected areas], the provincial government, environmental NGOs, and international aid missions with interests in the region."[163] In this arena, trade-offs between the competing goals of sustainable development are negotiated. For example, community members must collectively decide to what degree economic gains should outweigh ecological gains and vice versa. One proposal being looked at by the encompassing organization is to attain green certification for the wood from the project. "Selling these woods in the international markets would raise their price. Both the timber companies and the environmentalists support the proposal."[164] This project in Esmeraldas shows a process in which trade-offs among the competing goals of sustainable development can lead to better forest conservation and gains in economic growth. Contrary to radical accounts that external political and economic influences in ecological matters can be exploitative and destructive, Rudel argues that "outside intervention" helped create the encompassing organization in Esmeraldas that may form the basis of a more sustainable development. Communities around Esmeraldas have taken notice of the project's success and are taking steps to implement similar plans.

Different forms of organizing for sustainable development are taking place in the United States, as well. Weber explains the emergence of "hundreds of rural, place-based, grass-roots ecosystem management (GREM) efforts across the United States [that] constitutes a new environmental movement."[165] These groups are akin to the encompassing organizations that Rudel describes in Ecuador. GREM efforts attempt to gather stakeholders from communities to manage lands and recognizes that trade-offs are inevitable. Their goals are to promote environment, economy, and community; in sum, sustainable development. He provides examples of such groups from Willapa Bay, Washington; Applegate Valley, Oregon; and Blackfoot River Valley, Montana; among others, all communities that are dependent on "nature's bounty." Common among the author's writing of successful efforts to promote sustainable development is a locally based, democratic process that includes collaboration, participation, negotiation, and compromise.

Finally, we close this chapter with a quote from Leff who argues that Marxist thought and environmental thought can be reconciled. The quote summarizes the hopes of those who believe that alternatives are possible and necessary:

> Environmental thought can be inscribed within post-Marxist or post-modernist thought. It defends the specificity of local action (thinking globally, acting locally), the autonomy of social groups, and difference—difference in cultural values, and development styles, and options. It presents new ethical values and a new political culture, but at the same time poses the problem of political efficacy and of the real political power held by environmental groups. Although the defense of autonomy and local difference can be seen as part of the struggle against totalitarianism (against vertical and corporative power structures in traditional political organizations), it also rules out any universal demand other than one claiming the legitimacy of all local demands. Nevertheless, the demand for autonomous spaces as a starting point for the development of alternative local productive projects is part of the larger movement for another kind of material existence based on the integration of multiple development styles.[166]

CITATIONS AND NOTES

1. WCED, 1987.
2. Haedrich and Hamilton, 2000, p. 359.
3. Pearce et al., 1989.
4. Lélé, 1991, p. 610.
5. Meadows et al., 1972; Schumacher, 1973; Lovins, 1977, respectively.
6. For an overview, see Humphrey and Buttel, 1982, Chapter 5.
7. For a review, see Sandbach, 1978.
8. Buttel, 1998, p. 263.
9. Baker et al., 1997, pp. 2–3.
10. Cited in Ekins, 1993, p. 91.

11. Schumacher, 1973.

12. Schnaiberg, 1997, p. 74.

13. See Ekins, 1993; Lélé, 1991; Lohmann, 1990; Redclift and Woodgate, 1997, p. 55.

14. Ekins, 1993, p. 91.

15. Ekins, 1993, pp. 94–95.

16. Sachs, 1997, p. 73.

17. Baker et al., 1997; Sunderlin, 1995.

18. Dryzek, 1997, p. 123.

19. Richardson, 1997.

20. Rich, 1994, p. 196.

21. What is interesting about agriculture is that there has been a very conscious movement to include the "social" in sustainable agriculture. A significant amount of research has examined the connections between social and environmental issues around agriculture. See, for example, Allen, 1993; Clancy, 1993; Feenstra, 1997; and Kloppenburg, Hendrickson, and Stevenson, 1996. Sustainable agriculture has drawn the attention of sociology as is evidenced by several special issues of key journals over the last few years dedicated to the topic: *Society and Natural Resources'* (1997, vol. 10, no. 3) special issue on "The Politics and Policies of Sustainable Agriculture, *Monthly Review's* (1998, vol. 50) issue titled, "Hungry Profit: Agriculture, Food, and Ecology," and various issues of *Rural Sociology,* including a recent issue on restructuring agriculture in Australia and New Zealand (1999, vol. 64, no. 2).

22. U.S. Congress 1990 cited in Schwarzweller and Lyson, 1995, p. xi.

23. Sunderlin, 1995.

24. Baker et al., 1997, p. 17.

25. Lélé, 1991, p. 618.

26. Buttel, 1998, p. 265.

27. Barbier, 1987, p. 104. Barbier looks at the "biological and resource" system rather than the "ecological."

28. Sunderlin, 1995, pp. 487–88.

29. Costanza et al., 1995, p. 20.

30. Elgin, 1981.

31. Lewis, 2000.

32. Sunderlin, 1995, p. 82.

33. Brown et al., 1990.

34. Cernea, 1993; see also Cernea, 1991.

35. Mitlin, 1992.

36. Escobar, 1995; essays in O'Connor, 1994; Redclift, 1987; Sachs, 1993.

37. Redclift, 1987.

38. Escobar, 1995, preface.

39. Baker et al., 1997.

40. Schmidheiny, 1992.

41. Schmidheiny, 1992, p. xi.

42. Chatterjee and Finger, 1994, p. 116.

43. Information about 3M was gathered from its website.

44. See Mol and Sonnenfeld, 2000, for an overview of the history of research under the rubric of Ecological Modernization Theory-EM, pp. 3–14.

45. Mol and Spaargaren, 2000, p. 36.

46. Sonnenfeld, 2000, p. 236.

47. Mol and Spaargaren, 2000, pp. 22–23.

48. Schnaiberg, 1980; Schnaiberg and Gould, 1994.

49. O'Connor, 1994.

50. Buttel, 1998.

51. Buttel, 1998, p. 269.

52. Sonnenfeld, 1998, 2000.

53. Sonnenfeld, 2000, p. 237.

54. Sonnenfeld, 2000, p. 245.

55. Sonnenfeld, 2000, p. 250.

56. Sonnenfeld, 2000, p. 250.

57. Sonnenfeld, 2000, p. 251.

58. Sonnenfeld, 2000, p. 254.

59. Smith and Hawken website.

60. Ben & Jerry's website

61. Ben & Jerry's website.

62. CERES website.

63. ANSI website.

64. Northwest Earth Institute website.

65. Kruszewska, 1998, p. 76.

66. Kruszewska, 1998, p. 77.

67. Escobar, 1995; Sachs, 1993.

68. Richardson, 1997, p. 54.

69. Chatterjee and Finger, 1994, p. 119.

70. Kenya's NEAP website.

71. Larson, 1994; USAID website.

72. PCSD, 1992, p. iv.

73. USAID website.

74. USAID website.

75. USAID website.

76. Conroy and Litvinoff, 1988.

77. World Bank, 1992b.

78. World Bank, 1992b, p. 8.

79. World Bank, 1992b, Chapter 2.

80. World Bank, 1992a, pp. 12–13.

81. World Bank, 1992a.

82. World Bank website.

83. GEF, 1998.

84. World Bank, 1998.

85. Chatterjee and Finger, 1994; Seabrook, 1996.

86. Shiva, 1993.

87. Chatterjee and Finger, 1994.

88. Hayter, 1989; Rich, 1994.

89. Buttel and Taylor, 1994; Inman, 1992; Miller et al., 1991; Stoncich, 1989; Stone and Hamilton, 1991; Yearly, 1991.

90. Talbot, 1986.

91. Watson, 1999, p. 5.

92. World Bank, 1999, p. 256.

93. Rich, 1994.

94. Postel, 1989.

95. World Bank, 1991, p. 120.

96. Stone and Hamilton, 1991; Yearly, 1991.

97. Buttel and Taylor, 1994, p. 247.

98. Buttel and Taylor, 1994; Inman, 1992; Stoncich, 1989; Stone and Hamilton, 1991; Yearly, 1991.

99. Lewis, 1996.

100. Cheru, 1992.

101. Lewellen, 1995, p. 121.

102. UNDP, 1998, p. 100.

103. Cardoso and Faletto, 1979; Wallerstein, 1974.

104. Escobar, 1995, p. 195.

105. Another example of northern intervention is the creation of parks. See Foresta, 1991, for example.

106. Shiva, 1993, pp. 151–152.

107. Escobar, 1995.

108. Chatterjee and Finger, 1994; Sachs, 1997, p. 77.

109. Amanor, 1994; DeWalt, 1994; Hviding and Baines, 1994; Nazarea, 1998; Warren, Slikkerveer, and Brokensha, 1995.

110. Buttel and Taylor, 1994, p. 236.

111. Lohmann, 1993, p. 159.

112. Gudynas, 1993, p. 173.

113. Braidotti et al., 1994, p. 97.

114. Ekins, 1993, p. 99.

115. UNEP, 1992.

116. From a global perspective, radical critic, Vandana Shiva, contends that actions to preserve biodiversity serves the MDCs by providing them with control of the LDC's resources. In reference to the Convention on Biodiversity, Shiva (1993, p. 84) remarks, "The Biodiversity Convention started out primarily as an initiative of the North to 'globalise' the control, management and ownership [of biodiversity] (which due to ecological reasons lie primarily in the LDCs) so as to ensure free access to the biological resources which are needed as 'raw material' for the biotechnology industry" (1993, p. 151). Shiva argues that attempts to promote sustainable development ignore the North's role in the destruction of the environment and blames the locals for the destruction, then "places responsibility for conservation in the hands of the sources of destruction." She believes a focus on biodiversity is being used by the World Bank and other organizations to continue business as usual.

117. West and Brechin, 1991.

118. West and Brechin, 1991, p. 20.

119. Wheat, 1994.

120. deKadt, 1979.

121. Lafant and Graburn, 1992.

122. deKadt, 1992; Pigram, 1992.

123. Harmon, 1987, p. 152. See Akama, Lant, and Burnett, 1995; Ghimire, 1994; West and Brechin, 1991; and Marks, 1984, for other examples of negative effects of parks on resident peoples.

124. McNeely, 1991.

125. However, recent scholarship on the topic that focuses on nine projects administered by the Nature Conservancy through the Parks in Peril project (Brandon, Redford, and Sanderson, 1998) suggests that these projects are in fact too ambitious and that biodiversity conservation efforts might be best off focusing on the "nature" aspect of conservation rather than human welfare.

126. IUCN, 1997.

127. Conservation International (n.d.).

128. "ANWR Hypocrisies (Editorial)," 1995.

129. Bergman, 1995.

130. Chance and Andreeva, 1995.

131. Shiva, 1993.

132. Myers, 1994.

133. Redclift and Woodgate, 1997, p. 62.

134. Brown, 1990, p. 189.

135. UNDP, 1998, p. 16.

136. UNDP, 1998, p. 15.

137. UNDP, 1998, p. 15.

138. Hueting, Bosch, and DeBoer, 1992.

139. UNEP, 1992.

140. Goeteyn, 1996, p. 171.

141. Anderson, 1991.

142. Daly and Cobb, 1994.

143. Redefining Progress website.

144. Nordhaus and Kokkelenberg, 1999.

145. Stren, White and Whitney, 1992; Van der Ryn and Cowan, 1995; Wikan, 1995.

146. Goeteyn, 1996.

147. Sustainable Seattle website, see also Hatcher, 1996.

148. Miringoff and Miringoff, 1999.

149. Guha, 1989.

150. Escobar, 1995, p. 222.

151. Escobar, 1995.

152. Sunderlin, 1995, p. 485.

153. Sachs, 1997.

154. Barkin, 1998.

155. Sachs, 1997.

156. Nazarea, 1999.

157. For examples, see Baker et al., 1997; Braidotti et al., 1994; Friedmann and Rangan, 1993; Healy, 2001.

158. Braidotti et al., 1994, p. 15.

159. Taylor, 1995.

160. Rudel, 2000.

161. Rudel, 2000, p. 80.

162. Rudel, 2000, p. 78.

163. Rudel, 2000, p. 81.

164. Rudel, 2000, p. 81.

165. Weber, 2000, p. 237.

166. Leff, 1996, pp. 152–153.

8

Environmental Sociology and Alternative Environmental Futures

In Chapter 1 we stressed the fact that North American environmental sociology has been diverse in terms of theories employed and the issues that are explored. It is, nonetheless, important to note that, until about the early 1990s, most environmental sociology tended to share some common intellectual goals. Two such interrelated goals deserve mention here. The first was the commitment by most environmental sociologists to correct what they saw as the lack of attention to the biophysical environment in mainstream sociology.[1] Their aim was to show that the biophysical world was relevant to sociological analysis as both a causal factor shaping social change and as an outcome of social structures or social processes. Environmental sociologists' second goal was to explain the causes of environmental degradation or environmental problems.

Most major theories in environmental sociology, such as the ones we designate as the conservative, managerial, and radical paradigms, proceeded to focus on the task of explaining the powerful social forces leading to environmental destruction. In general, environmental degradation was seen as being an intrinsic or fairly automatic consequence of the key social dynamics of twentieth-century capitalist-industrial civilization. The most well-known theories in environmental sociology were those that posited a key factor (or a closely related set of factors) that had led to enduring environmental crisis; these well-known theories included Schnaiberg's theory of the "treadmill of production," Logan and Molotch's theory of the urban "growth machine," Catton and Dunlap's theory of the "dominant social paradigm" and of the "age of exuberance," and Murphy's theory of the irrationality of capitalist-industrial rationality.[2] Because of the stress placed on explaining theoretically why the United States and other advanced industrial societies were inexorably tending toward environmental crisis, North American environmental sociology found itself in an increasingly awkward position. Most environmental sociologists had given so much stress to explaining why environmental destruction and disruption were inevitable, given the major social institutions within which we live, that there remained little room for recognizing how a more sustainable society might be possible or how social arrangements could be changed to facilitate environmental improvements.

To be sure, many environmental sociologists—even those whose theories made environmental disruption sound essentially inevitable, and beyond the ability of groups and societies to deal with directly—began to devote attention to how societies could find their way out of the "iron cage" of environmental despair. Many of these attempts date from as early as the late 1970s and early 1980s. Allan Schnaiberg's acclaimed *The Environment,* published in 1980, contains a chapter on environmental movements that is still well worth reading today. While Schnaiberg's emphasis in his discussion of various types of environmental movements was on why they had serious shortcomings as vehicles for reversing the capitalist "treadmill" and its environmental destruction, he nonetheless argued that the mobilization of organized environmental movements was the only plausible way that the treadmill could be slowed or reversed. Likewise, while the conservative Durkheimian theoretical work of Riley Dunlap and William Catton[3] tended to stress the extraordinarily powerful momentum in the direction of environmental destruction, Dunlap in particular has remained strongly committed to the culturally oriented notion that the "new environmental paradigm" is compelling and likely to catalyze environmental citizens movements across the globe.[4] (Dunlap and Van Liere, 1984; Dunlap, 1993).

The need to continue to reassess the state of environmental sociology is not only an intellectual one. Ultimately, environmental sociology's contribution to the human community will be to help think through how humanity's environmental future can be enhanced. Until the late 1990s, however, environmental sociologists had made only modest contributions to identifying likely or possible mechanisms that can yield a positive environmental future. Most of the contributions by environmental sociologists to the way that environmental integrity can be sustained in the future had tended to revolve around the role of environmental social movements. In addition, the environmental-sociological logic behind emphasizing the role of environmental movements was also based on the presumption that they would ultimately catalyze national environmental management. But there are several reasons why many contemporary environmental sociologists have come to believe that there are strategies for environmental improvement other than mobilization of the kinds of environmental movements that currently predominate. There is also reason to argue that environmental mobilization does not necessarily lead to parallel national policy changes.

One factor in reconsidering the role of environmental movements in the future is the recognition that these movements, particularly the mainstream ones that focus on affecting managerial environmental policies of the U.S. federal government and of international organizations and regulatory bodies, are being increasingly challenged by environmental countermovements.[5] As Schnaiberg and Gould point out, one of the increasingly powerful types of "environmental movements" is that of the "anti-environmentalist movement."[6] The anti-environmentalist movement involves a range of organizations such as the Wise Use Movement, the Property Rights Movement, and several groups, such as the Climate Council, that are fighting to prevent the U.S. federal government from cooperating with the negotiations at the 1997 Kyoto Round and the 2000 Hague Round of the Framework Convention on Climate Change. The anti-environmentalist movement has developed a persuasive conservative ideological position—that the problem is environmental alarmists rather than environmental problems, and that the market is already doing a sound job of allocating resources—and has a well-funded network of think tanks and support groups (such as the Hudson Institute and the Cato Institute).

A second major reason for reevaluating the role of environmental movements is the observation by many environmental sociologists[7] that radical environmentalism, long viewed by many environmental sociologists as the type of social force needed to counter rampant environmental destruction,[8] has become increasingly irrelevant in dealing with modern environmental issues. These observers believe that environmentalists can be most effective if they engage in collaborative relationships with industrial corporations and other bureaucratic entities whose actions have an impact on the environment. More broadly, one of the strong tendencies among sociological observers of environmental movements over the past decade or so is for them to express reservations that one or another major segment of environmentalism is wrong-headed in its strategy and destined to fail. As noted in the chapter on movements, Gottlieb's *Forcing the Spring,* for example, is a hard-hitting critique of highly professionalized, managerially oriented East and West Coast environmental groups, and a brief on behalf of a more locally based, grassroots environmentalism.[9]

The third factor advanced by environmental sociologists and other scholars as a reason to look beyond conventional environmental movements as mechanisms for advancing the cause of environmental protection is that some of the most promising strategies for an environmentally sustainable future have little or no relationship to the movements. These strategies, which include industrial ecology, strategic environmental management, and "dematerialization" of production, will be discussed later in this chapter. These bring new and combined forms of conservative and managerial paradigms into the field.

The remainder of this chapter will focus primarily on exploring the changes that are now underway in environmental sociology as scholars have come to emphasize the explanation of environmental improvement (rather than mainly explaining environmental degradation), and as they have diversified their approaches to understanding ways that a sounder environmental future can be made possible. The portions of the chapter that follow will be organized around the four key mechanisms that environmental sociologists have identified as strategies or routes to environmental improvement. These are: (1) mobilization of environmental movements, particularly "new" or novel movements that expand upon mainstream environmentalism; (2) sustaining or enhancing the environmental managerial capacity of governments; (3) "ecological modernization," the notion that modern industrial societies can solve environmental problems through intensified development of innovative industrial technology, through ecological efficiencies in production and consumption, and through green marketing and other strategic environmental management practices; and (4) "environmental internationalism," the notion that due to the intrinsically global scale of environmental problems and the importance of "globalized" socioeconomic institutions, the most effective route to environmental protection is global managerialism through international environmental agreements, international environmental "regimes," and international intergovernmental organizations.

ENVIRONMENTAL MOVEMENTS, OLD AND "NEW": THE SOCIAL MOVEMENT IMAGE OF OUR ENVIRONMENTAL FUTURE

As we noted in Chapter 6, analyses of environmental movement organizations and of the movement as a whole have been the most longstanding focus of environmental sociology.[10] The sociological analysis of environmental movements has gone through tremendous shifts over the last decade. These shifts pertain to the role that environmental movements will play in shaping our environmental future. Although many of these themes were discussed in Chapter 6, here we underscore some of the most pressing changes as they relate to possible environmental futures.

One reason for change in the approach to environmental movements and movement organizations is that there is now much more attention to the specific mechanisms according to which environmental movement activities lead to environmental reforms or improvements. Early in the development of environmental sociology there was a presumption that, at least over the long term, there would be a relatively automatic tendency for environmental collective action to occur for one or more reasons. Many environmental sociologists presumed that evidence about and public awareness of environmental problems would eventually lead to citizen mobilization, as Robert Brulle has pointed out.[11] Other observers have suggested that as the United States and other industrial societies become increasingly affluent, the growth of the educated middle class would increase the base of support for environmental management.[12]

The linkages among affluence, environmental problems, and citizen environmental mobilization are by no means automatic, however. Consider, for example, the fact that the nature and extent of environmental problems are far better understood today than they were three decades ago, but that there has been little landmark environmental legislation passed in recent years, at least by comparison with the 1970s.[13] Thus, in addition to the need for scientific documentation—or a parallel process of popular or lay documentation of an environmental issue—in order to mobilize people to be concerned about an issue, these concerns need to be incorporated within environmental discourses or ideologies, and be seized upon by one or more environmental organizations. The attractiveness of an issue for media coverage is also a significant factor in shaping the extent to which the problem generates public interest and concern and becomes incorporated within the agenda of one or more environmental groups.[14]

A second reason why the role of environmental movements has come to be reassessed is that these movements are increasingly being challenged—and often overwhelmed—by anti- or counter-environmental groups. Andrew Rowell and Joe Thornton, for example, have each documented the growing trend toward well-funded anti-environmental organizations being formed to contest the efforts by environmental organizations to advocate for managerial environmentalism or reform policies.[15] Typically, these groups are funded by private corporations or by conservative philanthropies, though there are instances in which anti-environmental groups have emerged relatively spontaneously at the local level or are unaffiliated with conservative corporate interests.[16] Anti-environmental organizations are most prevalent and effective in the areas of land use regulation and in congressional and other domestic discussions of policies for controlling "greenhouse" gas emissions. Thus, one of the critical dimensions of the role played by environmental movement organizations and of the movement as a whole is the capacity of these groups to contend with anti-environmental groups at various levels.

A third significant change is that the environmental movement has undergone increasing differentiation. The movement is far more complex than it was at the dawn of environmental sociology as a recognized sociological specialty. In particular, there is now increased differentiation between the large Washington, D.C.- and New York-based national and international environmental groups on one hand, and much smaller local environmental groups on the other. Also, there has been continual ideological differentiation among these groups—witness, for example, the vast gulf between the relatively conventional, if not conservative, conservation groups such as the National Wildlife Federation and the Audubon Society compared to much more radical organizations such as the "deep ecology" group Earth First! and the relatively militant groups such as Friends of the Earth, Environmental Liberation Front, and Greenpeace.

Robert Brulle has studied the "discourses"—the competing paradigms with their major premises and claims—of major U.S. environmental groups and has based his research on the notion that studying common patterns in discourses can help to identify the major types of environmental groups that have existed over time.[17] Brulle has noted that from the mid-nineteenth century until the 1960s there were only two major types of pro-environmental paradigms and groups in the United States: "preservation" groups such as the Nature Conservancy and Wilderness Society advocating the preservation of wilderness and other natural areas, and "conservation" groups such as the National Wildlife Federation and Isaac Walton League advocating the reduction of resource waste through proper management and application of science to natural resource policymaking. Over the past 35 or so years, however, there have been four major new types of environmental movements that have emerged in the United States. These new types of environmental movement paradigms include the "ecocentric," political ecology, deep ecology, and ecofeminist discourses.

Ecocentric environmental groups—typified by the Natural Resources Defense Council, the Cousteau Society, and Zero Population Growth—adhere to the Malthusian view that natural systems are the basis of humanity, that human survival is linked to ecosystem survival, and that human ethics should be guided by ecological responsibility, or, moral restraint as discussed in Chapters 2 and 3. The political ecology paradigm is guided by a managerial view that the domination of humans by other humans leads to the domination of nature, and that political and economic power creates major environmental problems. Solutions to environmental problems require fundamental social change based on empowering subordinate groups such as local communities and poor people within these communities. Examples of well-known political ecology groups have included the Citizen's Clearinghouse for Hazardous Wastes during the 1980s and early 1990s, as discussed throughout the book, and the Government Accountability Project in recent years. Deep ecology groups' paradigm is based on the fundamental principle that the richness and diversity of *all* life—including nonhuman life forms—have value and should be protected, and that human life should be privileged only to the extent required to satisfy humans' vital needs.[18] The militant group Earth First! has been the classic deep ecology group, while Rainforest Action Network and Wild Earth are two more recent deep ecology environmental movement organizations. Finally, ecofeminism, also discussed throughout this book, is based primarily on the notion that ecosystem destruction is based on androcentric or patriarchical cultural values and institutions, and that eradication of androcentric institutions is the lynchpin of solving environmental and other social problems. World Women in Defense of the Environment and Women in Environment and Development are typical ecofeminist groups.

Fourth, the past decade has witnessed the rise of other new—and often highly innovative or provocative—environmental movement organizations and movements such as the environmental justice movement, the "grassroots" environmental movement, and radical ecological resistance movements in the developing world.[19] The closely related grassroots environmental movement and the environmental justice movement in the United States and "new social movements" in European countries are particularly notable instances of new types of environmental movements worth discussing here.[20]

There is a tendency when thinking about the environmental movement to focus largely on the major national and international environmental groups because of their visibility. But it is the case that Americans who are actually directly involved in environmental activism are much more likely to do so within local rather than nationally or globally focused environmental groups. The grassroots environmental movement is a highly activist component of the groups that operate mainly in particular communities or regions.

As discussed earlier, the principal impetus for the grassroots environmental movement was the discovery of widespread toxic chemical pollution in the Love Canal neighborhood in Niagara Falls, New York.[21] The grassroots environmental movement has continued to stress toxic chemical and related issues such as toxic waste dumps, contamination of water supplies, radioactive wastes, factory pollution, and siting of hazardous waste disposal facilities and garbage incinerators. Grassroots environmental groups also deal with broader issues of the protection of public health.

To some extent grassroots groups focus on issues that the more visible organizations in the environmental movement tend to ignore. Over the past 15 years the more visible parts of the environmental movement have tended to emphasize global-scale or transboundary environmental issues, and in so doing they have generally deemphasized relatively local kinds of problems such as toxic wastes, land use, and so on. Grassroots environmental groups fill the void created by mainstream groups having moved toward the national and international managerial policy arenas. Grassroots environmental groups differ from more mainstream ones in ways other than their stress on public health and toxic substance issues. While the large groups' members are mostly white and middle class, grassroots group members are from a broader cross-section of class backgrounds. Grassroots groups are especially likely to have women and volunteer leaders. Grassroots group members are also much more likely to distrust government and scientists, and to take strong or uncompromising stands, than are the national environmental groups. There are tendencies toward antagonism between the two groups, a good share of which comes from grassroots group members tending to "perceive the nationals as remote, overly legalistic, and too willing to accommodate to industry's concerns."[22]

The environmental justice movement is a particularly innovative and prominent form of the grassroots environmental movement with considerable potential to affect our environmental future.[23] The environmental justice movement was inspired by grassroots environmental mobilizations, but was catalyzed by the U.S. civil rights community, particularly the components of the faith community committed to social justice. It is thus a joint civil rights, social justice, and environmental movement. The environmental justice movement is based on the claim that many types of environmental destruction—particularly those involving toxics, pollution of the workplace, and polluting factories, waste dumps, and nuclear processing facilities—have their most adverse impacts on minority communities and the poor in general. Environmental protection is thus seen as a civil rights or social equity issue. Environmental reform and redirection of the processes for siting waste dumps and other polluting facilities have thus come to be redefined as social and racial justice concerns. What gives the environmental justice movement its force is the fact that it blends the themes of environmentalism and social and racial justice in a way that can bring forward an impressive level of mobilization around local and regional environmental issues. Environmental justice issues can also fall under civil rights and equal protection laws as well as under environmental laws.

Fifth, environmental movement organizations are changing as a result of new coalitions and alignments among various related movements. There are now, for example, increasingly close alliances between environmental movement organizations and other movements with which environmentalists were once thought to have very little in common. Environmental groups are now increasingly engaging in coalitions with organizations from movements such as the anti-trade movement, the labor movement, the sustainable agriculture movement, the consumer movement, the anti-biotechnology movement, genetic resources conservation movement, the human rights movement, and so on. A set of interrelated issues regarding globalization and trade has increasingly led environmental groups into unprecedented alliances with other movements.

The best illustration of these new patterns of coalition among movements is the role played by environmental social movement organizations both before and after the November 30, 1999 protest at the World Trade Organization (WTO) Seattle Ministerial Conference, that was held to kick off the Millennial Round of negotiations over extending the WTO. The "Battle in Seattle" was the culmination of a more than decade long tendency toward what Fred Buttel calls "environmentalization."[24] Environmentalization is the process by which a formerly nonenvironmental issue such as trade or human rights comes to be defined substantially as an environmental issue. During the 1990s, as trade liberalization policies such as the WTO and the North American Free Trade Agreement (NAFTA) were enacted and implemented, these policies' potential environmental consequences have been noted.

There has been considerable concern, for example, that trade liberalization policies such as WTO and NAFTA might have negative implications for the United States' ability to manage environmental problems, for two major reasons. First, trade liberalization policies are aimed at phasing out barriers to trade by defining certain types of trade restrictions to be illegal restraints on trade. WTO enables countries to challenge each others' laws if these laws can be seen to constitute a "non-tariff barrier to trade." Many of these types of newly forbidden restrictions on trade are environmental policies. A good example is the U.S. Marine Mammal Protection Act, which restricted imports of tuna that was not produced using "dolphin-safe" procedures minimizing dolphin deaths in the harvesting of tuna on the high seas. Mexico filed a complaint with the WTO, which ruled in favor of Mexico and forced the repeal of the U.S. import restriction.

A second and closely related concern is that trade agreements such as WTO and NAFTA may result in a "downward harmonization" of regulations across the world, or, "the race to the bottom." WTO and NAFTA, in other words, could result more often in countries watering down their regulations on imports than they are to countries increasing their health, safety, and environmental standards on imported goods. As an example, in 1997 the WTO overturned part of the U.S. Clean Air Act—the part that prevented the import of low-quality gasoline with a high potential for air pollution.

When NAFTA and the World Trade Organization were originally considered for ratification by the U.S. Congress in 1993 and 1994, respectively, essentially all major U.S. environmental groups either supported or were neutral about NAFTA, and only five opposed ratifying the WTO agreement. Since that time, however, virtually all major U.S. environmental groups have come to have grave reservations over conservative "free trade" policies because of their potential environmental impacts or their implications for effectively repealing U.S. environmental legislation. The potentially negative impacts of trade liberalization on the environment have proved to be critical in stitching together the surprisingly broad coalition of movements that joined the Battle in Seattle. Trade, along with certain other issues such as opposition to genetically modified food products, has proven to be a "bridging issue" that serves to help cement a far broader coalition than might otherwise be possible. The wide range of environmental, labor, consumer, farmer, international development, human rights, anti-biotechnology, and related groups that joined forces in Seattle often had little in common before. Their opposition to trade liberalization, as well as their opposition or ambivalence toward genetic engineering, served to unite these groups into a relatively harmonious coalition that has had a decisive impact on the politics of international trade liberalization. While it was largely taken for granted in 1998 and early 1999 that the WTO was well established and, if anything, would be strengthened in the Millennial Round negotiations, the strengthening of WTO during the early years of the twenty-first century now appears to be problematic.

Finally, developing country environmental movements, some of which are very radical and closely aligned with social and international justice concerns, are a further instance of innovative and potentially transformative environmental movements. Some developing country environmental movements are relatively similar in their membership characteristics and goals to preservationist or ecocentric movements in the developed countries of the North. The most innovative and dynamic developing country movements, however, are those that have their origins as much or more in social justice concerns as in the impulse to preserve biodiversity or sensitive ecozones. Many of these developing country environmental movements, for example, have been organized around advocacy of the rights of indigenous peoples and peasants, particularly in the context of struggles over access to land. The tragic death of Chico Mendes over the right of the rubber tappers to protect trees in the

Brazilian Amazon exemplifies this kind of environmentalism in developing countries.

Other developing country movements have been highly involved in the international processes of negotiating treaties, protocols, and other international agreements relating to biodiversity, forest conservation, and control of greenhouse gas emissions. Developing country environmental groups usually weigh in on these discussions by advocating agreements that involve fairness to the developing world and to poor people and peasants.[25] It should also be noted that developing country environmental movements very often have close organizational and financial linkages to counterparts in the North, leading not only to coalitions but also to cross-fertilization of ideas.

The social movement image of our environmental future is largely a managerial one of political mobilization and political pressure placed on state officials and private decision makers. The essence of this image of the future is that as justifiable and rational as environmental protection might seem in the abstract, there is such a strong tendency for private interests to favor expansion of production and consumption. There must be constant political pressure from mobilized citizenries to keep public as well as private decision makers environmentally accountable. There clearly is no necessary relationship between environmental movement mobilization and pro-environmental outcomes. The recent history of environmental movements around the globe, nonetheless, suggests an innovative diversification of approach and tactics, and perhaps a reasonably well-functioning division of labor, that will be at the heart of any environmental progress to be made in the future.

THE MANAGERIAL PARADIGM, THE ENVIRONMENTAL REGULATORY STATE, AND OUR ENVIRONMENTAL FUTURE

The notion that government or state regulation of environmentally related private decision making, particularly by industrial corporations, would be central to a promising environmental future is an old one. Numerous histories of the early origins of environmentalism and environmental protection success stories in the United States and elsewhere point to the fact that the quest for resource conservation was, more often than not, very closely associated with supportive, if not catalytic, managerial actions from government agencies and officials. At the turn of the twentieth century in the United States, for example, much of the thrust behind what we now would call environmental protection came from nascent government agencies such as the Forest Service and Department of Interior. The Progressive Era conservation movement at the turn of the twentieth century was as much a federal government-sponsored movement among government-agency-based resource managers armed with new developments in the sciences such as forestry, fisheries, and agriculture as it was a social movement among citizens. To be sure, conservation organizations in civil society such as the Sierra Club and Isaac Walton League, along with professional resource management associations such as the American Forestry Association, played major roles in encouraging government officials to take steps to improve the conservation performance of America's natural resource sectors. But it has been repeatedly documented that the impetus for conservation programs often came from *within* government circles.[26]

Stephen Skowronek, in his now-classic study of the development of the American federal government, has noted that the rise of the natural resource management agencies and of the regulatory apparatus that went along with them was one of the most critical changes in the "modernization" of the American state.[27] As recently as the late nineteenth century, the American state was a government of "courts and parties," in which a conservative Congress strongly protected states rights and blocked attempts to have a stronger federal role in the economy and society, while a conservative court system staunchly protected the prerogatives of property. At the time there was little impetus or mechanism for collective interests or concerns to be reflected in national governmental policies, especially if doing so might involve significant public expenditure, reduction of "states rights," or federal intervention in the decision making of private capital. Ultimately, however, many of the accumulating excesses of the United States' highly decentralized governmental system created massive social and natural resource problems that could not be dealt with through the traditional governmental order of "courts and parties." Farmers, among many other groups, agitated for protection from railroad, farm machinery, and other monopolies that had been permitted to develop to an extraordinary degree under protection of a conservative property-protecting judiciary. Middle-class reformers clamored for federal laws that would protect the young and the working class from the problems of an unregulated workplace. Most significantly for present purposes, there was a growing voice in support of the need for federal regulation of the activities of loggers, miners, and others who were seen to be despoilers of the country's natural bounty and patrimony.[28] Many scholars thus take what we now call the rise of the environmental regulatory state, a reflection of the managerial paradigm, to be one of the central and defining features of the development of the modern form of liberal democratic government in the Western countries.

There can be little doubt that the environmental regulatory state in the United States has contributed richly to environmental protection in our country. Table 8.1, taken from Michael Kraft, identifies the major federal environmental laws in the United States since 1964.[29] Essentially all of these laws, particularly the truly landmark laws passed in the 1970s, involved "nationalizing" environmental policy. Kraft notes that

> Environmental policy was "nationalized" by adopting federal standards for the regulation of environmental pollutants, action-forcing provisions to compel the use of particular technologies by specified deadlines, and tough sanctions for noncompliance. Congress could no longer tolerate the

Table 8.1 Major Federal Environmental Laws, 1964–1999

1964	Wilderness Act, PL 88-577
1968	Wild and Scenic Rivers Act, PL 90-542
1969	National Environmental Policy Act, PL 91-190
1970	Clean Air Act Amendments, PL 91-604
1972	Federal Water Pollution Control Act Amendments (Clean Water Act), PL 92-500
	Federal Environmental Pesticides Control Act of 1972 [amended the Federal Insecticide, Fungicide and Rodenticide Act (FIFRA) of 1947, PL 92-516]
	Marine Protection, Research, and Sanctuaries Act of 1972, PL 92-532
	Marine Mammal Protection Act, PL 92-522
	Coastal Zone Management Act, PL 92-583
	Noise Control Act, PL 92-574
1973	Endangered Species Act, PL 93-205
1974	Safe Drinking Water Act, PL 93-523
1976	Resource Conservation and Recovery Act (RCRA), PL 94-580
	Toxic Substances Control Act, PL 94-469
	Federal Land Policy and Management Act, PL 94-579
	National Forest Management Act, PL 94-588
1977	Clean Air Act Amendments, PL 95-95
	Clean Water Act (CWA), PL 95-217
	Surface Mining Control and Reclamation Act, PL 95-87
1980	Comprehensive Environment Response, Compensation, and Liability Act (Superfund), PL 96-510
1982	Nuclear Waste Policy Act of 1982, PL 97-425 (amended in 1987 by the Nuclear Waste Policy Amendments Act of 1987, PL 100-203
1984	Hazardous and Solid Waste Amendments (RCRA amendments), PL 98-616
1986	Safe Drinking Water Act Amendments, PL 99-339
	Superfund Amendments and Reauthorization Act, PL 99-499
1987	Water Quality Act (CWA amendments), PL 100-4
1988	Ocean Dumping Act of 1988, PL 100-688
1990	Clean Air Act Amendments of 1990, PL 101-549

NOTE: A fuller list with a description of the key features of each act can be found in Vig and Kraft (2000), Appendix 1. Chapters 5 and 6 provide brief descriptions of the major environmental protection and natural resource policies.

SOURCE: Kraft, 2001. Used by permission of Pearson Education.

cumbersome and ineffective pollution control procedures used by state and local governments (especially evident in water pollution control). Nor was it prepared to allow unreasonable competition among the states created by variable environmental standards.[30]

The nationalization of environmental regulation depicted by Kraft clearly bore fruit. Virtually all of the national-level legislation in Table 8.1 has yielded significant results. In particular, there has been great progress during the twentieth century in adding and protecting wilderness, forests, and sensitive habitats within nature reserves such as those of the U.S. Park Service. Also, since the late 1960s, there has been considerable progress in air and water pollution control (particularly relative to what could have been the state of air and water quality if the trends in production, consumption, and pollution after World War II would have continued until the end of the century) and in workplace health and safety. It is often noted, in fact, that the 1970s were a kind of Golden Age of American environmental protection policy, in that this was the most significant epoch of environmental policy innovation in our history.

Why is it that political systems such as the U.S. federal government have become increasingly involved in environmental regulation over the twentieth century? Many scholars argue that the federal regulatory role resulted partly from the pressures placed on government by the environmental movement.[31] But it is also apparent that there was a definite momentum behind the nationalization of responsibility for environmental control and protection well before the mobilization of the late 1960s and early 1970s environmental movement. Thus, it must be the case that, at least to some degree, there has been some impulse toward federal environmental regulation that originated independently of environmental movement pressures.

In the industrialized countries of North America, Europe, and Oceania the modern form of institutionalizing environmental tasks in state policies and politics generally dates back to the 1960s. From the 1960s through the 1980s the state rapidly expanded the span of managerial activities and powers in environmental protection and occupied a "comfortable" and unquestioned position in dealing with environmental problems. In the United States, for example, the expansion of the scope of federal responsibility for environmental protection coincided with the establishment of the Environmental Protection Agency by Executive Order in 1970.

Scholars and other observers, of course, pointed out repeatedly at the time that government responses to environmental problems, challenges, and crises were very uneven and often inadequate.[32] But during the heyday of national government environmental regulation, concerns about the limited successes of government natural resource and environmental protection policies invariably led to calls for *more rather than less* state management activity and intervention in the economic processes of investment, production, and even consumption. The nearly universal reaction among environmental management professionals and scholars was that there was no realistic alternative to assigning to the state the key role in ensuring environmental "public goods." There was also broad consensus that the only way to realize societal demands for high environmental quality and minimized environmental risks was for a stronger state to better counterbalance the power of corporate capital. Even where the capitalist economy was seen as one of the major causes of environmental deterioration, more active intervention of the nation-state in the essential economic decisions in the private sector was believed to be the only plausible remedy. This consensus behind the necessity of an interventionist environmental state was cemented further by the more general view in the 1960s and 1970s, particularly in Europe, of the desirability of developing an activist welfare state.

A useful perspective on the origins of national-governmental environmental regulatory capacity and the historic consensus about its role can be based on the notion that one of the intrinsic roles of governments in a societal division of labor is to rationalize social arrangements in the interest of order and efficiency. The government or political system can be distinguished from other social institutions in that the government or state is the only institution with the ability—and thus ultimately the responsibility—to make possible what might be called the rationalization of society. Most other institutions in society, particularly economic institutions, are based on private incentives or group interests, many of which are represented before the branches of government in pursuit of (narrow) group benefit. The state, by contrast, is constituted with the expectation (or, at a minimum, with an ideology) of providing collective benefits, and with the prerogative to foster changes in society that make social and economic life function more smoothly, efficiently, or rationally.[33] Environmental protection can be thought of as the textbook case of a policy arena in which government agencies and officials are in a distinctive position of being able to take steps to rationalize institutional rules and societal behaviors in order to create a level of ecosystem protection that benefits citizens as a whole (at least from their perspective).[34] Thus, many sociologists suggest that the national and other levels of government tends to take on the role of environmental protection because government is the only institutional sphere that has the capacity and the potential legitimacy to do so. Responsibility for ensuring environmental protection (or at least some modicum of it) is inherent in the state's role in a societal division of labor. Environmental movements clearly can increase the level of demand and pressure on the state to increase its commitment to environmental protection, but movement pressure cannot itself establish the fact that state managers and agencies can legitimately take steps to intervene on behalf of resource conservation and maintenance of environmental quality.

A recent spate of impressive histories of twentieth-century American environmental policies has converged on the notion that a society's ability to make possible environmental protection is essentially a function of the national-state's capacity to manage—to enact and to implement regulations of private behaviors.[35] Thus, it is not surprising that, as a result of the enactment of 1970s and subsequent environmental regulation policies, and following on ever more conclusive evidence that these policies have more or less worked, many social scientists have felt that the development and maintenance of the state's capacity to regulate private resource decision making comprise the critical factor in our environmental future. There has thus tended to be a presumption in many quarters that environmental protection can go only so far as there is capacity of government resource management and environmental agencies to implement an environmental regulatory and control agenda.

In the 1980s, however, the comfortable state of affairs of the continually expanding responsibility of national governments to enact and implement regulations to protect the environment came under serious scrutiny and pressure for the first time. The conservative regimes of the 1980s—especially those in the United States (Reagan) and the United Kingdom (Thatcher)—were heavily inspired by neoliberal scholars who argued for and legitimized strong deregulation and privatization programs. These neoliberal tendencies affected a wide range of policy fields, including the environment.[36] In addition, Reaganism and Thatcherism indirectly but substantially influenced the political cultures and regimes of a variety of countries around the world and led to similar deregulatory demands elsewhere.

Thus, while there is still considerable recognition that the government's managerial role in environmental protection remains critical, in the late twentieth and early twenty-first centuries the future of environmental regulation—particularly in the U.S. federal government's key environmental regulatory agency, the Environmental Protection Agency (EPA)—has come under considerable doubt. The expansion of the American government's responsibility in environmental regulation and protection, for one thing, was never complete. Many historical analyses of the rise of the American regulatory state, especially its environmental regulatory apparatus, have noted that the rise of this regulatory system did not involve displacing the previous system of subsidy to resource consumption and protection of private property rights. Thus, the regulatory system was superimposed on our decentralized governmental system, leading to endemic conflict among federal agencies and among the three branches of government over the implementation of environmental policies.[37] The Environmental Protection Agency, which has become the most important U.S. federal environmental protection bureaucracy, has never acquired cabinet status and tends to have far less influence in the federal government than do agencies such as the Departments of Defense, Treasury, and State. Thus, many critics of the traditional environmental regulatory state feel that the subordinate stature of the environmental agencies means that they can only be reactive.

Perhaps the most unsettling area of doubt about the future role of government environmental regulation is that the conventional form of environmental control—what is often referred to as "command-and-control" regulation—is increasingly seen as being outmoded. Raymond Murphy, for example, has argued that command-and-control regulation tends not to be very innovative or dynamic because it cannot escape the limits of "end-of-the-pipe" control.[38] Conventional environmental regulation, that is, cannot go beyond setting standards for regulating corporate behavior in terms of the levels of emissions of pollutants of various kinds and litigating when these standards are not met. This style of regulation, accordingly, presumes that firms will continue to pollute, albeit less so. Mandating use of specific pollution control structures will often merely shift contaminants from one place to another such as from water to the land. This form of regulation, in Murphy's view, provides little or no incentive for firms to make innovative changes in their production practices that could result simultaneously in reduced costs and reduced resource usage.

Another commonly expressed variant of the conventional critique of command-and-control regulation is that it is inflexible and inefficient. Some of this criticism is based on empirical studies showing, for example, that most command-and-control regulation sends inefficient signals "at the margin" so that firms often respond to regulations in ways that reduce employment or national income.[39] Command-and-control regulation can be cost inefficient because it often mandates costly pollution abatement equipment when overall pollution levels could be reduced more cheaply through some market mechanism such as pollution trading permits. Regulations can become obsolete very rapidly in industries where there is a brisk pace of technological innovation. There has also been a trend toward rising government outlays and privately incurred costs associated with environmental regulation.[40] Command-and-control regulation is also argued to foster more adversarial relations between agency staff managers and private decision makers than is necessary or desirable.

It should also be noted that much of the hesitation about national government environmental regulation has had to do with the fact that conservative think tanks and related groups have felt that their future lies with a more globalized world in which national regulations, as well as other government interferences such as corporate taxation and restrictions on investment and trade, play a decreased role. Thus, the title of Fred Block's book, *The Vampire State,* indicates how there has been a general "demonization" of the role of government on the part of many corporate officials, conservative think tanks, and associated intellectuals.[41] The anti-environmental movement has been a major voice contributing to the demonization of the state as a whole, and to the demonization of centralized environmental rule making in particular.

Beginning in the 1980s, these various criticisms of national environmental regulation led to attempts by conservative governments to achieve "regulatory relief" for their supporters and clients—or, in other words, to reduce the role of the state in managing the environmental performance of the productive sectors—with varying degrees of success. In the United States, for example, the first Reagan administration achieved a very substantial rollback of environmental state activities and influence, as indicated by the fact that EPA expenditures declined in real (1997) dollars from $24.4 billion in 1980 to $18.2 billion in 1984; as late as 1998 environmental agency spending in constant 1997 dollars remained below the 1980 pre-Reagan expenditure level.[42] While the environmental deregulatory impulse was particularly strong in the United States and the United Kingdom and had considerable influence abroad, its impacts were quite variable internationally. Environmental policies in the Netherlands, for example, were hardly affected by the wave of deregulation and privatization of the 1980s.

Over the past several years there has been an intensifying debate in the United States over whether centralized or nationalized command-and-control regulation is desirable for environmental protection—or, in other words, over whether state environmental management should be thought of as the necessary centerpiece of a desirable environmental future. On one hand, it is now clearly established in the United States that the trend in environmental policy is toward deregulation. Walter Rosenbaum, for example, argues that the EPA has acquired a "battered agency syndrome," in that it has been the target of a range of interest groups—as well as the bulk of Congress and the Republican Party—and has become halting and indecisive in its role, and on the defensive about command-and-control regulation and centralized environmental rule making.[43] The EPA has thus taken on what Rosenbaum refers to as the "gamble with regulatory reinvention." EPA's regulatory reinvention has included steps such as developing industry-specific standards—rather than standards being applied to all industries, giving the states more responsibility in environmental protection, and developing market incentives for pollution control as, for example, administering a national market for sulfur oxide emissions through the 1990 Clean Air Act Amendments, and promoting a new type of "market enviornmentalism." The EPA has also widely implemented "risk assessment" procedures that replace mandatory pollution control with a cost-benefit assessment of regulatory decisions and standards. There have been some notable successes associated with these reforms. There are significant concerns, however, that these shifts are not supported by the EPA staff and will reduce the long-term capacity of the government to control major environmental problems.[44]

There are emerging economic analyses of American environmental regulations, on the other hand, suggesting that claims of the cost-inefficiency and inflexibility of national command-and-control regulation are exaggerated if not incorrect.[45] Many sociologists have also suggested that it is unrealistic to expect that the environmental role of governments can become as consensual, efficient, and innovative as is implied in many of the critiques of government regulation. There is now a growing tendency to think of government environmental regulation by employing the terminology of the "environmental state."[46] The notion of the environmental state means not only that the government is the key agent of environmental control and rationalization but that there also is a corresponding tendency for governments to take on the major responsibility for ensuring environmental protection. The notion of the environmental state also suggests an expectation that, as the state's responsibility for environmental protection grows, it becomes inevitable that state activities will involve conflict and contradictory responsibilities. The essence of these contradictory responsibilities is that, on one side, states face strong pressures to expand production, consumption, and living standards, and thereby the state is implicated in *causing environmental destruction,* mostly indirectly but sometimes directly through its public works and other programs. On the other side, and just as fundamentally, the state is being expected by citizens and various social groups to ultimately be the key entity *ensuring environmental conservation*. There is thus something of an inescapable contradiction between causing and being responsible for the ameliorative management of environmental problems, and this contradiction leads to an environmental state that functions in an indefinite pattern of ambivalence and internal struggle.

Defenders of the national-state role in ensuring a promising environmental future do not confine their advocacy to defending command-and-control regulation. There has been particular enthusiasm in recent years for innovations in regulatory practice such as applying the legal "Precautionary Principle"

(PP) to regulatory decision making. The PP is now being looked to by most environmental groups and many in environmental regulatory agencies around the world as playing a particularly important role regarding regulation of the approval and introduction of new foods, drugs, and chemicals.

In January 1998 a group of scientists, government officials, lawyers, and activists met at the Wingspread Conference Center in Racine, Wisconsin, to develop a "Wingspread Statement on the Precautionary Principle." The PP, which is relatively widely recognized as the guiding principle for regulation of chemicals and potentially hazardous practices in the European Union, has two major components. First, it involves a shift in the burden of proof from government regulatory agencies to private firms; thus, under the PP, it is not the obligation of government to prove that a new product or production practice is harmful, but rather an obligation of private firms to prove that the practice is safe. Second, the scientific standard for implementing the PP in regulatory decision making is a more encompassing one. Products or practices could be disapproved if there is evidence of any harm and/or if there it is a plausible scientific rationale that approval could lead to negative health or environmental effects. In addition, the Wingspread conferees generally supported the notion advanced by the ecological economist Robert Costanza of the University of Maryland that firms introducing new technologies, chemicals, and production practices should be required to provide "assurance bonds," a procedure he calls the "4P approach to scientific uncertainty."[47] Assurance bonds would be based on a worst-case scenario of the costs of a new technology, process, or chemical, and would be forfeited, at least partly, if there are eventually found to be damages associated with the practice. Bonds would be returned to firms with interest if and when harmlessness was proven over time.[48]

Conceptualization and advocacy of the PP comprise is but one of a large number of strategic and policy innovations that have been pursued within the governmental and civil-society communities interested in enhancing the role of environmental regulation. Another exciting frontier of environmental policy thought is that of the growing interest in national environmental accounting and in providing information on corporate environmental performance through modalities such as the Toxic Release Inventory (TRI) and the CERES principles, as discussed in Chapter 7.[49] It is useful to note in this regard that, while environmental agencies in a number of world nations (particularly those in Europe) have been receptive to the PP, the impetus for institutionalizing the PP in national law and international agreements has come largely from environmental movements. These innovations suggest that, while a reevaluation of theoretical presumptions about the role of environmental movements and environmental regulation has long been overdue in the field, there exists an extraordinary vitality and dynamism in environmental policy thought. Further, the critical role played by environmental movements is apparent. Accordingly, researching the nature of the relations between environmentalism and regulatory practice represents a high priority for environmental sociologists interested in exploring our environmental future and the role of the managerial paradigm in our future.

THE CONSERVATIVE PARADIGM, ECOLOGICAL MODERNIZATION, AND THE IMAGE OF OUR ENVIRONMENTAL FUTURE

In contrast to those who believe that either environmental mobilization and movements or national governmental environmental management are the ultimate guarantors of a secure environmental future are ecological modernizationists. They have far less faith in either movements or governments as agents of a sounder environmental future. Ecological modernization theorists believe that as much as environmental problems in the past have been caused by an industrially driven process of expanded production and consumption, the solution to environmental problems cannot be found in radical movements that seek to restore the lower levels of output and consumption that prevailed years ago, or in centralized, managerial command-and-control regulation.

As discussed in earlier chapters, ecological modernizations—as proponents of the conservative paradigm—are critical of both radical environmentalism and conventional environmental regulation for several reasons. First, ecological modernizationists have observed that such radical environmental movements aimed at reversing the process of modernization—what they often call "countermodernity" movements—have not been very successful. These radical movements have attracted very little public support, and are mostly ignored by government officials and private decision makers. Second, ecological modernizationists suggest that going back to some imagined utopia of a less-industrialized past is infeasible. Most people—those in the industrial countries, and many of the privileged in the developing world—will be unwilling to reduce their living standards significantly even if doing so might make possible major improvements in the health and sustainability of the environment. Also, most of the processes that have led to ecological deterioration such as capital intensive industrial expansion, corporate competition, and international competition are so powerful that they are not likely to be restrained or reversed even if there was a broad consensus in favor of environmental protection. Industrialization, for example, has led to such extraordinary advantages in terms of life expectancy, safety, comfort, and so on that rolling back the industrialization process around the world seems inconceivable. Finally, the ecological modernizationists raise many of the concerns about the inflexibility and inefficiency of command-and-control regulation that we previously discussed.

In particular, the ecological modernization image of our environmental future is based on the observation that some of the core features of a more environmentally secure tomorrow are already emerging or already in place though they are less visible than radical environmentalism or government standard setting. Ecological modernizationists see several hopeful trends and processes. One such process, which we will discuss at greater length later, is that there are areas of production and consumption in which improvements are being made that are resulting in reduced use of resources and lower levels of pollution. These areas of improvement are best typified by "industrial ecology" advances in manufacturing sectors such as in European chemical and paper production.[50] Industrial ecology practices go far beyond reduction of pollution emissions at the "end of the pipe." Industrial ecology practices involve drastic restructuring of production processes in order to tighten recycling loops inside and outside of the factory. These tight or closed loops are such that "waste" in any given production process—by-products in making paper, for example—becomes a valuable input in another production process. In addition to the increasingly widespread use of industrial ecology practices and other eco-efficiency measures, ecological modernizationists also see the global spread of "green marketing" and strategic environmental management as evidence that there is an ongoing process of environmentally friendly "modernization."

The ecological modernization image of the future has elements of both "automaticity" and political specificity. That is to say, on one hand there are some respects in which ecological modernizationists believe that the tendency toward a more environmentally friendly future is essentially an extension of well-established institutional patterns of social change. Private industry, for example, has an interest in efficiency. Industrial eco-efficiency can be achieved by being more sparing in the use of resources and raw materials; or, in other words, by minimizing production costs and by minimizing the ancillary costs of production, such as pollution control expenditures. Likewise, being able to market "green products" such as organic foods, recycled paper, or dolphin-safe tuna gives firms an advantage in the marketplace. Further, industrial-ecological production practices, cultivation of a positive pro-environmental image, and associating the corporate or brand name with environmentally friendly practices may serve to build brand loyalty and reduce expenses associated with lawsuits, liability, and litigation. The continual competition faced by private industry—reasoning that relies on the conservative paradigm—provides an ongoing incentive that can reinforce the incentives for pro-environmental decision making.

But if we accept that there are some sound management reasons why environmentally friendly corporate behavior can occur, why is it that corporate environmental accountability and improved eco-performance are far from universal or far from the norm? Here ecological modernizationists have observed that it is not simply corporate competition that is a necessary and sufficient condition for more positive environmental outcomes in the future. Ecological modernization thus involves political specificity; there must be a "modernization" of politics that reshapes the competitive corporate environment in order to make the pursuit of environmentally friendly production and management decisions more rational and more likely. One of the conditions for the modernization of politics is the persistent presence of a strong and effective environmental movement.[51]

Managerial environmentalists often believe that a positive environmental future is contingent upon environmental movements becoming more radical in their demands and more comprehensive in their vision. Ecological modernizationists, however, believe that a strong movement—measured by the number of supporters and the degree to which their claims are strident and their demands are uncompromising—may not yield significant ecological improvements. Radical, uncompromising movements may catalyze active corporate opposition or reinforce counterproductive regulatory practices by government agencies. If aggressive, uncompromising environmental movement groups, for example, force governments to increase their "command-and-control" regulations, corporate behavior may shift more toward evading regulation. Corporate decision makers might move to "pollution havens" or engage in litigation rather than making positive moves toward compliance with environmental regulations.

Ecological modernizationists have observed that the region of the world in which the most positive changes are occurring in environmental policy and performance is that of Northern and Northwest Europe. The ecological modernization image of the future is that two interrelated institutional changes, both of which have occurred most extensively in Northern Europe, are needed to ensure that environmental sentiments and the impulse of governments to regulate and rationalize will have positive consequences. First, Arthur Mol has observed that in European countries such as the Netherlands and Germany, governments have modernized their managerial efforts by moving away from command and control, and toward more collaborative relationships with industry.[52] Government regulatory officials thus devote more of their efforts to collaborating with private corporations and to bringing more ecologically efficient, less risky, and more profitable alternatives to the attention of corporate officials than they do to setting, monitoring, and litigating over "end-of-the-pipe" standards. Second, the ecological modernizationists believe that environmental groups will be more effective to the degree that they work with industry to achieve environmental goals rather than putting the bulk of their effort into inducing government agencies to take stronger regulatory action.

Note, though, that the ecological modernization image of our environmental future, while very hopeful and optimistic, is not a naïve one. Ecological modernizationists do not assume that corporate, government, and environmental movement decision makers will normally be in complete agreement or that collaboration and compromise are easy to achieve. Even as ecological modernization processes proceed, for example, environmental groups will reserve the right to "go public" and organize campaigns against firms that are recalcitrant in improving their environmental performance. Thus, it is presumed that the public-private collaboration process is contested and partly conflictual, at least beneath the surface. The overall argument, however, is that a modernized government "oversight" and "guidance" process is more likely to create an atmosphere of corporate innovation and environmental citizenship than the largely adversarial relations that characterize command-and-control structures.

Ecological modernizationist thought not only supports selective "deregulation"—away from command-and-control management—but it also embraces another notion that has tended to be the antithesis of environmental-regulationist thought. Environmental modernization thinking has involved endorsing the concept of increased reliance on market mechanisms of environmental protection. It was noted earlier that the United States and other Western governments have taken steps to introduce market mechanisms such as the national market for sulfur oxide emissions that was created through the 1990 Clean Air Act Amendments. The intention behind a market in pollution permits is to ensure that reductions in pollution at any given time will tend to be in areas, or on the part of firms, where the costs of pollution control are least.

But the type of market mechanism that ecological modernizationists are most enthusiastic about is that of "green taxes." Green taxes are government levies on the extraction of raw materials or on the emission of pollution. Green taxes are thus designed to "internalize" what would otherwise be the "external" costs of environmental degradation and resource depletion. The types of green taxes that are given most attention as we enter the twenty-first century are taxes on the sulfur or BTU content of fossil fuels. Fossil fuel taxes are an attractive focus for green taxes because reduced use of fossil fuels induced by higher prices can aid in reducing "greenhouse" gases. Because fossil energy is implicated in most production and consumption activities, fossil fuel taxes will create an incentive to spend less on environmentally destructive activities of all types, and to allocate incomes and funds toward nonenvironmentally destructive areas such as services, leisure, and acquisition of information.

While fossil fuel taxes and taxes on pollution emissions are the most direct and efficacious kinds of environmental taxes, ecological modernizationists generally support consumption taxes such as the consumer sales tax imposed by most state and some city governments in the United States, and particularly value-added taxes such as those in Canada and most European countries. Not only will broad consumption taxes tend to dampen consumption below what it otherwise might be, but the logic behind consumption taxes is that these taxes are or can be substituted for taxes on labor income. Thus, not only will consumption taxes tend to reduce consumption; they will also reallocate income and other benefits toward workers and be beneficial on social equity grounds.

Ecological modernization is in one sense a quite specific perspective on social and environmental change, reflecting both the conservative and managerial paradigms. But there are also several prominent variants on the theme of ecological modernization that employ similar assumptions or concepts. One parallel terminology is that of "dematerialization." Those who hold an ecological modernizationist image of our environmental future believe that there can be, and in many ways there already is, an overall tendency toward production processes being "dematerialized." Dematerialization means that for each unit of output—for each automobile, each ton of steel, or each container of breakfast cereal—there will be progressively fewer environmental resources required as inputs into production. Dematerialization thus implies that "environmental inputs" such as energy and raw materials become replaced by other inputs such as industrial-ecological factory designs, by more use of labor, or by better information, organization, and management skill. At a highly aggregated level, dematerialization of production leads to societies and economies becoming "decoupled" from resource use.[53] Decoupling involves income growth and improvements in living standards becoming less and less dependent on inputs of natural resources and environmental services. Optimistic assessments of dematerialization suggest not only that ecological modernization, decoupling, and dematerialization are feasible, but also that they can be a source of economic growth over the long term.[54] Other advocates envision the possibility of a future "ecological service economy" that is equitable as well as environmentally friendly.[55]

One of the most interesting arenas of research for ecological modernizationists, and their critics, is that of estimating "environmental Kuznets curves." Environmental Kuznets curves involve exploring empirically whether social change and development across the world tends, on balance, to have environmentally positive, dematerializing, or decoupling effects. This question, of course, is the opposite of the question of radical environmental sociologists who look for reasons why modern industrial societies are inherently environmentally destructive. A particularly critical issue is how the lower-income regions of the world will fit into an overarching environmental future. If, for example, shifts toward improved living standards and increased development in the lower-income countries of the South will inevitably involve substantially increased pressure on global environmental resources, an environmentally concerned person, group, or nation might have reservations about whether to promote development in the South or to have any optimism about the prospects for an environmentally sustainable future for the world as a whole.

Environmental Kuznets curves are a variant of the notion of Kuznets curves, which have been a common concept in the sociology of international development. The term "Kuznets curve" reflects the notion, named after the famous Nobel-Prize-winning development economist Simon Kuznets, that as very low-income societies begin to develop and become "middle-income" countries, their levels of income and wealth inequality will become temporarily more pronounced; that is, the degree of income inequality will generally increase as developing countries' incomes rise from $300 to $500 per capita per annum—the level that many of the poorest countries in the world are at as we enter the new millennium—to $3,000 to $8,000 per capita per year. Kuznets received his Nobel Prize, in part, for documenting the fact that there was a general pattern, among both the highly industrial countries as well as late industrializing countries in the South, that continued per capita income growth after "middle-income" status had been achieved tended to be followed by lower levels of income inequality. A Kuznets curve, then, is a graphical depiction of the hypothesis that income inequality takes on an inverted-U shape over the course that a society takes as it moves from a low level of development, to middle income, and ultimately to high income.

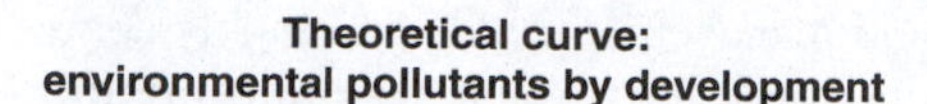

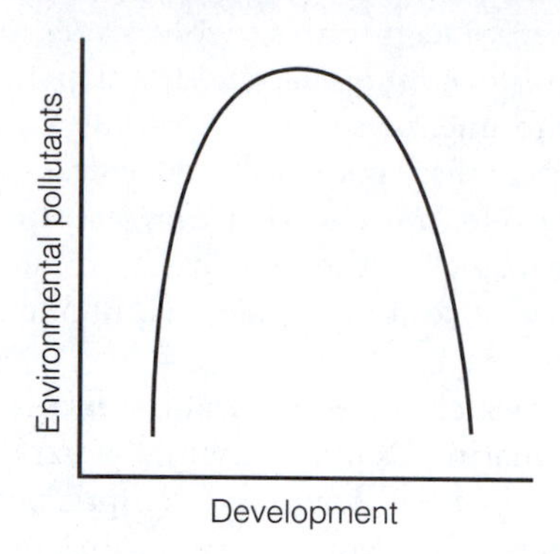

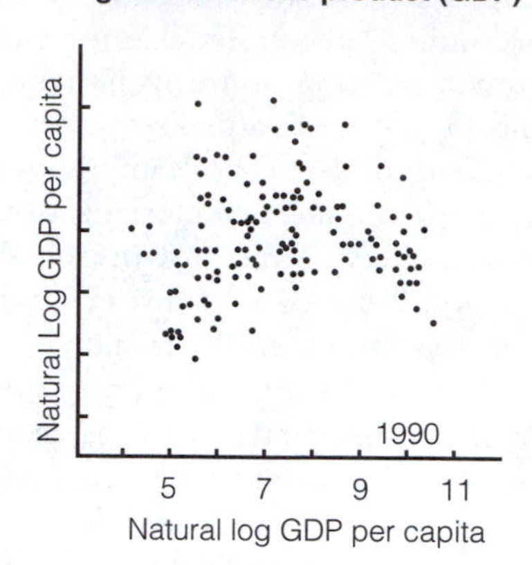

FIGURE 8.1 Environmental Kuznets Curves

SOURCE: Figure at right from Roberts and Grimes, 1997. Reproduced by permission of Elsevier Science.

An environmental Kuznets curve involves the same type of graphical portrayal as the more standard notion of Kuznets curve except that it is a hypothesis about societies' environmental impacts rather than their levels of income inequality (see Figure 8.1). Thus, an environmental Kuznets curve is a graphical depiction of the notion that as countries move from low income to middle income, their impacts on the environment will become increasingly negative or destructive because of the expansion of production, their still-growing populations, and the inefficiencies associated with obsolete production practices and equipment. As the development process proceeds to a high level of per capita income, however, one can expect that environmental performance will progressively improve due to private incentives to dematerialize, to the fact that environmental movements will increasingly organize to address environmental concerns and risks, and because governments will improve their capacity to militate against environmental degradation. High-income societies are considered to be the best able to allocate scarce public and private investments to environmental protection. The environmental Kuznets curve image of our environmental future is thus a doubly optimistic one. Not only does the notion of Kuznets curve imply that higher living standards can be environmentally positive. The argument also suggests that there will be environmental benefits if steps are taken to facilitate development in the low-income countries of the South.

While the notion or future image of ecological modernization does not necessary rise or fall along with empirical evidence on whether the environmental Kuznets curve hypothesis has empirical support, it is worth noting that the evidence available thus far has been mixed. There has been evidence that there is an "environmental Kuznets effect" with regard to pollutants such as

CO_2 and other greenhouse gases.[56] CO_2 emissions per unit of national income have been found to increase as national income rises from low- to "middle-income" levels, and then decrease as nations grow further to higher income levels.[57] Timmons Roberts and Peter Grimes (see Figure 8.1) argue that the environment Kuznets Curve "is the result not of individual countries passing through stages of development, but of a relatively small number of wealthy ones becoming more efficient since 1970 while the average for the rest worsens." But J. B. Opschoor notes that the process of delinking economic and income growth from the demand on the biosphere for materials and services in the industrial countries is currently too slow to yield a Kuznets curve-type response.[58] Opschoor, in fact, suggests that the most recent evidence suggests that there may have even been a *relinking* process between income, pollution, and resource consumption among rich nations in the early 1990s. Opschoor concludes his review article on environmental Kuznets formulations by suggesting, contrary to the notion widely embraced by ecological modernizationists, that more rather than less state managerial intervention and environmental regulation will be required to induce this pattern of change in our environmental future.[59]

INTERNATIONAL ENVIRONMENTAL AGREEMENTS AND OUR ENVIRONMENTAL FUTURE

There can be little doubt that the most significant shift in environmental thought in the late twentieth century was that of viewing environmental problems and their potential solutions in global context. Seldom does a major scholarly book in environmental science or environmental studies fail to make note of the fact that *global* environmental problems, such as global warming, atmospheric ozone depletion, loss of biodiversity, and transboundary movement of toxic wastes, are among the most serious challenges to face humanity. Since the late 1980s, accordingly, the notion that global environmental problems are the most significant, serious, and challenging ones has become commonplace in the social sciences, including but not limited to sociology.[60] There has thus been a clear trend in recent decades toward seeing our environmental future as being premised on our ability to manage these global-scale ecological processes and concerns.

It should be noted, however, that the notion that environmental problems—particularly, our most pressing or challenging ecological concerns—are essentially global in nature is hardly new. Ever since the rise of the modern environmental movement beginning in the late 1960s, the mainstream environmental movement has premised most of its thinking and strategy on global conceptions of environmental problems. Paul Ehrlich's famous book, *The Population Bomb,* as discussed in Chapter 3, was perhaps the single most important inspiration and guide for environmentalism in the late 1960s and early 1970s.[61] In *The Population Bomb* Ehrlich popularized the notion that there exists a global population, with its own global dynamics, and that the essence of the human role on the earth is that this global population is threatening planet-wide environmental catastrophe. The strongly Malthusian flavor of the environmental movement at the time was due in no small measure to the great influence that Ehrlich's notion of "population bomb" had on movement leaders.

During the early and mid-1970s, another global conception of environmental problems that developed in the Donella Meadows and associates book, *The Limits to Growth,* came to be even more prominent in academic and activist environmental thought.[62] Meadows and associates argued that because of the strong tendency for economic expansion to lead to insoluble pollution and resource depletion problems, there was a need to adopt "limits to growth" policies at a global level. The search for feasible strategies to limit global growth, and thereby to reduce the degree to which humans were affecting the integrity of the natural world, came to be the overaching goal of the movement. The reasoning of Meadows and associates about the limits to growth also played a significant role in the discussions at the 1972 Stockholm Conference on the Human Environment that strongly framed environmental discussions, in the developed industrial countries at least, during the 1970s.

Note, though, that despite the longstanding tendency for environmental thought to have a significant global dimension, it was the case that both Ehrlich's notion of population bomb and the Meadows and associates notion of limits to growth failed to ignite durable environmental mobilization. Global notions of environmental problems and their solutions have long been associated with North-South tensions. These tensions were manifest at the 1972 Stockholm Conference, and particularly at the 1974 World Population Conference at Bucharest, Romania. In large part these tensions emerged because the notions of "population bomb" and "limits to growth" implied that the developing countries of the South were major causes of environmental problems and/or that their aspirations for the levels of living standards enjoyed in the North would need to be restrained if global environmental problems were to be solved. In addition, there was considerable opposition, particularly among industrial corporations, and general public ambivalence about population control and the imperative to constrain growth and increased living standards.

The general lack of enthusiasm for modern environmentalism's early forays into global thinking, in fact, led to the movement having lost much of its momentum during the late 1970s and early 1980s. What would change all this would be the appointment of the World Commission on Environment and Development (the Commission) by the Secretary-General of the United Nations in the early 1980s. The Commission was charged with developing new ideas about how South and North could come to agreement on ways to make progress in solving environmental and human problems. As noted earlier in this book, the Commission played a highly influential role in popularizing the notions of sustainability and sustainable development.[63] Most significantly, the Commission's work led to some measure of compromise among representatives of various world governments, environmental organizations, international development NGOs, and development agencies. The essence of the compromise worked out within the Commission was that the contradiction between economic growth and development could be diminished very substantially if new growth was harnessed in a sustainable development framework. Equally importantly, Commission's *Our Common Future* also argued that the major ecological problems that sustainable development were to address were essentially global-scale ecological problems. *Our Common Future,* for example, was perhaps the first globally circulated book in which the "greenhouse" problem was portrayed as the master or overarching global environmental issue. Most of the other ecological problems that the Commission argued must be addressed through sustainable development programs and policies were global-scale problems such as deforestation, loss of biological diversity, desertification, soil and land degradation, and so on.

The Commission's *Our Common Future,* and the 1992 Rio "Earth Summit" that it paved the way for, represented a hopeful pattern of international collaboration and agreement, which has subsequently become one of the pillars of modern thought about how a more promising environmental future can be made possible. In addition to the pioneering work of the Commission, by the time of the Earth Summit it was becoming well known that the 1987 Montreal Protocol had begun to make major accomplishments in reducing the introduction of chlorofluorcarbons (CFCs) into the stratosphere, and in making possible a reduced depletion rate of the stratospheric ozone layer. The relatively nonconflictual and effective process of agreeing to and implementing the Montreal Protocol suggested that international treaties and agreements, and the international organizations and "regimes" that are formed in association with these agreements, would be the logical course to take in creating a better environmental future.[64]

The general impulse that led to enthusiasm about and fascination with global environmental policymaking also had some precedent in the modern environmental movement. From an environmental movement standpoint, the logic behind an international approach is fairly compelling. For one thing, focusing on global-scale problems, particularly if these problems could be connected with suggestions that future global-scale environmental disasters might occur, could be an effective strategy for environmental groups to obtain media attention and to multiply their impact.[65] Thus, there has been some association between international environmental claims making and cultivation of an atmosphere of imminent crisis in environmental thought and strategy. Global strategies also provide a way for environmental group to multiply their impacts on policy. Instead of environmental groups having to contest policy decisions in every capital city across the world, successful passage of a global-scale agreement could, in one fell swoop, leverage governments across the world to implement new environmentally friendly managerial policies. Third, as noted earlier, there was growing disillusion with and opposition to standard command-and-control national-level regulation, and the international arena promised a fresh and possibly more comprehensive approach to environmental reform. Finally, international negotiations promised more access by civil-society groups to policymaking than was often the case with regulatory implementation in the United States and other industrial countries.

As a result of the late 1980s and 1990s spurt of interest in global environmental problems and in the global frameworks for solving these problems, there have been some significant changes in how many organizations and governments think about a desirable environmental future. As the work of the Commission and activities leading up the the Rio de Janeiro Earth Summit moved forward and as global environmental problems were propelled into the spotlight, there was a tendency for most large environmental organizations on the North American coasts and across the major cities of Western Europe to become increasingly global in their discourses, focus, and strategies. Second, prompted by the activities leading up to the Earth Summit, prospective environmental "conventions" and "protocols" were put into place (and will be discussed a more specific discussion of terminology later.). Most of the critical international environmental negotiations that have occurred over the last decade have been those connected in some way to the 1980s and 1990s work of the Commission and its successor the United Nations Commission on Environment and Development. Major examples of these frameworks for international negotiations include the Convention on Biological Diversity and the Framework Convention on Climate Change. A large share of the work of the major environmental organizations continues to focus on global environmental arenas such as these.

There are several different vehicles for such an international approach to improving the future of the global environment. The most common mechanism is typically referred to as "international environmental regimes." The regimes are systems of norms and rules specified in a multilateral agreement among signatory states to regulate actions on a specific issue or set of issues.[66] Regimes generally involve some binding legal agreement or instrument, the most typical of which is a "convention." A convention is a legal instrument that contains all the binding obligations that have been negotiated, and a detailed legal inventory of norms and rules. A framework convention such as the Framework Convention on Climate Change was laid down in advance of the Rio Earth Summit. That Framework continues to be negotiated in the new Millennium and is a very general or formal agreement negotiated in anticipation of additional texts to be agreed to later that specify rules and obligations of the parties, called "protocols." Some agreements are "soft," or in other words nonbinding, an example of which is the Agenda 21 Plan of Action agreed to at the 1992 UN Conference on Environment and Development Earth Summit. Nonbinding agreements, however, tend to have minimal impacts. Such soft agreements typically lead to efforts to create a legally binding agreement.

The second major component of international environmental policymaking consists of international governmental organizations (IGOs). Environmental IGOs are intergovernmental organizations formed for some specific purpose in relation to the environment. Important environmental IGOs include the UN Commission for Environment and Development, the United Nations Environment Programme, and the United Nations Commission on Sustainable Development. Also, many other IGOs, such as the World Bank, the Food and Agriculture Organization of the UN, the United Nations Development Programme, and the World Health Organization, play crucial roles in global environmental policymaking because their mandates relate closely to the environment in one or more ways. The UN Environmental Programme has played a particularly prominent role in international environmental policymaking. The Programme, for example, was responsible for convening a group of experts who adopted a World Plan for Action on the Ozone Layer. Five years later, in 1982, negotiations leading to the Montreal Protocol began. The Montreal Protocol on Substances that Deplete the Ozone Layer, usually referred to as the "Montreal Protocol," was ultimately adopted in 1987. But some environmental IGOs, such as the Intergovernmental Panel on Forests, largely fail and become nonfunctional. The Framework Convention on Climate Change has yet to develop a protocol with concrete agreed-upon norms and rules for implementation. The latest round of that Framework Convention, at The Hague in November 2000, essentially ended in failure.

While international regimes and international governmental organizations have some similarities, their roles should be recognized as being quite distinct. Environmental IGOs themselves are not empowered to formulate international agreements, whereas the raison d'être of international regimes is to establish norms, rules, and sanctions relating to environmentally related conduct of signatory countries and their agents. Some international IGOs, however, have very substantial funding programs that have a great deal of impact on the environment globally. The UN Development Programme has a particularly large development assistance grant fund—of approximately $1.5 billion annually—and is the major grant, as opposed to loan-based, funder of sustainable development and integrated conservation and development programs in the world today. The Programme is one of the three implementers of the Global Environmental Facility that grew out of the Earth Summit.[67] The World Bank is the largest international development finance agency and has an even greater global environmental impact, historically a substantially negative one.[68]

There are some differences of view about how the international level of environmental policymaking and policy development operates or ought to operate. The most common view is that the most straightforward and immediate route to solving international environmental problems is to engage in international negotiations with an eye to securing an agreement for an international regime and legal protocol. Though there have been some recent successes, such as the Categena Protocol on Biosafety under the umbrella of the Convention on Biological Diversity—initiated in conjunction with the 1992 Rio Earth Summit, most recent attempts at making breakthroughs on protocols involving major global environmental issues have largely failed. The essential failure of the 1997 Kyoto Protocol of the Framework Convention on

Climate Change to secure ratification in a critical mass of signatory states, particularly the United States, and the complete failure of the Hague Round to resolve the impasse make it quite likely that there will be no significant agreement on greenhouse gases and climate change for the foreseeable future.

The shortcomings of the past decade or so of international environmental policymaking notwithstanding, there have recently been more optimistic assessments of the constructive role that world society plays in environmental policymaking. David Frank and his colleagues have argued that the most significant role played by world society is through IGOs,[69] rather than only or primarily through international regimes.[70] IGOs serve to diffuse the shared pro-environment and environmental-scientific cultures of their "epistemic (expert) communities" down to the government agencies and officials of nation-states. Thus, for Frank and his associates, the existence of the large, prominent environmental IGOs such as UN Environmental Programme and UN Commission on Sustainable Development, plus the many smaller environmental IGOs such as the International Union for the Conservation of Nature, has been critical over time in inducing the governments of world nations to take positive steps toward environmental control.[71] The IGOs establish autonomous environmental ministries, become signatories to international environmental protocols, and set aside land and other natural areas for conservation and preservation.[72]

Despite the widespread interest in the environment at the level of international policymaking, its status as the focal point for ensuring a desirable environmental future is by no means clear. As noted earlier, the concrete accomplishments at the level of international regime and protocol negotiation since the Rio Earth Summit have been very modest. It is also useful to evaluate the success of the international strategy of environmental policymaking by bearing in mind that, to some degree, global environmental change is a "frame" or social construction in which preexisting problems or concerns have essentially undergone "repackaging."[73] Global environmental change, or atmospheric disruption, essentially tends to boil down to two key longstanding issues: air pollution and energy conservation. It is therefore important to ask whether the decade or so of reframing air pollution and energy conservation as global environmental change—or reframing the case for controlling air pollution and conserving energy being that of staving off atmospheric or climate disruption—has given us any greater leverage on the problem.

Again, it is not clear that environmental globalization has had advantages in making possible significant environmental reforms in the managerial arenas of air pollution control and energy conservation. In fact, in the United States the framing or social construction of air pollution control and energy efficiency as being global issues requiring global-scale policy action has clearly energized right-wing think tanks and conservative corporations to fight creeping internationalism. Prior to Rio, there was seldom a concerted right-wing movement to counter air pollution control and energy efficiency improvement programs. Now, however, the U.S. Congress is very unlikely to ratify an international regime and protocol that appears to partially exonerate developing countries from complying with greenhouse standards. Environmental internationalism has so catalyzed the right-wing countermovement by 2000 that there actually occurred an unprecedented failure by Congress to pass amendments to the Clean Air Act.

In addition to environmental internationalism catalyzing right-wing opposition and heading off regulatory improvements that might have been possible if particular issues had been defined as national ones, there are other actual or potential shortcomings of international environmental agreements and environmental IGOs. First and most important is that there are very definite North-South differences of interest in coming to agreement on protocols relating to fundamental global environmental concerns. The industrial countries of the North now account for the bulk of resource-consumption-related impacts on the global environment, but these countries are strongly divided on whether they are willing to sacrifice growth or jeopardize their international economic stature in pursuit of international environmental public or collective goods such as healthier forests, a more stable climate, conservation of scarce land and soil resources, and so on. Many of the developing countries of the South are unwilling to enter into agreements unless they are essentially exonerated from major commitments over the short to medium term, and unless they can expect to receive "green foreign aid" to help finance a transition to a more sustainable pattern of natural resource use. With the end of the Cold War, however, most industrial countries, particularly the United States, are beginning to pare foreign aid budgets that were originally established after the Korean War in a climate of East-West rivalry over the hearts and minds of developing country governments and their peoples. The end of East-West Cold War rivalry has led to declining foreign aid outlays and to a decreased likelihood that the South will receive subsidies to invest in new "green technology" or in sustainable development programs. In effect, then, despite the allusion in Commission's *Our Common Future* that all of humankind has a common stake in international environmental protection,[74] the apparent reality is that different countries perceive very different domestic interests in international cooperation over the environment.[75]

Differences of national interest in international environmental policy discussions have clearly been the major factor preventing new landmark international agreements. Social research on international environmental regimes is, in addition, increasingly discovering that national political and regulatory styles remain very resistant to external pressures for change deriving from international agreements. The fact that an international environmental agreement has been secured by no means ensures that fundamental changes in national styles of regulation and policymaking will result. Recent controversies over the role of the World Trade Organization—particularly over how the Organization's policies relate to trade in hormone-treated meat and genetically modified foods—suggest that international negotiators are increasingly discovering the limits on how much global negotiations can override national differences in regulatory cultures and practices.[76]

WHAT ROLE FOR GREEN CONSUMERISM?

In this chapter we have suggested that in the world today there are four major alternative images of the route to a more promising environmental future—or four major mechanisms that are now being emphasized in theories of environmental improvement. It is useful to note, however, that our list of the four major mechanisms of environmental improvement omitted what is perhaps the single most common environmental reform strategy advocated by environmentalists and environmental scientists—that of stressing cultural changes to reduce consumption on the part of individual consumers.[77] While changes in consumer culture have been a focal point of environmental group doctrine for many years,[78] the attention to consumption strategies has, if anything, increased in recent years as a result of the attention now given to "green" products and "green" marketing.[79] How should we understand the future role that changes in consumer culture, behavior, and green marketing will play with regard to our environmental future?

Consumption and consumptionism occupy a curious position in contemporary social science and environmental sociology. The predominant theoretical tradition in the social sciences has been to regard consumption phenomena as being of relatively minor importance, save for how consumption styles and symbols serve as "markers" of class or status position, or for how humans construct meaning around the activities of their daily lives. Environmental sociology is one of the few branches of the social sciences in which consumption is taken seriously. Even so, consumption phenomena are somewhat controversial in environmental sociology as well. Thus, on one hand, there are a number of environmental sociologists who have argued strongly about the need to take consumption and individual consumption processes seriously.[80] On the other hand, a number of environmental sociologists have argued that the exhortation of individuals to consume less or to buy "green" products such as organic food or recyled paper is an ineffective means of securing environmentally friendly social changes.[81]

Environmental sociologists have raised four major concerns about placing major theoretical stress on a green-consumption-driven environmental future. One of the major points that has been stressed is one with deep roots in the neo-Marxian, radical tradition. This is the notion that production institutions, particularly the laws and tendencies of capitalist competition and capital accumulation, have causal priority over consumption institutions. Thus Allan Schnaiberg, in his influential book, *The Environment,* devotes an entire chapter (Chapter 4) to the issue of whether environmental deterioration is primarily a consumption-driven problem.[82] Schnaiberg by and large agrees with the neo-Marxist position that, in capitalist societies, the principal decisions made about production processes and consumption choices are made within production institutions—that is, mainly by corporations and other production groups—rather than by individual consumers. Accordingly, Schnaiberg has stressed that environmental movements need to focus on changing the rules of production institutions rather than trying to influence consumer purchasing decisions. In addition, Schnaiberg has stressed that most of the energy and materials consumption in the U.S. economy is accounted for outside the sphere of household consumption—by corporations and governments, and in the transportation and circulation of nonconsumer goods—and is thus largely beyond the purview of the individual consumer to influence.

A second and related criticism of seeing environmental problems and solutions to these problems mainly in consumption terms has also been made by Schnaiberg, as well as by a number of other environmental sociologists.[83] This criticism is that the choices of products and services available to consumers are shaped very substantially by a society's physical infrastructure. For example, Chapter 5 showed that individual consumption is shaped heavily by whether a society's transportation system dictates individual dependence on the private automobile, and by whether city-planning designs and other public policies encourage a predominance of automobile-dependent shopping and of large, detached, single-family residences. Thus, it is held that individuals have relatively little latitude in making fundamental reductions in their consumption of energy and materials, given the physical infrastructure within which they must make consumer decisions. These environmental sociological critics of the green consumerism strategy thus suggest that voluntary limitations on consumption or green consumerism would involve little more than a drop in the bucket compared to the huge flows of resources that are shaped by public policy through its effects on the transportation system, the urban infrastructure, and the character of the built environment.

A third issue raised by environmental sociologists about consumption is that of the potential "greenwashing" effects of green consumption. Schnaiberg and Gould, for example, have noted that, although many consumers would welcome new products that involve major reductions in the use of resources or in pollution levels, green products are typically more matters of corporate public relations than they are significant improvements in an ecological sense.[84] Thus, they observe that the significance of green consumerism lies more in corporate image enhancement—or, in other words, in "greenwashing"—than in reduction of resource consumption or pollution.

Finally, Tellegen and Wolsink have stressed that in most instances there appear to be only small relationships between individual attitudes and green consumption behaviors.[85] They thus suggest that the emergence of more pro-environmental public attitudes are not likely to yield major changes in consumption of energy and materials. Further, while they acknowledge that it is possible for public policies to affect attitudes that are very specific to a given type of consumption, such as energy consumption in autos or in the home, and these programs can have some limited effectiveness, it remains the case that influencing environmental behaviors through cultural changes of values and attitudes will have only modest results.

Each of these limits to the role that green consumerism might play in the future has a strong element of truth, and suggests that consumption and green production strategies do not constitute another important category of images of how to make for a sounder environmental future. It would be premature, however, to argue that individual consumption matters are unimportant for environmental sociology. In particular, individual consumption behaviors and green consumerism can be quite relevant to several of the four major images of our environmental future that we have stressed in this chapter. For example, green consumerism is a key issue that relates to environmental movements. Several European environmental groups such as Friends of the Earth have made major gains by targeting specific industrial corporations that play strategic roles in making possible beneficial industry-wide changes; typically these environmental groups target an industry leader such as Siemens in electronics and raise the threat of mobilizing consumer boycotts as a strategy to secure voluntary corporate agreement to undertake an environmentally friendly change in production practices.

Another critical arena of state managerialism are environmental regulations, rules, and incentives that affect the society's transportation, urban, and built environmental infrastructures. Clearly, we know from Chapter 5 that one of the major factors that has stimulated growth in energy and materials consumption has been the "automobilization" of the United States. Similarly, many observers have pointed to the constructive conservation role that the U.S. government's strict CAFE (Corporate Average Fuel Economy) standards had in the late 1970s and early 1980s. Beginning in 1975 federal CAFE standards mandated ongoing improvements in automobile gas mileage by auto manufacturers. CAFE standards had very positive impacts on automobile fuel economy in the United States until they began to be relaxed during the Reagan administration in the mid-1980s. Similarly, government policy changes that would make cities more compact and encourage greater use of mass transportation could make possible major decreases in resource consumption. Further, green consumerism is now one of the most important areas of research in the ecological modernization tradition,[86] particularly with regard to areas in which environmental group and corporate collaboration can yield improvements in environmental efficiency. Thus, while there are limits to the role that individual- or household-level green consumerism can play in creating a sounder environmental future, consumption will become an increasingly important component of environmental improvement in the future.

WHICH WAY FORWARD INTO THE FUTURE?

North American environmental sociology found its way in the 1970s and 1980s by emphasizing the understanding of how ongoing patterns of environmental problems have been shaped by social factors and social changes. The ultimate promise of environmental sociology, however, is arguably that of helping to chart the way forward to more socially secure and environmentally sustainable arrangements. Recent work by environmental sociologists on how a better environmental future can be made possible has been extremely exciting, but not yet very conclusive. Which of the four focal processes of environmental improvement—environmental movement mobilization, national government environmental regulation, ecological modernization, or international environmental policymaking—is the most enduring or promising in this regard?

One potential answer to this question is to say that all four processes are important and that choices among them really need not be made. Indeed, a reasonable case could be made for such a judgment. On several occasions in this chapter, for example, we have noted significant connections among the four overarching images of our environmental future as, for example, the hypotheses by ecological modernizationists about the kinds of environmental movement strategies that tend to be most constructive. But it remains worthwhile, nonetheless, to ponder the issue of which of the four basic processes is the most fundamental and which ought to attract the most research and application. In this concluding section we will make a few tentative and highly speculative remarks that scholars and students might want to consider.

This chapter began by noting that the conventional wisdom in environmental sociology has traditionally been that the ultimate way out of the "iron cage" of persistent environmental crisis is that of citizen mobilization through environmental movements. There is good reason to scrutinize the assumptions environmental sociologists have made about the future role that environmental movements will play. This is the case if for no reason other than the fact that the paradigms, strategies, support bases, and alliances of environmental movements are now enormously different than was the case when sociologists began speculating about and doing research on the role of these movements in the 1970s. It is also clear that environmental sociologists, in their recent quest to think broadly about what our environmental future might hold, have actively considered a range of alternatives to the conventional notion that environmental challenges lead more or less directly to environmental mobilization, and ultimately to institutional changes of a pro-environmental nature.

In this chapter, however, we have moved toward—or actually back to—an overall view that if there is a particular social force that ought to be considered most fundamental to a sounder environmental future, this force is probably that of environmental movements and environmental mobilization. When all is said and done, the pressures for an environmentally problematic business-as-usual—if not the environmental backsliding and deregulation that so many anti-environmental countermovements now seek—is now so strong that public environmental mobilization is now the ultimate guarantor that public responsibility is taken to ensure environmental protection. We do not wish to imply that there is no room for concern about the directions that various environmental movements are taking. Global environmental movements are in some senses becoming more splintered and fractionalized as they increasingly embrace very different discourses and strategies, as noted by Brulle.[87] The chasm between internationally oriented and grassroots movements is a particular concern, as is the fact that so many of the largest and best-funded movement organizations have stressed the global arena and thus far have little to show for it. Internationally oriented environmental movements are increasingly encountering the problem that emphasizing problems that cannot be directly experienced such as greenhouse warming (three or so generations hence) or goals that will not be achievable in citizens' lifetimes, or climate stability at the end of the twenty-first century, is problematic for sustaining mobilization.

Second, we would suggest that the national-state system of sovereign countries is so entrenched in the world today, and reflects such different regulatory cultures and styles, that there is no substitute for working out sounder environmental management policies at the national governmental level. Successful international agreements, if they are to be broadly implemented, need to accommodate differing national interests and need to be adopted, implemented, and enforced by national governments. In addition, nationally constituted governments remain the master institutional complex with an ultimate managerial authority to ensure that steps are taken to make changes in society to accomplish environmental goals. Some of the prevailing concerns about conventional command-and-control regulation—particularly those about inflexibility and the obstacles posed to private and public innovation—have an element of truth and need to be addressed in the future. At the same time there is evidence that a national environmental regulatory system will remain critical in dealing with environmental challenges in the future.[88]

Third, there is much to be said for the possibilities that ecological modernization processes might emerge somewhat autonomously from the actions of social movements and the (de)regulatory activities of national governments. It is ultimately the case that successful environmental movement initiatives and more effective national environmental oversight will be needed to harness the capacity of private firms to respond to new incentives, and to stimulate the creativity and accountability of government officials in designing and implementing innovative new environmental control incentives. Ecological modernization perspectives can be of particular help in conceptualizing the possibilities for comprehensive, effective, and socially benign systems of environmental incentives.

It is particularly noteworthy that there has emerged sound socioeconomic modeling evidence that green taxation, particularly CO_2 or BTU taxes, would be enormously effective in environmental protection and, through replacement of income taxes, would ensure that such a policy regime would be reasonably equitable.[89] There have been promising steps taken toward green taxation in several European countries,[90] and the European Union has taken a forward-looking position on green taxation as a means of making possible an effective agreement on a protocol for the Framework Convention on Climate Change. At the same time, the fact that even modest advances in energy-related green taxation are not on the U.S. political agenda at the beginning of the twenty-first century—and that there is no small amount of resistance to green taxes in some European countries—is one of several realities suggesting that the potentials of ecological modernization processes ought not to become a reason for complacency. Ecological modernization, to the degree that such a process begins to emerge across the globe, will no doubt be as conflictual, contested, and subject to backsliding as more conventional environmental regulation has been.

There is, in particular, growing evidence that the incentives and institutional innovations that make ecological modernization possible will need to be anchored in a supportive political climate, which can only be created by environmental movements. David Sonnenfeld, in his comprehensive empirical test of ecological modernization theory in the context of the transformation of pulp and paper manufacturing, has found that environmental social movements and activism were critical in inducing the ecological upgrading that has occurred in this sector across the globe.[91] The results of Sonnenfeld's study were consistent with several hypotheses of the ecological modernizationists such as Mol and Spaargaren, such as the notion that movements will tend to move away from mainly contesting government policy, and move toward working more directly with manufacturing industry.[92] But Sonnenfeld has also concluded that there was little support for the ecological modernization hypothesis that successful environmental activism would be moderate or reformist, as opposed to radical environmentalism; he found that both radical and managerial environmentalism was to be successful in helping to prompt pulp and paper industries into more environmentally responsible behavior. Further, some environmental groups moved back and forth from radical demands to reformist, managerial-oriented, direct relations with government agencies and private firms.

Fourth, the track record of the past decade or so of internationalized environmental policymaking is not particularly encouraging. It seems unlikely that the pursuit of international agreements, or that the role of environmental IGOs in the diffusion of environmental protection norms to national governments, has sufficient potential to justify seeing the international policy strategy as the "lead horse" for achieving a sounder environmental future. This notion is in some sense still in conflict with the conventional wisdom, though.[93] It also seems somewhat in conflict with the idea that environmental progress in this era of globalization would not necessarily be achieved most effectively and comprehensively at the global level. But perhaps there is no contradiction here because of the fact that, rightfully or not, pro-environmental social movements are becoming increasingly opposed to the trade liberalization and international monetary policies that are increasingly creating the conditions for offshore corporate veto of environmental laws.

The environmental movement at the turn of the beginning of the twenty-first century shows every sign of becoming part of a larger social movement complex that is increasingly anti-globalization and anti-international in its focus. Increasingly, a wide range of environmental groups has joined hands with other movements to pursue agendas about anti-globalization, pro-human rights, pro-consumer, pro-family farming and sustainable agriculture, and opposition to genetically engineered foods in tandem. The anti-globalization agenda seems ironic given the emphasis of the movement on international environmental policymaking over the past decade and a half—though this is perhaps not so surprising when one considers the difficulties that have been encountered at the international levels. And perhaps it remains the case as well that the essence of the human relationship to the environment is that the relationship occurs somewhere—at a particular location on the globe, in particular communities or countries, and particular institutions and social circles—and that there is something irreducibly local about the human experience with the biophysical environment.[94]

While we have our doubts about whether international regimes or organizations are likely to prompt global-scale sustainability improvements over and above those that environmental movements, government regulatory agencies, and ecologically modernizing firms will be able to achieve, there is no doubt that the international arena will be critical to the shape of our environmental future. The German scholar, Wolfgang Sachs has usefully summarized the promise and peril of global economic and social integration for the future of sustainability.[95] Global competition, for example, can be a powerful force in ferreting out inefficiencies in production organizations, resource use, state policies, and other areas. Similarly, the Dutch scholar, Arthur Mol, in a creative linking of his ecological modernization perspective with that of the environmental internationalists, has noted that international organizations and processes of negotiating global environmental agreements will be a key mechanism in diffusing new conservation ideas and technologies around the world.[96] But Sachs notes that the outcome of global competition and efficiency improvements will undoubtedly be a pattern of economic expansion that requires increased use of energy, minerals, raw materials, and ecological services; efficiency, in other words, tends to lead to growth, which may negate the ecological benefits of producing goods and services more efficiently.[97] A related contradictory pattern identified by Sachs is that governments and environmental movements have good reasons to want to get on board the turn-of-the-century globalization trajectory at the same time that trade liberalization and the shift of economic, social, and environmental regulation to the transnational levels will be unsettling to nation-states and their citizens. Sachs has entitled his discussion of the contradictory roles and ambivalent views about globalization "Planet Dialectics." By "dialectics," Sachs means that instead of the modern world changing according to readily foreseeable linear trends such as those toward modernization, efficiency, development, or sustainability, global change is characterized more by ambivalences, ironies, contradictory forces, and counter-trends. Achieving environmental security in the turbulent social environment of the twenty-first century is the epitome of a dialectical vision. The new century will very possibly be one in which the master trends that created environmental disruption and degradation—industrialization, global economic integration, and unfathomable technological transformations—will be reshaped in ways that help to solve the problems they created in the last century. Sociological vision has much to contribute to this, perhaps the most important social as well as biophysical project of this new century.

CITATIONS AND NOTES

1. See, for example, Catton and Dunlap, 1978; Goldblatt, 1996; Martell, 1994; Murphy, 1994.

2. Schnaiberg, 1980; Logan and Molotch, 1987; Catton and Dunlap, 1980; Murphy, 1994.

3. Catton, 1976, 1980; Catton and Dunlap, 1978; Dunlap and Catton, 1994.

4. Dunlap and VanLiere, 1984; Dunlap, 1993.

5. See Thornton (2000) for a fascinating case study of the "Chlorine Chemistry Council," an association of corporations that make and sell potentially dangerous chlorinated chemicals. Thornton identifies the diverse ways that the Council is an omnipresent lobbying force at the same time that the organization has remained almost completely invisible to the public.

6. Schnaiberg and Gould, 1994, p. 148.

7. For example, Mol, 1995.

8. See Schnaiberg, 1980.

9. Gottlieb, 1993.

10. Also note that the analysis of environmental attitudes and values on one hand, and of environmental movements on the other, has traditionally been presumed to be largely the same subject matter. For purposes of this chapter we will take the matter of environmental values, attitudes, and movements—or, in other words, the matter of environmental attitude change as a component of a more promising environmental future—to be mostly synonymous. We do so while recognizing that environmental attitude research (see, for example, the summary in Tellegen and Wolsink, 1998, Chapter 6) has a definite literature of its own, and that the translation of pro-environmental values into concrete social movements is highly problematic (Hannigan, 1995; Brulle, 2001). Also, we would suggest that the value change-behavior connection is sufficiently small (Tellegen and Wolsink, 1998) so that environmental value change ought not to be considered an environmental futures mechanism apart from environmental movements.

11. Brulle, 2001, p. 234.

12. It is useful to note that Inglehart's postmaterialism hypothesis—the notion that growth in income leads social groups, and presumably countries as a whole, to come to embrace "postmaterial" values such as environmentalism—has been subjected to a number of cross-national tests and found to have major shortcomings. The postmaterialism hypothesis, for example, is inconsistent with the fact, discussed later in this chapter, that the past decade and a half or so has witnessed the mobilization of numerous environmental movements in the developing countries of the South (Gadgil and Guha, 1995). Brechin and Kempton (1994) and Dunlap and Mertig (2001) have also found that there is little association between the level of affluence of nations and the degree to which their citizens express environmental concerns.

13. Kraft, 2001, Chapter 4.

14. See also Hannigan, 1995.

15. Rowell, 1996; Thornton, 2000.

16. McCarthy, 1998.

17. Brulle, 2001.

18. Kraft (2001, pp. 89–90) uses the term "greens" or "radical greens" to encompass Brulle's (2001) categories of ecocentric and deep ecology discourses and groups.

19. Taylor, 1995; Peet and Watts, 1996.

20. Scott, 1996; Beck, 1992.

21. See Levine, 1982; Szasz, 1994.

22. Freudenburg and Steinsapir, 1992, p. 33.

23. See, for example, Bullard, 2001; Szasz, 1994.

24. Buttel, 1992.

25. It is worth noting, though, that while various developing country groups and governments have been a base of strong support for pro-environmental agendas (such as supporting the Precautionary Principle [discussed later this chapter] in the discussions that led to the Categena Protocol of the Conventional on Biological

Diversity), this is not always the case. Many developing country governments, for example, have objected to the environmental protection and conservation procedures that the World Bank is increasingly insisting on. Also, most developing countries have opposed the World Trade Organization moving toward permitting environmental restrictions on trade, or insisting on particular environmental regulatory practices as a condition of membership.

26. Hays, 1987, 2000; Andrews, 1999; Kraft, 2001.

27. Skowronek, 1990.

28. Hays, 1959, 1987, 2000.

29. Kraft, 2001, p. 87.

30. Kraft, 2001, p. 87.

31. See, for example, Kraft, 2001.

32. For example, Rosenbaum, 1973.

33. As we will note later, some sociologists stress that the referent for "efficiency" or "rationalization" tends to be international in nature—in other words, that state managers tend to be influenced in their sense of the possibilities and responsibilities of government by the standards exhibited in other countries. Frank and associates (2000) have argued that it has generally been the case at the global level that the major factor that shapes national government's views and perspectives of the proper role of government in relation to the environment is the expectations in this regard that are transmitted through international organizations such as the Intergovernmental Panel on Climate Change and negotiations of the Convention on Biological Diversity. We will suggest later, however, that at least for the bulk of the developed industrial countries of the North, the tendency is that the blueprint for the state's environmental role tends to be shaped by domestic politics. Thus, state agencies and managers, acting in relation to environmental and other groups in civil society and in relation to competing forces within the state, develop conceptions of the appropriate or possible roles of government in a manner that is not decisively shaped by international organizations and forces.

34. Buttel, 1998.

35. Kraft, 2001; Hays, 2000; Andrews, 1999.

36. Compare Simon, 1982; Vig and Kraft, 1984.

37. Hays, 1987; Kraft, 2001.

38. Murphy, 1997.

39. See, for example, Freeman, 2000.

40. Rosenbaum, 2000, p. 175.

41. Block, 1996.

42. Vig and Kraft, 2000, p. 396.

43. Rosenbaum, 2000.

44. Rosenbaum, 2000, Andrews, 1999.

45. Grossman and Cole, 1999.

46. Mol and Buttel, 2000.

47. The Precautionary Principle is sometimes referred to as the Precationary Polluter Pays Principle (or "4P") in recognition of the fact that it is premised on the notion that firms should be directly responsible—including the payment of fines to the government and compensating victims—if a technology or production practice proves to be harmful.

48. Note that the WTO in 1998 ruled against the application of the Precautionary Principle by the European Union in banning imports of hormone-treated beef.

49. Murphy, 1994; Sachs et al., 1998; Milani, 2000.

50. Mol, 1995.

51. See also Sonnenfeld, 2000.

52. Mol, 1995, 1997.

53. See, for example, Sachs et al., 1998.

54. Weale, 1992, p. 76.

55. Milani, 2000, p. 86.

56. Selden and Song, 1994.

57. It should be stressed, however, that it essentially never occurs that a single nation moves from low-income to high-income status over a period as brief as a few generations. Kuznets curve formulations are based mainly on cross-sectional differences among nations as arrayed by their per capita income levels.

58. Opschoor, 1997; for another critique, see Roberts and Grimes, 1997.

59. Opschoor, 1997, p. 284.

60. Taylor and Buttel, 1992; Redclift and Benton, 1994.

61. Ehrlich, 1968.

62. Meadows et al., 1972.

63. World Commission on Environment and Development, 1987.

64. Note, however, that many of the most successful international environmental regimes and agreements were negotiated well before the current era of "global environmental change" and fascination with the global level of international policymaking. In fact, most international environmental regimes currently "on the books" involved protocols agreed to prior to 1990. An excellent example of a preexisting international environmental regime and agreement was the 1973 Convention on International Trade in Endangered Species of Wild Fauna and Flora.

65. Mol, 2000; Taylor and Buttel, 1992.

66. See Porter and Brown, 1996, Chapter 1.

67. The Global Environmental Fund plays a major role in funding sustainable development and biodiversity protection projects in the tropical developing countries.

68. The World Bank and its tendency to direct loan funds to the development and expansion of resource and raw material extraction sectors continue to be a major focal point for criticism by international environmental groups. At the same time, the World Bank has taken steps to establish a substantial Environment Department and to insist on greater environmental protections than many developing country governments are comfortable with.

69. Frank, 1997; Frank et al., 2000.

70. See also Mol, 2000.

71. Frank et al., 2000.

72. Frank (1997) and Frank and colleagues (2000) stress the role that scientifically driven conceptions of environmental problems play in environmental IGOs. It should be noted, however, that the increasing knowledge-based nature of international environmental policymaking has both advantages and disadvantages. On one hand, scientific documentation of the global character of environmental problems and of the threats and risks that are resulting from these problems has largely given rise to new opportunities to make use of international channels and processes for environmental protection. But it is now increasingly apparent than the application of Western science, rather than being able to bring closure to environmental policy discussions, typically has the opposite effect (Yearley, 1996). Controversies or disagreements within science become seized upon by various national governments and nongovernmental parties. The likelihood that matters of politics and social choice will become transformed into technocratic exercises is almost certain to generate opposition as well.

73. Taylor and Buttel, 1992.

74. World Commission on Environment and Development, 1987.

75. Yearley, 1996.

76. O'Neill, 2000; Weale et al., 1996.

77. See, for example, Ehrlich and Ehrlich, 1990, Chapter 12.

78. See, for example, Ehrlich, 1968.

79. Milani, 2000.

80. For example, Catton, 1976; Redclift, 1996; Redclift and Woodgate, 1997; Shove, 1997; Murphy, 1994.

81. Schnaiberg and Gould, 1994.

82. Schnaiberg, 1980, Chapter 4.

83. Schnaiberg, 1980; see also Milani, 2000; Schnaiberg and Gould, 1994; Tellegen and Wolsink, 1998.

84. Schnaiberg and Gould, 1994.

85. Tellegen and Wolsink, 1998, p. 138.

86. Spaargaren, 1996.

87. Brulle, 2001.

88. Grossman and Cole, 1999.

89. Buttel, 1998; Milani, 1999; Sachs et al., 1998; Weale, 1992.

90. Milani, 1999.

91. Sonnenfeld, 2000.

92. Mol, 1995; Spaargaren, 1996.

93. See, for example, Frank et al., 2000; Buttel, 2000.

94. Macnaghten and Urry, 1998.

95. Sachs, 1999, Chapter 8.

96. Mol, 2000.

97. Sachs, 1999.

References

Aber, J., and J. Melello. 1991. *Terrestrial Ecosystems.* Philadelphia, PA: Saunders.

Adelman, M.A. 1995. *The Genie Out of the Bottle: World Oil Since 1970.* Cambridge, MA: MIT Press.

Adeola, Francis O. 2000. "Cross-national environmental injustice and human rights issues." *American Behavioral Scientist* 43(4):686–706.

Agarwal, Bina. 1986. *Cold Hearths and Barren Slopes: The Wood Fuel Crisis in the Third World.* Riverdale, MD: Riverdale Co.

———. 1992. "The gender and environment debate: Lessons from India." *Feminist Studies* 18:119–58.

———. 1994. *A Field of One's Own: Gender and Land Rights in South Asia.* Cambridge: Cambridge University Press.

Agbo, Valentin, Nestor Sokpon, John Hough, and Patrick C. West. 1993. "Population-environment dynamics in a constrained ecosystem in Northern Benin." Pp. 283–300 in Gayl D. Ness, William D. Drake, and Steven R. Brechin (eds.) *Population-Environment Dynamics: Ideas and Observations.* Ann Arbor: University of Michigan Press.

Akama, John S., Christopher L. Lant, and G. Wesley Burnett. 1995. "Conflicting attitudes toward state wildlife conservation programs in Kenya." *Society and Natural Resources* 8:133–44.

Alexandratos, Nikos, (ed.). 1995. *World Agriculture: Towards 2010, an FAO Study.* Rome, *IT:* Food and Agriculture Organization of the United Nations.

Alford, Robert R., and Roger Friedland. 1985. *Powers of Theory: Capitalism, the State, and Democracy.* New York: Cambridge University Press.

Allen, Patricia. 1993. "Connecting the social and the ecological in sustainable agriculture." Pp. 1–16 in P. Allen (ed.) *Food for the Future: Conditions and Contradictions of Sustainability.* New York: John Wiley & Sons, Inc.

Alter, George. 1992. "Theories of fertility decline: A nonspecialists guide to the current debate." Pp. 13–27 in John R. Gillis, Louise A. Tilly, and David Levine (eds.) *The European Experience of Declining Fertility, 1850–1970: The Quiet Revolution*. Cambridge, MA: Basil Blackwell.

Altieri, M. 1995. *Agroecology*. Boulder, CO: Westview Press.

Amanor, Kojo. 1994. "Ecological knowledge and the regional economy: Environmental management in the Asesewa District of Ghana." Pp. 41–68 in D. Ghai (ed.) *Development & Environment: Sustaining People and Nature*. Oxford: Blackwell.

Anderson, V. 1991. *Alternative Economic Indicators*. London: Routledge.

Anderton, Douglas L., Andy B. Anderson, John Michael Oakes, and Michael R. Fraser. 1994. "Environmental equity: The demographics of dumping." *Demography* 31:229–48.

Andrews, R.N.L. 1999. *Managing the Environment, Managing Ourselves*. New Haven: Yale University Press.

ANSI website, *www. ansi.org*.

"ANWR hypocrisies" (Editorial). 1995. *Oil and Gas Journal* 93(45):13.

Ashford, Lori S. 1995. "New perspectives on population: Lessons from Cairo." *Population Bulletin* 50(1):1–44.

Baker, Susan, Maria Kousis, Dick Richardson, and Steven Young. 1997. *The Politics of Sustainable Development: Theory, Policy and Practice within the European Union*. London: Routledge.

Banks, J.A., and Olive Banks. 1964. *Feminism & Family Planning in Victorian England*. New York: Schocken Books.

Barbier, Edward B. 1987. "The concept of sustainable economic development." *Environmental Conservation* 14(2):101–10.

Barkin, David. 1987. "The end of food self-sufficiency in Mexico." *Latin American Perspectives* 14:271–97.

———. 1998. "Sustainability: The political economy of autonomous development." *Organization and Environment* 11(1):5–32.

Barkin, David, Rosemary Batt, and Billie DeWalt. 1990. *Food Crops vs. Feed Crops: Global Substitution of Grains in Production*. Boulder, CO: Lynne Rienner.

Barney, Gerald O. 1980. *Global 2000 Report to the President*. Washington, DC: U.S. Government Printing Office.

Bates, Marston. 1968. "Environment." Pp. 91–93 in D. L. Sills (ed.) *International Encyclopedia of the Social Sciences*. New York: Macmillan.

Beck, U. 1992. *Risk Society*. Beverly Hills, CA: Sage.

Bell, Daniel. 1976a. *The Coming of Post-Industrial Society: A Venture in Social Forecasting*. New York: Basic Books.

———. 1976b. *The Cultural Contradictions of Capitalism*. New York: Basic Books.

Bell, Michael M. 1998. *An Invitation to Environmental Sociology*. Thousand Oaks, CA: Pine Forge Press.

Ben & Jerry's website. *www.benjerry.com*.

Benton, Thomas. 1989. "Marxism and natural limits: An ecological critique and reconstruction." *New Left Review* 178:51–86.

Bergman, B.J. 1995. "Leader of the pack." *Sierra* 80(6):50+.

Bidwell, Charles E., and John D. Kasarda. 1998. "An ecological theory of organizational structuring." Pp. 85–116 in Michael Micklin and Dudley L. Poston, Jr., (eds.) *Continuities in Sociological Human Ecology*. New York: Plenum Press.

Block, F. 1996. *The Vampire State*. New York: New Press.

Bluestone, Barry, and Bennett Harrison. 1982. *The Deindustrialization of America: Plant Closings, Community Abandonment, and the Dismantling of Basic Industry*. New York: Basic Books.

Bonar, James. 1966. *Malthus and His Work*. London: F. Cass.

Bongaarts, John. 1994. "Population policy options in the developing world." *Science* 263:771–76.

———. 1996. "Population pressure and the food supply system in the developing world." *Population and Development Review* 22:483–503.

Bongaarts, John, W.P. Mauldin, and J. Phillips. 1990. "The demographic impact of family planning programs." *Studies in Family Planning* 21:299–310.

Booth, Roberta. 1998. "Personal communication."

Bormann, F. Herbert, Diana Balmori, and Gordon T. Geballe. 1993. *Redesigning the American Lawn.* New Haven, CT: Yale University Press.

Boserup, Ester. 1965. *The Conditions of Agricultural Growth: the Economics of Agrarian Change under Population Pressure.* Chicago: Aldine Publishing.

———. 1981. *Population and Technical Change: A Study of Long-Term Change.* Chicago: University of Chicago Press.

———. 1985. "The impact of scarcity and plenty on development." Pp. 185–210 in Robert I. Rothberg and Theodore K. Rabb (eds.) *Hunger and History.* Cambridge: Cambridge University Press.

Botkin, Daniel R. 1990. *Discordant Harmonies: A New Ecology for the Twenty-First Century.* New York: Oxford University Press.

Bradley, P.N., and S.E. Carter. 1989. "Food production and distribution—and hunger." Pp. 101–24 in R.J. Johnson and P.J. Taylor (eds.) *A World in Crisis? Geographical Perspectives.* Oxford: Basil Blackwell.

Braidotti, Rosi, Ewa Charkiewicz, Sabine Hausler, and Saskia Wieringa. 1994. *Women, the Environment and Sustainable Development: Towards a Theoretical Synthesis.* London: Zed Books in association with INSTRAW.

Brandon, Katrina, Kent H. Redford, and Steven E. Sanderson. 1998. *Parks in Peril: People, Politics, and Protected Areas.* Washington, DC: Island Press.

Bray, Francesca, 1986. *The Rice Economics: Technology and Development in Asian Societies.* New York: Blackwell.

Brechin, Steven R., and Willett Kempton. 1994. "Global environmentalism: A challenge to the postmaterial thesis?" *Social Science Quarterly* 75:245–69.

Brechin, Steven R., Peter R. Wilshusen, Crystal L. Fortwangler, and Patrick C. West. 2000. "Reinventing a square wheel: A critique of the new protectionist paradigm in international biodiversity conservation." Under review.

Broadbent, Jeffrey. 1998. *Environmental Politics in Japan: Networks of Power and Protest.* New York: Cambridge University Press.

Bromley, Simon. 1991. *American Hegemony and World Oil.* University Park: The Pennsylvania State University Press.

Brosius, J. Peter. 1997. "Prior transcripts, divergent paths: Resistance and acquiescence to logging in Sarawak, East Malaysia." *Comparative Studies in Society and History* 39:468–510.

Brown, Harrison. 1954. *The Challenge of Man's Future.* New York: Compass Books, Viking Press.

Brown, Lester. 1990. *State of the World 1990.* New York: W.W. Norton.

———. 1991a. "Fertilizer engine losing steam." *Worldwatch Paper* no. 85. Washington, DC: Worldwatch Institute.

———. 1991b. "The new world order." Pp. 3–20 in Lester Brown (ed.) *State of the World 1991.* London: Earthscan Publications.

———. 1995. *Who Will Feed China? Wake-up Call for a Small Planet.* New York: W.W. Norton.

Brown, Lester R. 1987. "Analyzing the demographic trap." Pp. 20–37 in Lester R. Brown (ed.) *State of the World 1987.* New York: W. W. Norton.

Brown, Lester R., Christopher Flavin, and Sandra Postel. 1990. "Picturing a sustainable society." Pp. 173–90 in Brown et al., *State of The World.* New York: W.W. Norton.

Brown, Lester., and Hal Kane. 1994. *Full House: Reassessing the Earth's Carrying Capacity.* New York: W.W. Norton.

Brown, Lester., and E. Wolfe. 1984. "Soil erosion: Quiet crisis in the world economy." *Worldwatch Paper no. 60.* Washington, DC: Worldwatch Institute.

Brulle, Robert J. 1995. "Environmental discourse and social movement organizations: A historical and rhetorical perspective on the development of U.S. environmental organizations." *Sociological Inquiry* 65.

———. 1995. "Environmentalism and human emancipation." Pp. 309–28 in S. M. Lyman (ed.) *Social Movements.* London: Macmillan.

Brulle, Robert J. 2001. "Environmental discourse and social movement organizations: A historical and rhetorical perspective on the development of U.S. environmental organizations." Pp. 217–37 in R.S. Frey (ed.) *The Environment and Society Reader.* Needham Heights, MA: Allan and Bacon.

Bryant, Bunyan. 1995. *Environmental Justice: Issues, Policies, and Solutions.* Washington, DC: Island Press.

Bugbee, B., and O. Monje. 1992. "The limits of crop productivity." *Bioscience* 42(7):494–502.

Bullard, R. 2001. "Anatomy of environmental racism and the environmental justice movement." Pp. 97–105 in R. S. Frey (ed.) *The Environment and Society Reader.* Needham Heights, MA: Allan and Bacon.

Bullard, Robert D.1993. *Confronting Environmental Racism: Voices from the Grassroots.* Boston, MA: South End Press.

———. 1994. *Dumping in Dixie.* Boulder, CO: Westview Press.

Bullard, Robert D., and Beverly Hendrix Wright. 1986. "The politics of pollution: Implications for the black community." *Phylon* 46:71–78.

———. 1992. "The quest for environmental equity: Mobilizing the African-American community for social change." Pp. 39–50 in R. E. Dunlap and A. G. Mertig (eds.) *American Environmentalism.* Philadelphia PA: Taylor and Francis.

Burch, William R., Jr. 1971. *Daydreams and Nightmares: A Sociological Essay on the American Environment.* New York: Harper & Row.

Burgess, Ernest W. 1925. "The growth of the city: An introduction to a research project." Pp. 47–62 in Robert E. Park, Ernest W. Burgess, and Roderick D. McKenzie (eds.). *The City.* Chicago: University of Chicago Press.

Busch, Lawrence, William B. Lacy, Jeffrey Burkhardt, and Laura R. Lacy. 1991. *Plants, Power, and Profit: Social, Economic, and Ethical Consequences of the New Biotechnologies.* Cambridge, MA: Basil Blackwell.

Buttel, Frederick H. 1978. "Environmental sociology: A new paradigm?" *The American Sociologist* 13:252–56.

———. 1986. "Sociology and the environment: The winding road toward human ecology." *International Social Science Journal* 109:337–56.

———.1987. "New directions in environmental sociology." *Annual Review of Sociology* 13:465–88.

———. 1992. "Environmenralization: Origins, processes, and implications for rural social change." *Rural Sociology* 57:1–27.

———. 1996. "Environmental and resource sociology: Theoretical issues and opportunities for synthesis." Rural Sociology 61:56–76.

———. 1998. "Some observations on states, world orders, and the politics of sustainability." *Organization & Environment* 11(3):261–86.

———. 2000 "World society, the nation-state, and environmental protection." *American Sociological Review* 65 (2000):117–21.

Buttel, Frederick H., and Laura Raynolds. 1989. "Population growth, agrarian structure, food production, and food distribution in the Third World." *Food and Natural Resources.* Pp. 325–61.

Buttel, Frederick H., and Peter J. Taylor. 1992. "Environmental sociology and global environmental change: A critical assessment." *Society and Natural Resources* 5:211–30.

———. 1994. "Environmental sociology and global environmental change: A critical assessment." In M. Redclift and T. Benton (eds.) *Social Theory and the Global Environment.* London: Routledge.

Cable, Sherry, and Charles Cable. 1995. *Environmental Problems Grassroots Solutions.* New York: St. Martin's Press.

Caldwell, John. 1976. "Toward a restatement of demographic transition theory." *Population and Development Review* 2(3–4):321–66.

———. 1980. "Mass education as a determinant of the timing of fertility decline." *Population and Development Review* 6(2):225–55.

———. 1985. Educational Transition in Rural South India." *Population and Development Review* 11:29–51.

Campbell, Colin, and Jean H. Leherrere. 1998. "Mideast oil forever?" *Scientific American* 278:78–83.

Capek, Stella M. 1993. "The 'Environmental justice' frame: A conceptual discussion and an application." *Social Problems* 41:5–24.

Cardoso, Fernando Henrique, and Enzo Faletto (translation). 1979. *Dependency and Development in Latin America.* Berkeley: University of California Press.

Carpenter, Ted G. 1991. "Introduction." Pp. 1–3 in Ted G. Carpenter (ed.) *America Entangled: The Persian Gulf Crisis and Its Consequences.* Washington, DC: CATO Institute.

Carson, Rachel. 1962. *Silent Spring.* New York: Houghton-Mifflin.

Catton, William R. 1976. "Why the future isn't what it used to be (and how it could be made worse than it has to be)." *Social Science Quarterly* 57:276–91.

———. 1980. *Overshoot: The Ecological Basis of Revolutionary Change.* Urbana: University of Illinois Press.

Catton, W. R., Jr., and R. E. Dunlap. 1978. "Environmental sociology: A new paradigm." *The American Sociologist* 13:41–49.

Catton, Willliam R., and Riley E. Dunlap. 1978. "Environmental sociology: A new paradigm?" *The American Sociologist* 13:307–23.

———. 1980. "A new ecological paradigm for post-exuberant sociology." *American Behavioral Scientist* 24:15–48.

CERES website, *www.ceres.org.*

Cernea, Michael M., (ed.) 1991. *Putting People First-Sociological Variables in Development.* New York: Oxford University Press.

———. 1993. "The sociologist's approach to sustainable development." *Finance & Development,* December: 11–13.

Chance, Norman A., and Elena N. Andreeva. 1995. "Sustainability, equity, and natural resource development in Northwest Siberia and Arctic Alaska." *Human Ecology* 23(2): 217–31.

Chatterjee, Pratap, and Matthias Finger. 1994. *The Earth Brokers: Power, Politics and World Development.* London: Routledge.

Checkoway, Barry. 1984. "Large builders, federal housing programs, and postwar suburbanization." Pp. 152–73 in William K. Tabb and Larry Sawers (eds.) *Marxism and the Metropolis: New Perspectives in Urban Political Economy,* Second Edition. New York: Oxford University Press.

Cheru, Fantu. 1992. "Structural adjustment, primary resource trade and sustainable development in Sub-Saharan Africa." *World Development* 20(4):497–512.

Chesler, Ellen. 1992. *Women of Valor: Margaret Sanger and the Birth Control Movement in America.* New York: Simon and Schuster.

Clancy, Katherine L. 1993. "Sustainable agriculture and domestic hunger: Rethinking a link between production and consumption." Pp. 251–93 in P. Allen (ed.) *Food for the Future: Conditions and Contradictions of Sustainability.* New York: John Wiley & Sons, Inc.

Clements, Frederick E. 1905. *Research Methods in Ecology.* Lincoln: University of Nebraska.

Coale, Ansley J., and Susan Cotts Watkins, (eds.) 1986. *The Decline of Fertility in Europe.* Princeton, NJ: Princeton University Press.

Coates, Peter. 1991. *The Trans-Alaska Pipeline Controversy: Technology, Conservation, and the Frontier.* Bethlehem, PA: Lehigh University Press.

Cohen, Maurie. 1997. *Sustainable Development and Ecological Modernization,* Research Paper no. 14. Oxford Centre for Environment, Ethics, and Society. Oxford: Oxford University.

Cohen, Roger. 1990. "A man's fight for the rain forest." *New York Times,* August 14: Section C, p. 15.

Collinson, Helen. 1996. *Green Guerrillas: Environmental Conflicts and Initiatives in Latin America and the Caribbean.* London: Latin American Bureau.

Commission for Racial Justice. 1987. *Toxic Wastes and Race in the United States.* New York: United Church of Christ.

Commoner, Barry. 1963. *Science and Survival.* New York: Viking Press.

———. 1966. *Science and Survival.* New York: Viking Press.

———. 1971. *The Closing Circle.* New York: Knopf.

———. 1979. *The Politics of Energy.* New York: Alfred Knopf.

———. 1990. *Making Peace with the Planet.* New York: Pantheon Books.

Conroy, Czech, and Miles Litvinoff (eds.) 1988. *The Greening of Aid.* London: Earthscan Publications Ltd.

Conservation International. n.d. *The Tagua Initiative.* Washington, DC.

Conway, G., and Jules Pretty. 1991. *Unwelcome Harvest: Agriculture and Pollution.* London: Earthscan Publications.

Costanza, Robert, John Audley, Richard Borden, Paul Ekins, Carl Folke, Silvio O. Funtowicz, and Jonathan Harris. 1995. "Sustainable trade: A new paradigm for world welfare." *Environment* 37(5):16–44.

Crane, Barbara B. 1994. "The transnational politics of abortion." in Jason L. Finkle and C. Alison McIntosh (eds.) *The New Politics of Population.* Oxford: Oxford University Press.

Crosson, Pierre, and Jock R. Anderson. 1992. "Resources and global food prospects: supply and demand for cereals to 2030." *World Bank Technical Paper* no. 184. Washington, DC: The World Bank.

Daly, Herman E., and John B. Cobb, Jr. 1994. *For the Common Good: Redirecting the Economy towards Community, the Environment and a Sustainable Future,* 2d ed. Boston: Beacon Press.

Dankelman, Irene, and Joan Davidson. 1988. *Women and Environment in the Third World: Alliance for the Future.* London: Earthscan and the International Union for the Conservation of Nature.

Deal, Carl. 1993. *The Greenpeace Guide to Anti-environmental Organizations.* Berkeley, CA: Odonian Press.

D'Eaubonne, Francoise. 1974. "Feminism or death." Pp. 64–67 in Elaine Marks and Isaballe de Courtivron (eds.) *New French Feminisms.* New York: Schocken Books.

Defenders of Wildlife. 1999. "Conservationists file federal court challenge to Vail Ski expansion, press release, 11 June 1998." *www.defenders.org/pr611c98.html.*

DeKadt, Emanuel. 1979. *Tourism: Passport to Development?* New York: Oxford University Press.

———. 1992. "Making the alternative sustainable: Lessons from development for tourism." In V.L. Smith and W.R. Eadington (eds.) *Tourism Alternatives: Potentials and Problems in the Development of Tourism.* Philadelphia: University of Pennsylvania Press.

Devall, Bill. 1991. "Deep ecology and radical environmentalism." *Society and Natural Resources* 4:247–58.

———. 1992. "Deep ecology and radical environmentalism." Pp. 51–62 in R. E. Dunlap and A. G. Mertig (eds.) *American Environmentalism.* Philadelphia, PA: Taylor and Francis.

Devall, Bill and George Sessions. 1985. *Deep Ecology: Living As If Nature Mattered.* Salt Lake City, UT: Gibb M. Smith, Inc.

DeWalt, Billie. 1994. "Using indigenous knowledge to improve agriculture and natural resource management." *Human Organization* 53(2):123–30.

Dewees, Donald. 1970. "The decline of the American street railways." *Traffic Quarterly* 24:563–82.

Dickens, Peter. 1992. *Society and Nature: Toward a Green Social Theory.* Philadelphia, PA: Temple University Press.

Dietz, Thomas, and Eugene A. Rosa. 1994. "Rethinking the environmental impacts of population, affluence, and technology." *Human Ecology Review* 1:277–300.

Dixon-Mueller, Ruth. 1993. *Population Policy & Women's Rights: Transforming Reproductive Choice.* Westport, CT: Praeger.

Donaldson, Peter S. 1990. *Nature Against Us, the United States and the World Population Crisis.* Chapel Hill: University of North Carolina Press.

Donaldson, Peter J., and Amy Ong Tsui. 1990. "The international family planning movement." *Population Bulletin* 45(3):3–44.

Donaldson, Scott. 1969. *The Suburban Myth.* New York: Columbia University Press.

Dowie, Mark. 1995. *Losing Ground: American Environmentalism at the Close of the Twentieth Century.* Cambridge, MA: The MIT Press.

Dryzek, John S. 1997. *The Politics of the Earth: Environmental Discourses.* Oxford: Oxford University Press.

Dumont, C. 1996. "The demise of community and ecology in the Pacific Northwest." *Sociological Perspectives* 39:277.

Duncan, Otis D. 1959. "Human ecology and population studies." Pp. 678–716 in Philip M. Hauser and O.D. Duncan (ed.) *The Study of Population.* Chicago: University of Chicago Press.

———. 1961. "From social system to ecosystem." *Sociological Inquiry* 30:140–49.

Duncan, Otis D., and Leo F. Schnore. 1959. "Culture, behavioral, and ecological perspectives in the study of social organization." *American Journal of Sociology* 65:132–46.

Duncan, Otis D., W.R. Scott, Stanley Lieberson, Beverly Duncan, and Hal H. Winsborough. 1960. *Metropolis and Region.* Baltimore, MD: Johns Hopkins Press.

Dunlap, Riley E. 1992. "Trends in Public Opinion Toward Environmental Issues: 1965–1990." Pp. 89–116 in R. E. Dunlap and A. G. Mertig (eds.) *American Environmentalism. The U.S. Environmental Movements, 1970–1990.* Philadelphia, PA: Taylor and Francis.

Dunlap, Riley E., 1993. "From environmental to ecological problems." In C. Calhoun and G. Ritzer (eds.) *Social Problems.* New York: McGraw-Hill.

Dunlap, Riley E., and William R. Catton, Jr. 1979a. "Environmental sociology." *Annual Review of Sociology* 5:243–73.

———. 1979b. "Environmental sociology: A framework for analysis." Pp. 57–85 in T. O'Riordan and R.C. d'Arge (eds.) *Progress in Resource Management and Environmental Planning,* vol. 1. Chichester: Wiley.

———. 1980. "Paradigmatic change in the social sciences: From human exemptionalism to an ecological paradigm." *American Behavioral Scientist* 24:5–14.

———. 1984. "Commitment to the dominant social paradigm and concern for environmental quality." *Social Science Quarterly* 65:1013–28.

———. 1994. "Struggling with human exemptionalism: The rise, decline, and revitalization of environmental sociology." *The American Sociologist* 21:5–30.

Dunlap, Riley E., George H. Gallup Jr., and Alec M. Gallup. 1993. *Health of the Planet.* Princeton, NJ: Gallup International Institute.

Dunlap, Riley E., and Angela G. Mertig. 1992. "American environmentalism: The U.S. environmental movement, 1970–1990." Philadelphia, PA: Taylor and Francis.

———. 2001. "Global concern for the environment: Is affluence a prerequisite?" Pp. 202–15 in R. S. Frey (ed.)

The Environment and Society Reader. Needham Heights, MA: Allan and Bacon.

Dunlap, Riley E., and Kent D. Van Liere. 1978. "The 'new environmental paradigm': A proposed measuring instrument and preliminary results." *Journal of Environmental Education* 9:10–19.

———. 1984. "Commitment to the dominant social paradigm and concern for environmental quality." *Social Science Quarterly* 65:1013–1028.

Durkheim, Emile. [1893] 1960. *The Division of Labor in Society.* Glencoe, IL: The Free Press of Glencoe.

———. [1893] 1964. *The Division of Labor in Society.* New York: Free Press.

Durning, Alan T. 1989. *Action at the Grassroots: Fighting Poverty and Environmental Decline.* Washington, DC: Worldwatch Institute.

Dyson, Tim. 1994. "World population growth and food supplies." *International Social Science Journal* 46:361–65.

———. 1996. *Population and Food: Global Trends and Future Prospects.* New York: Routledge.

Easterlin, Richard. 1983. "Modernization and Fertility: A critical essay." Pp. 562–86 in R. Bulatao and R.D. Lee (eds.) *Determinants of Fertility in Developing Countries.* New York: Academic Press.

———. 1987. *Birth and Fortune: The Impact of Numbers on Personal Welfare.* Chicago: University of Chicago Press.

Echeverria, John D., and Raymond Booth Eby. 1995. *Let the People Judge: Wise Use and the Private Property Rights Movement.* Washington, DC: Island Press.

Eckholm, Erik P. 1976. *Losing Ground: Environmental Stress and World Food Prospects.* New York: W.W. Norton.

Eddy, Mark. 1998. "Environmental groups condemn 'heinous act'." *The Denver Post* 20 October, A1.

Edwards, Bob. 1995. "With liberty and environmental justice for all: The emergence and challenge of grassroots environmentalism." In B. Taylor (ed.) *Grassroots Ecological Resistance.* Albany: State University of New York Press.

Ehrlich, Anne. 1988. "Development and agriculture." Pp. in Paul R. Ehrlich and John P. Holdren (eds.) *The Cassandra Conference.* College Station: University of Texas Press.

Ehrlich, Paul R. 1968. *The Population Bomb.* New York: Ballantine Books.

Ehrlich, Paul R., and A. H. Ehrlich. 1990. *The Population Explosion.* New York: Simon and Schuster.

Ehrlich, Paul, Anne Ehrlich, and Gretchen Daily. 1993. "Food security, population, and the environment." *Population and Development Review* 19:1–32.

———. 1991. *Healing the Planet.* Boston, MA: Addison-Wesley.

Eicher, Carl K. 1986. "Strategic issues in combating hunger and poverty in Africa." Pp. 242–78 in Robert J. Berg and Jennifer F. Whitaker (eds.) *Strategies for African Development.* Berkeley: University of California Press.

Ekins, Paul. 1993. "Making Development Sustainable." Pp. 91–103 in W. Sachs (ed.) *Global Ecology.* London: Zed Books.

Elgin, Duane. 1981. *Voluntary Simplicity: Toward a Way of Life That is Outwardly Simple, Inwardly Rich.* New York: Morrow.

Engels, Frederick. 1892. *The Conditions of the Working Class in England in 1844.* London: Allen and Unwin.

Escobar, Arturo. 1995. *Encountering Development: The Making and Unmaking of the Third World.* Princeton, NJ: Princeton University Press.

Escobar, Arturo. Forthcoming. "Gender, place and networks: A political ecology of cyberculture." In W. Harcourt. (ed.) *Women @ Internet: Creating New Cultures in Cyberspace.* London: Zed Books/UNESCO.

Faber, Daniel. 1998. *The Struggle for Ecological Democracy: Environmental Justice Movements in the United States.* New York: The Guilford Press.

Faber, Daniel, and James O'Connor. 1989. "The struggle for nature: Environmental crises and the crisis of environmentalism in the United States." *Capitalism, Nature, Socialism* 2:12–38.

Falkenmark, M., and C. Widstrand. 1992. "Population and water resources: A delicate balance." *Population Bulletin* 47(3).

FAO-Tech-5. 1995 (Nov.). "Overall socio-political and economic environment for food security at national, regional, and global levels." *World Food Survey* 96/TECH/5 Rome, IT.

Farrell, John G. and Miguel A. Altieri. 1995. "Agroforestry systems" Pp. 247–63 in Miguel Altieri (ed.) *Agroecology: The Science of Sustainable Agriculture.* Boulder, CO: Westview.

Feenstra, Gail W. 1997. "Local food systems and sustainable communities." *American Journal of Alternative Agriculture* 12(1):28–36.

Field, Donald R., and William R. Burch, Jr. 1988. *Rural Sociology and the Environment.* New York: Greenwood Press.

Finkle, Jason L., and C. Alison McIntosh. 1994. *The New Politics of Population: Conflict and Consensus in Family Planning.* Oxford: Oxford University Press.

Flavin, Christopher. 1990. "Beyond the Gulf crisis: an energy strategy for the '90's." *Challenge* 33:4–10.

Flavin, Christopher, and Nicholas Lennsen. 1994. *Power Surge: Guide to the Coming Energy Revolution.* New York: W.W. Norton.

Food and Agriculture Organization of the United Nations (FAO). 1994. *The State of Food and Agriculture 1994.* Rome: FAO Publication.

———. 1995. *Assessment of the Current World Food Security Situation and Medium Term Review.* Item II of the Provisional Agenda, 20th Session, Committee on World Food Security. Rome: FAO Publication.

Foreman, Dave, and Bill Haywood. 1987. *Ecodefense: A Field Guide to Monkeywrenching,* Second Edition. Tuscon, AZ: A Ned Ludd Book.

Foresta, Ronald A. 1991. *Amazon Conservation in the Age of Development.* Gainesville: University of Florida Press.

Fornos, Werner 1991. "Population politics." *Technology Review* 94:42–51.

Foster, John Bellamy. 1997. "The crisis of the earth." *Organization and Environment* 10:278–95.

———. 1999. "Marx's theory of metabolic rift: Classical foundations for environmental sociology." *American Journal of Sociology* 105:366–405.

Francis C., C. Flora, and L. King. 1990. *Sustainable Agriculture in Temperate Zones.* New York: Wiley-Interscience.

Frank, David John. 1997. "Science, nature, and the globalization of the environment, 1870–1990." *Social Forces* 76:409–35.

Frank, David John, Ann Hironaka, and Evan Schofer. 2000. "The nation-state and the natural environment over the twentieth century." *American Sociological Review* 65:96–116.

Frank, David John, Ann Hironaka, John W. Meyer, Evan Schofer, and Nancy Brandon Tuma. 1999. "The Rationalization and Organization of Nature in World Culture." Pp. 81–99 in John Boli and George M. Thomas (eds.) *Constructing World Culture.* Stanford, CA: Stanford University Press.

Freebairn, Donald K. 1973. "Income disparities in the agricultural sector: regional and institutional stresses." Pp. 97–119 in T.T. Poleman and D. K. Freebairn (eds.) *Food, Population, and Employment: The Impact of the Green Revolution.* New York: Praeger.

Freeman, A.M. III. 2000. "Economics, incentives, and environmental regulation." Pp. 190–209 in N.J. Vig and M. E. Kraft (eds.) *Environmental Policy,* 4th ed. Washington, DC: CQ Press.

Freidmann, Harriet. 1993. "The political economy of food: a global crisis." *The New Left Review* 197:29–57.

Friedmann, John, and Haripriya Rangan, (eds.) 1993. *In Defense of Livelihood.* West Hartford, CT: Kumarian Press.

Freudenberg, N., and C. Steinsapir. 1992. "Not in our backyards: The grassroots environmental movement." Pp. 27–37 in R.E. Dunlap and A.G. Mertig, *American Environmentalism*. Philadelphia, PA: Taylor and Francis.

Freudenburg, William. 1989. "The emergence of environmental sociology; contributions of Riley E. Dunlop and William R. Catton, Jr." *Sociological Inquiry* 59:439–44.

Freudenburg, William, and Robert Gramling. 1994. *Oil Over Troubled Waters: Perception, Politics and the Battle Over Offshore Drilling.* Albany: State University Press of New York.

Gadgil, M., and R. Guha. 1995. *Ecology and Equity.* London: Routledge.

Gamson, William A. 1975. *The Strategy of Social Protest*. Homewood, IL: The Dorsey Press.

Gamson, William A., and Andre Modigliani. 1989. "Media discourse and public opinion on nuclear power." *American Journal of Sociology* 95:1–38.

Gans, Herbert J. 1962. *The Urban Villagers: Group and Class in the Life of Italian-Americans.* New York: The Free Press.

———. 1967. *The Levittowners: Ways of Life and Politics in a New Suburban Community.* New York: Random House.

Gear, A., (ed.). 1983. *The Organic Food Guide.* Essex: Longman, Harlow.

GEF (Global Environment Facility). 1998. *Operational Report on GEF Programs.* Washington, DC: The GEF Secretariat.

Gezon, Lisa. 1997. "Institutional structure and the effectiveness of integrated conservation and development projects: Case study from Madagascar." *Human Organization* 56(4):462–70.

Ghai, Dharam, (ed.). 1994. *Development and Environment: Sustaining People and Nature.* Cambridge: Blackwell Publishers.

Ghimire, Krishna B. 1994. "Parks and People: Livelihood Issues in National Parks Management in Thailand and Madagascar." Pp. 195–229 in D. Ghai, (ed.) *Development and Environment: Sustaining People and Nature.* Oxford: Blackwell Publishers/UNRISD.

Gibbs, Jack P., and W.T. Martin. 1962. "Urbanization, technology, and the division of labor: International patterns." *American Sociological Review* 27:667–77.

Glass, D. V. 1973. *Numbering The People.* Farnborough: Saxon House Publishers.

Gleik, Peter H. 1993. "An introduction to global fresh water issues." Pp. 3–12 in Peter H. Gleik (ed.) *Water in Crisis.* New York: Oxford University Press.

Goeteyn, Luc. 1996. "Measuring sustainable development at the national and international level." Pp. 161–80 in B. Nath et al., (eds.) *Sustainable Development*. Brussels: VUB University Press.

Goldblatt, D. 1996. *Social Theory and the Environment*. Boulder, CO: Westview Press.

Goldstone, Jack A. 1980. "The weakness of organization: A look at Gamson's *The Strategy of Social Protest*." *American Journal of Sociology* 5:1017–42.

Gottlieb, Robert. 1993. *Forcing the Spring: The Transformation of the American Environmental Movement*. Washington, DC: Island Press.

Gould, Kenneth A. 1991. "The sweet smell of money: Economic dependency and local environmental political mobilization." *Society & Natural Resources* 4:133–50.

Gould, Kenneth A., Allan Schnaiberg, and Adam S. Weinberg. 1996. *Local Environmental Struggles: Citizen Activism in the Treadmill of Production*. New York: Cambridge University Press.

Gould, Kenneth A., Adam S. Weinberg, and Allan Schnaiberg. 1993. "Legitimating impotence pyrrhic victories of the modern environmental movement." *Qualitative Sociology* 16:207–46.

Greer, Edward. 1991. "The hidden history of the Iraq war." *Monthly Review* 43:1–14.

Griffin, Keith B. 1974. *The Political Economy of Agrarian Change: An Essay on the Green Revolution*. Cambridge, MA: Harvard University Press.

Grossman, P.Z., and D.H. Cole. 1999. "When is command and control efficient?" *Wisconsin Law Review* 5:887–938.

Grove, Richard. 1992. "Origins of Western environmentalism." *Scientific American* 267:42–47.

———. 1992a. *Green Imperialism: Colonial Expansion, Tropical Island Edens, and the Origins of Environmentalism, 1600–1860.* Cambridge: Cambridge University Press.

Grueso, Libia, Carlos Rosero, and Arturo Escobar. 1998. "The process of black community organizing in the southern Pacific coast region of Colombia." Pp. 196–219 in S. E. Alvarez, E. Dagnino, and A. Escobar (eds.) *Cultures of Politics/Politics of Cultures.* Boulder, CO: Westview Press.

Gudynas, Eduardo. 1993. "The fallacy of ecomessianism: Observations from Latin America." Pp. 170–78 in Wolfgang Sachs (ed.) Global Ecology. *A New Avenue of Political Conflict.* London: Zed Books.

Guha, Ramachandra. 1989. *The Unquiet Woods: Ecological Change and Peasant Resistance in the Himalaya.* Berkeley: University of California Press.

Guha, Ramachandra, and Juan Martinez-Alier. 1997. *Varieties of Environmentalism: Essays North and South.* London: Earthscan Publications, Ltd.

Hadden, Jeffery K., and Josef J. Barton. 1973. "An image that will not die: thoughts on the history of anti-urban ideology." Pp. 79–119 in Louis H. Masotti and Jeffery K. Hadden (eds.) *The Urbanization of Suburbs.* Beverly Hills, CA: Sage.

Haedrich, Richard L., and Lawrence C. Hamilton. 2000. "The fall and future of Newfoundland's cod fishery." *Society & Natural Resources* 13:359–72.

Hall, D.H. Mynick, and R. Williams. 1991. "Alternative roles for biomass in coping with greenhouse warming." *Science and Global Security* 2:1–39.

Hannigan, John A. 1995. *Environmental Sociology: A Social Constructionist Perspective.* London: Routledge.

Harmon, David. 1987. "Cultural diversity, human subsistence, and the national park ideal." *Environmental Ethics* 9:147–58.

Harper, Charles L. 1996. *Environment and Society.* Saddle River, NJ: Prentice-Hall.

———. 2001. *Environment and Society. Human Perspectives on Environmental Issues.* Upper Saddle River, NJ. Prentice Hall.

Harrison, Bennett, and Barry Bluestone. 1988. *The Great U-Turn: Corporate Restructuring and the Polarizing of America.* New York: Basic Books.

Hartmann, Betsy. 1987. *Reproductive Rights and Wrongs: The Global Politics of Population Control and Contraceptive Choice.* New York: Harper and Row.

———. 1994. "Consensus and contradiction on the road to Cairo." Pp. 15–26 in Rosiska Darcey de Oliveira (ed.) *Terra Femina.* Rio, Brazil: Institute of Cultural Action.

Hatcher, R. Lee. 1996. "Local Indicators of Sustainability: Measuring the Human Ecosystem." Pp. 181–203 in B. Nath et al., (eds.) *Sustainable Development.* Brussels: VUB University Press.

Hauser, Philip M., and A. J. Jaffee. 1947. "The extent of the housing shortage." *Law and Contemporary Problems* 12:3–15.

Havens, Eugene A., and William J. Flinn. 1975. "Green revolution technology and community development: the limits of action programs." *Economic Development and Cultural Change* 23:469–81.

Hawley, Amos H. 1950. *Human Ecology: A Theory of Community Structure.* New York: The Ronald Press.

———. 1971. *Urban Society: An Ecological Approach.* New York: Ronald Press.

———. 1986. *Human Ecology: A Theoretical Essay.* Chicago: University of Chicago Press.

Hawley, T.M. 1992. *Against the Fires of Hell: Environmental Disaster of the Gulf War.* New York: Harcourt, Brace Jovanovich, Publishers.

Hayami, Yujiro, and Vernon W. Ruttan. 1985. *Agricultural Development: An International Perspective.* Baltimore, MD: John Hopkins University Press.

———. 1987. "Population growth and agricultural productivity." Pp. 57–100 in D. G. Johnson and R. L. Lee (eds.) *Population Growth and Economic Development: Issues and Evidence.* Madison: University of Wisconsin Press.

Hays, Samuel P. 1959. *Conservation and the Gospel of Efficiency: The Progressive Conservation Movement, 1980–1920.* Cambridge: Harvard University Press.

———. 1987. Beauty, *Health and Permanence.* New York: Cambridge University Press.

———. 2000. *A History of Environmental Politics Since 1945.* Pittsburgh, PA: University of Pittsburgh Press.

Hayter, Teresa. 1989. *Exploited Earth: Britain's Aid and the Environment.* London: Earthscan Publishers.

Healy, Kevin. 2001. *Llamas, Weavings and Organic Chocolate: Multi-cultural Grassroots Development in the Andes and Amazon of Bolivia.* Notre Dame, IN: University of Notre Dame Press.

Hecht, Susanna, and Alexander Cockburn. 1990. *The Fate of the Forest: Developers, Destroyers and Defenders of the Amazon.* New York: Harper.

Henderson, David R. 1991. "The myth of Saddam's oil stranglehold." Pp. 41–45 in Ted G. Carpenter (ed.) *America Entangled: The Persian Gulf Crisis and Its Consequences.* Washington, DC: CATO Institute.

Hicks, James F., et al. 1990. *Ecuador's Amazon Region: Development Issues and Options.* Washington, DC: The World Bank.

Himmelfarb, Gertrude, (ed.). 1960. *On Population by Thomas Malthus.* New York: Modern Library.

Hobsbawm, Eric J. 1959. *Primitive Rebels.* New York: W.W. Norton.

Homer-Dixon, Thomas F., Jeffrey H. Boutwell, and George W. Rathjens. 1994. "Environmental Change and Violent Conflict." Pp. 391–400 in Laurie Ann Mazur (ed.) *Beyond the Numbers.* Washington, DC: Island Press.

Houghton, R.A. 1992. "Tropical deforestation and atmospheric carbon dioxide." Pp. 99–118 in Norman Myers (ed.) *Tropical Forests and Climate.* Dordrecht, ND: Kluwer Academic Publishers.

Hoyt, E. 1988. *Conserving the Wild Relatives of Crops.* Rome, IT and Gland, SW: International Board for Plant Genetic Resources, International Union for the Conservation of Nature, and Worldwide Fund for Nature.

Hueting, R.P., P. Bosch, and B. De Boer. 1992. *Methodology for the Calculation of Sustainable National Income.* Gland, Switzerland: WWF International Publication.

Hull, Terence H. 1990. "Recent trends in sex ratios at birth in China." *Population and Development Review* 16(1):63–84.

Humphrey, Craig R., and Frederick H. Buttel. 1982. *Environment, Energy, and Society.* Belmont, CA: Wadsworth Publishing.

Hurley, Andrew. 1995. *Environmental Inequalities: Race, Class, and Industrial Pollution in Gary, Indiana, 1945–1980.* Chapel Hill: The University of North Carolina Press.

Hviding, Edvard, and Graham B.K. Baines. 1994. "Community-based fisheries management, tradition and the challenges of development in Marovo, Solomon Islands." Pp. 13–40 in D. Ghai (ed.) *Development & Environment: Sustaining People and Nature.* Oxford: Blackwell.

Inglehart, Ronald. 1977. *The Silent Revolution: Changing Values and Political Styles among Western Publics.* Princeton, NJ: Princeton University Press.

———. 1990. *Culture Shift in Advanced Industrial Society.* Princeton, NJ: Princeton University Press.

———. 1995. "Public support for environmental protection: Objective problems and subjective values in 43 societies." *PS: Political Science & Politics* 28:57–72.

Inman, Katherine. 1992. "Fueling expansion in the Third World: Population, development, debt, and the global decline of forests." *Society and Natural Resources* 6:17–39.

International Energy Agency. 1996. *World Energy Outlook*. Paris, FR: Organization for Economic Cooperation and Development.

IUCN, Inter-Commission Task Force on Indigenous Peoples. 1997. *Indigenous Peoples and Sustainability.* (Prepared by Darrell A. Posey and Graham Dutfield.) Utrecht: International Books.

Jaggar, Alison M. 1983. *Feminist Politics and Human Nature.* Totowa, NJ: Rownam and Allanheld.

Jenkins, J. Craig, and Charles Perrow. 1977. "Insurgency of the powerless: Farm worker movements (1946–1972)." *American Sociological Review* 42:249–68.

Jiggins, J. 1986. "Gender-related impacts and the work of the international agricultural research centers." *Consultative Group on International Agriculture Research Study Paper* no. 17. Washington, DC: The World Bank.

Kamieniecki, Sheldon. 1993. *Environmental Politics in the International Arena.* Albany: State University of New York Press.

Kanagy, Conrad L., Craig R. Humphrey, and Glenn Firebaugh. 1994. "Surging environmentalism: Changing public opinion or changing publics?" *Social Science Quarterly* 75:804–19.

Kasarda, John D. and Charles E. Bidwell. 1984. "A human ecological theory of organizational structuring." Pp. 179–236 in M. Micklin and H. Choldin (eds.) *Sociological Human Ecology.* Boulder, CO: Westview Press.

Kaufman, Joan, Zhang Zhirong, Qjao XinJian, and Zhang Yang. 1989. "Family planning policy and practice in China: A study of four rural communities." *Population and Development Review* 15(4):707–30.

Kelly, Barbara M. 1993. *Expanding the American Dream: Building and Rebuilding Levittown.* Albany: State University of New York Press.

Kempton, Willett, James S. Boster, and Jennifer A. Hartley. 1995. *Environmental Values in American Culture.* Cambridge, MA: The MIT Press.

Kennedy, Paul. 1988. *First the Seed: The Political Economy of Plant Biotechnology.* New York: Cambridge University Press.

———. 1993. *Preparing for the Twenty-First Century.* New York: Vintage Press.

Kenney, Martin. 1986. *Biotechnology: The University-Industrial Complex.* New Haven, CT: Yale University Press.

Kenya's NEAP website, *www.rri.org/envatlas/africa/kenya/ke-sum.html.*

Keyfitz, Nathan. 1991. "Population and sustainable development: Distinguishing fact and preference concerning the future human population and environment." *Population and the Environment* 14:441–62.

Kidd, Quentin, and Aie-Rie Lee. 1997. "Postmaterialist values and the environment: A critique and reappraisal." *Social Science Quarterly* 78:1–15.

Kimerling, Judith, 1991. *Amazon Crude.* New York: Natural Resources Defense Council.

Kitschelt, Herbert P. 1986. "Political opportunity structures and political protest: Anti-nuclear movements in four democracies." *British Journal of Political Science 16:57–85.*

Klandermans, Bert. 1992. "The social construction of protest and multiorganizational fields." Pp. 77–103 in A. D. Morris and C. M. Mueller (eds.) *Frontiers in Social Movement Theory.* New Haven: Yale University Press.

Klandermans, Bert, and Sidney Tarrow. 1988. "Mobilizing into social movements: Synthesizing European and American approaches." *International Social Movements Research* 1:1–40.

Kloppenburg, Jack. 1988. *First the Seed: The Political Economy of Plant Biotechnology.* New York: Cambridge University Press.

Kloppenburg, Jack Jr., John Hendrickson, and G.W. Stevenson. 1996. "Coming in to the foodshed." Pp. 113–23 in W. Vitek and W. Jackson (eds.) *Rooted in the Land: Essays on Community and Place.* New Haven: Yale University Press.

Knodel, John, and Etienne van de Walle. 1986. "Lessons from the past: Policy implications of historical fertility studies." Pp. 390–419 in Ansley J. Coale and Susan Cotts Watkins (eds.) *The Decline of Fertility in Europe.* Princeton, NJ: Princeton University Press.

Knudsen, O., and J. Nash. 1990. "Redefining government's role in agriculture in the nineties." Washington, DC: The World Bank.

Kraft, M.E. 2001. *Environmental Policy and Politics.* 2nd ed. New York: Longman.

Kriesi, H. 1989. "New social movements and the new class in the Netherlands." *American Journal of Sociology* 94:1078–116.

Kruszewska, Iza. 1998. "Changing consumption and production patterns in the North." *Development* 41(1):74–80.

Lafant, Marie-Françoise, and Nelson H. Graburn. 1992. "International Tourism Reconsidered: The Principle of the Alternative." In V.L. Smith and W.R. Eadington (eds.) *Tourism Alternatives: Potentials and Problems in the Development of Tourism.* Philadelphia: University of Pennsylvania Press.

Lappé, Frances M. 1986. *World Hunger.* San Francisco, CA: Institute for Food and Development Policy.

Lappé, Frances M., and Joseph Collins. 1978. *Food First* New York: Ballantine.

Laraña, Enrique, Hank Johnston, and Joseph R. Gusfield. 1994. *New Social Movements: From Ideology to Identity.* Philadelphia, PA: Temple University Press.

Larson, Bruce A. 1994. "Changing the economics of environmental degradation in Madagascar: Lessons from the national environmental action plan process." *World Development* 22(5):671–89.

Layne, Christopher. 1991. "Why the Gulf War was not in the national interest." *The Atlantic Monthly* 268:55, 65–81.

Leff, Enrique. 1996. Marxism and the environmental question: From the critical theory of production to an environmental rationality for sustainable development. Pp. 137–56 in T. Bento (ed.) *The Greening of Marxism.* New York: The Guilford Press.

Lélé, Sharachandra M. 1991. "Sustainable development: A critical review." *World Development* 19(6):607–21.

Lesthaeghe, Ron J. 1977. *The Decline of Belgian Fertility, 1800–1970.* Princeton, NJ: Princeton University Press.

———. 1983. "A century of demographic and cultural change in Western Europe: An exploitation of underlying dimensions." *Population and Development Review* 9(3):411–36.

Levi-Bacci, Massimo. 1986. "Social-group forerunners of fertility control in Europe." Pp. 182–200 in Ansley J. Coale and Susan Cotts Watkins (eds.) *The Decline of Fertility in Europe.* Princeton, NJ: Princeton University Press.

———. 1992. *A Concise History of World Population.* Translated by Carl Ipsen. Cambridge, MA: Blackwell.

Levine, Adeline G. 1982. *Love Canal: Science, Politics, and People.* Lexington, MA: Lexington Books.

Lewellen, Ted C. 1995. *Dependency & Development: An Introduction to the Third World.* Westport, CT: Bergin & Garvin.

Lewis, Tammy L. 1996. *The land protection strategy of conserving global biodiversity: Northern influences on Southern sustainability.* Ph.D. dissertation, Department of Sociology, University of California-Davis.

———. 2000. "Media representations of 'sustainable development': Sustaining the status quo?" *Science Communication* 21(3):244–73.

Logan, J.R., and H. Molotch. 1987. *Urban Fortunes.* Berkeley: University of California Press.

Lohmann, Larry. 1990. "Whose common future?" *The Ecologist* 20:82–84.

———. 1993. "Resisting green globalism." Pp. 159–69 in Wolfgang Sachs (ed.) *Global Ecology.* London: Zed Books.

Lovelock, J.E. 1979. *Gaia: A New Look at Life on Earth.* Oxford: Oxford University Press.

Lovins, Amory. 1976. "Energy policy: The road not taken?" *Foreign Affairs* (fall): 65–96.

———. 1977. *Soft Energy Paths.* Cambridge, MA: Ballinger.

———. 1996. "Negawatts: Twelve transitions, eight improvements and one distraction." *Energy Policy* 24:331–43.

Lovins, Amory, and L. Hunter Lovins. 1991. "Winning the peace." *Whole Earth Review* 73 (winter): 60–62.

Lovins, Amory B., L. Hunter Lovins, and Paul Hawken. 1999. "A road map for natural capitalism." *Harvard Business Review* 77:145–58.

Luciani, Giacomo. 1989. "Oil and instability: A political economy of petroleum and the Gulf war." Pp. 17–31 in Hanns W. Maull and Otto Pick (eds.) *The Gulf War: Regional and International Dimensions.* New York: St. Martin's Press.

Macnaghten, P., and J. Urry. 1998. *Contested Natures.* London: Sage.

Maisel, S. J. 1953. *House Building in Transition.* Berkeley: University of California Press.

Marks, Stewart A. 1984. *The Imperial Lion: Human Dimensions of Wildlife Management in Central Africa.* Boulder, CO: Westview Press.

Martell, L. 1994. *Ecology and Society.* Amherst: University of Massachusetts Press.

Martínez-Alier, Juan. 1997. "Environmental justice (local and global)." *Capitalism Nature Socialism* 8:91–107.

Marx, Karl. 1890. *Capital: A Critical Analysis of Capitalist Production.* New York: Humboldt Publishing Co.

Massey, Douglas S., and Nancy A. Denton. 1993. *American Apartheid: Segregation and the Making of the Underclass.* Cambridge, MA: Harvard University Press.

Matthiesen, Paul C., and James McCann. 1978. "The effects of infant and child mortality on fertility." Pp. 47–68 in Samuel Preston (ed.) *The Effect of Infant and Child Mortality on Fertility.* New York: Academic Press.

Mayer, Tom. 1991. "Imperialism and the Gulf." *Monthly Review* 42:1–11.

Mazur, Allan. 1994. "How does population growth contribute to rising energy consumption in America?" *Population and Environment* 15:371–78.

McAdam, Doug. 1982. *Political Process and the Development of Black Insurgency, 1930–1970.* Chicago: University of Chicago Press.

McAdam, Doug, and David A. Snow. 1997. *Social Movements: Readings on Their Emergence, Mobilization, and Dynamics.* Los Angeles: Roxbury Publishing Company.

McCarthy, J. 1998. "Environmentalism, wise use, and the nature of accumulation in the rural West." Pp. 126–49 in B. Braun and N. Castree (eds.) *Remaking Reality.* London: Routledge.

McCarthy, John D., and Mark Wolfson. 1992. "Consensus movements, conflict movements, and the cooptation of civic and state infrastructures." In A. D. Morris and C. M. Mueller (eds.) *Frontiers in Social Movement Theory.* New Haven: Yale University Press.

McCormick, John. 1989. *Reclaiming Paradise: The Global Environmental Movement.* Bloomington: Indiana University Press.

McIntosh, Alison C., and Jason L. Finkle. 1995. "The Cairo Conference on Population and Development: A new paradigm." *Population and Development Review* 21(2):223–60.

McKeown, Thomas. 1976. *The Modern Rise of Population.* New York: Academic Press.

McLaren, Angus. 1992. "The sexual politics of reproduction in Britain." Pp. 85–100 in John R. Gillis, Louise A. Tilly, and David Levine (eds.) *The European Experience of Declining Fertility, 1850–1970: The Quiet Revolution.* Cambridge, MA: Basil Blackwell.

McLaughlin, Martin M. 1996. "Food security in a globalizing economy." *Development* no. 4:51–55.

McMichael, Philip, and Laura Raynolds. 1994. "Capitalism, agriculture, and the world economy." Pp. 316–38 in Leslie Sklair (ed.) *Capitalism and Development.* New York: Routledge.

McNeely, Jeffrey. 1991. "Bio-diversity: The economics of conservation and management." Pp. 145–55 in J.T. Winpenny (ed.) *Development Research: The Environmental Challenge.* London: Overseas Development Institute.

Meadows, Donella H., Dennis L. Meadows, Jørgen Randers, and William Behrens III. 1972. *The Limits to Growth.* New York: Universe Books.

Meadows, Donella H., Dennis L Meadows, and Jørgen Randers. 1992. *Beyond the Limits: Confronting Global Collapse, Envisioning a Sustainable Future.* Post Mills, VT: Chelsea Green Publishing Company.

Merchant, Carolyn. 1984. "Women of the Progressive Conservation Movement: 1900–1916." *Environmental Review* 8:57–85.

———. 1992. *Radical Ecology: The Search for a Livable World.* New York: Routledge.

Merrick, Thomas W. with the Population Reference Bureau staff. 1991. "World population in transition." *Population Bulletin* 41: no. 2.

Michaelson, Marc. 1994. "Wangari Maathai and Kenya's Green Belt Movement: Exploring the evolution and potentialities of consensus movement mobilization." *Social Problems* 41:540–61.

Michelson, William. 1970. *Man and His Urban Environment: A Sociological Approach.* Reading, MA: Addison-Wesley Publishing Company.

Micklin, Michael, and David F. Sly. 1998. "The ecological complex: a conceptual elaboration." Pp. 51–66 in Michael Micklin and Dudley L. Poston, Jr., (eds.) *Continuities in Sociological Human Ecology.* New York: Plenum Press.

Mies, Maria. 1986. *Patriarchy and Accumulation on a World Scale.* London: Zed Books.

———. 1994. "People or population." Pp. 41–64 in Rosiska Darcey de Oliveira (ed.) *Terra Femina.* Brazil: Institute of Cultural Action.

Milani, B. 2000. *Designing the Green Economy.* Lanham, MD: Rowman and Littlefield.

Milbraith, Lester. 1984. *Environmentalists: Vanguard for a New Society.* Albany: State University of New York Press.

Miller, Kenton R., Walter V. Reid, and Charles V. Barber. 1991. "Deforestation and species loss: Responding to the crisis." In J.T. Mathews (ed.) *Preserving the Global Environment: The Challenge of Shared Leadership.* New York: W.W. Norton & Company.

Miringoff, Marc, and Marque-Luisa Miringoff. 1999. *The Social Health of the Nation: How America Is Really Doing.* New York: Oxford University Press.

Mitchell, Donald O., and Merlinda D. Ingko. 1993. *The World Food Outlook.* Washington, DC: The World Bank.

Mitchell, Robert Cameron, Angela G. Mertig, and Riley E. Dunlap. 1992. "Twenty years of environmental mobilization: Trends among national environmental organizations." Pp. 11–26 in R. E. Dunlap and A. G. Mertig (eds.) *American Environmentalism.* Philadelphia, PA: Taylor and Francis.

Mitlin, Diana. 1992. "Sustainable development: A guide to the literature." *Environment and Urbanization* 4(1):111–24.

Mol, A. P. J. 1995. *The Refinement of Production*. Utrecht: Van Arkel.

———. 1997. "Ecological modernization: Industrial transformations and environmental reform." Pp. 138–49 in M. Redclift and G. Woodgate (eds.) *The International Handbook of Environmental Sociology.* Northhampton, MA: Edward Elgar.

———. 2000. "Globalization and environment: Between apocalypse-blindness and ecological modernization." Pp. 121–50 in G. Spaargaren et al., (eds.) *Environment and Global Modernity.* London: Sage.

Mol, A. P. J., and F. H. Buttel. 2000. "The environmental state under pressure." Unpublished manuscript, Wageningen University, Netherlands.

Mol, Arthur P. J., and David A. Sonnenfeld. 2000. "Ecological modernisation around the world: An introduction." *Environmental Politics* 9(1):3–14.

Mol, Arthur P. J., and Gert Spaargaren. 2000. "Ecological modernisation theory in debate: A review." *Environmental Politics* 9(1):17–49.

Mollenkopf, John H. 1983. *The Contested City.* Princeton, NJ: Princeton University Press.

Molotch, Harvey. 1970. "Oil in Santa Barbara and power in America." *Sociological Inquiry* 40:131–44.

Molotch, Harvey, and Marilyn Lester. 1975. "Accidental news: The great oil spill as local occurrence and national event." *American Journal of Sociology* 81:235–60.

———. 1976. "The city as a growth machine." *American Journal of Sociology* 82:309–30.

Moore, Raymond A. 1993. "The case for war." Pp. 91–110 in Marcia Lynn Whicker, James Pfiffner, and Raymond A. Moore (eds.) *The Presidency and the Persian Gulf War.* Westport, CT: Praeger.

Morris, Aldon D., and Carol McClurg Mueller. 1992. *Frontiers in Social Movement Theory.* New Haven: Yale University Press.

Mueller-Dixon, Ruth. 1993. *Population Policy & Women's Rights: Transforming Reproductive Choice.* Westport, CT: Praeger.

Murdock, Steve H., and Don E. Albrecht. 1998. "The human ecology of agriculture in the United States." Pp. 131–56 in Michael Micklin and Dudley L. Poston, Jr., (eds.) *Continuities in Sociological Human Ecology.* New York: Plenum Press.

Murphy, Raymond. 1994. *Rationality & Nature: A Sociological Inquiry into a Changing Relationship.* Boulder, CO: Westview Press.

———.1997. *Sociology and Nature.* Boulder, CO: Westview Press.

Myers, Norman. 1992. *The Primary Source: Tropical Forests and Our Future.* New York: WW Norton.

———. 1992. Tropical forests: Present status and future outlook. Pp. 3–32 in Norman Myers (ed.) Tropical Forests and Climate. Dordrecht, ND: Kluwer Academic Publishers.

———. 1994. "Population and biodiversity." In S.F. Graham-Smith (ed.) *Population: The Complex Reality.* London: The Royal Society.

Naess, Arne. 1973. "The shallow and the deep, long-range ecology movement." *Inquiry* 16:95–100.

Nash, Roderick Frazier. 1989. *The Rights of Nature: A History of Environmental Ethics.* Madison: University of Wisconsin Press.

National Research Council, Board on Agriculture. 1991. *Managing Global Genetic Resources.* Washington, DC: National Academy Press.

Nazarea, Virginia. 1999. *Cultural Memory and Biodiversity.* Tucson: University of Arizona Press.

Nelkin, Dorothy, and Jane Poulsen. 1990. "When do social movements win? Three campaigns against animal experiments." In 85th Annual Meeting of the American Sociological Association, Washington, DC

Newman, Oscar. 1973. *Defensible Space: Crime Prevention Through Urban Design.* New York: Macmillan.

Nordhaus, William D., and Edward C. Kokkeleberg, (eds.). 1999. *Nature's Numbers: Expanding the National Economic Accounts to Include the Environment.* Washington, DC: National Academy Press.

Northwest Earth Institute website. *www.nwei.org.*

Notestein, Frank W. 1945. "Population, the long view." Pp. 36–57 in T. Schultz (ed.) *Food for the World.* Chicago: University of Chicago Press.

———. 1953. "Economic problems of population change." Pp. 47–68 in *8th International Conference of Agricultural Economists.* London: Oxford University Press.

O'Connor, James. 1973. *The Fiscal Crisis of the State.* New York: St. Martin's Press.

———. 1988. "Capitalism, nature, socialism." *Capitalism, Nature, Socialism* 1:11–38.

———. 1994. "Is sustainable capitalism possible?" Pp. 152–75 in Martin O'Connor (ed.) *Is Capitalism Sustainable?* New York: Guilford Press.

———. 1998. "Murder on the Orient Express: The political economy of the Gulf War." Pp. 212–26 in James O'Connor (ed.) *Natural Causes: Essays in Ecological Marxism.* New York: Guilford Press.

O'Connor, Martin, (ed.). 1994. *Is Capitalism Sustainable? Political Economy and the Politics of Ecology.* New York: The Guilford Press.

Oldeman, L., V. Van Engelen, and J. Pulles. 1990. "The extent of human-induced soil degradation." Annex 5 of L. Oldeman et al., *World Map of the Status of Human-Induced Soil Degradation: An Explanatory Note,* rev. 2nd ed. Waginengen, Netherlands: International Soil Reference and Information Center (ISRIC).

Oliveira, Rosiska Darcy De. 1994. "Memories of Planeta Femea." Pp. 87–99 in Alessandra Alde (ed.) *Terra Femina: Women and Human Rights.* Brazil: Institute of Cultural Action.

Olofsson, G. 1988. "After the working-class movement? An essay on what's 'new' and what's 'social' in the new social movements." *Acta* 31:15–34.

Olsen, Marvin E., Dora G. Lodwick, and Riley E. Dunlap. 1992. *Viewing the World Ecologically.* Boulder, CO: Westview Press.

O'Neill, K. 2000. *Waste Trading Among Rich Nations: Building a New Theory of Environmental Regulation.* Cambridge, MA: MIT Press.

Opschoor, J.B. 1997. "Industrial metabolism, economic growth and institutional change." Pp. 274–86 in M. Redclift and G. Woodgate (eds.) *The International Handbook of Environmental Sociology.* Northhampton, MA: Edward Elgar.

Park, Robert E., and Ernest W. Burgess. 1921. *Introduction to the Science of Sociology.* Chicago: University of Chicago.

Parry, M. 1990. *Climate Change and World Agriculture.* London: Earthscan Publications.

Parsons, Howard L. 1977. *Marx and Engels on Ecology.* Westport, CT: Greenwood Press.

Partridge, Eric. 1966. *Origins. A Short Etymological Dictionary of Modern English.* New York: Macmillan.

Paulino, Leonardo A. 1987. "The evolving food situation." Pp. 23–38 in John Mellor, Christopher L. Delgado, and Malcolm J. Blackie (eds.) *Accelerating Food Production in Sub-Saharan Africa.* Baltimore, MD: Johns Hopkins University Press.

PCSD (The President's Council on Sustainable Development). 1996. *Sustainable America: A New Consensus.* Washington, DC: USGPO.

———. 1999. *Towards A Sustainable America.* Washington, DC: USGPO.

Pearce, D.W., Markandya, A., and Barbier, E. B. 1989. *Blueprint for a Green Economy.* London: Earthscan.

Pearse, A. 1980. *Seeds of Plenty, Seeds of Want.* Oxford: Clarendon Press.

Peel, John and Malcom Potts. 1969. *Textbook of Contraceptive Practice.* Cambridge: University Printing House.

Peet, R., and M. Watts. 1996. "Liberation ecology: Development, sustainability, and environment in an age of market triumphalism." Pp. 1–45 in R. Peet and M. Watts (eds.) *Liberation Ecologies.* London: Routledge.

Pellow, David N. 2000 "Environmental inequality formation." *American Behavioral Scientist* 43(4):581–601.

Peluso, Nancy L. 1991. *Rich Forests, Poor People: Resource Control and Resistance in Java.* Berkeley: University of California Press.

Perelman, Michael. 1976. "The Green Revolution: American agriculture in the Third World." Pp. 111–26 in R. Merrill (ed.) *Radical Agriculture.* New York: Harper Colophon.

Perry, William J. 1991. "Desert Storm and deterrence." *Foreign Affairs* 70:66–83.

Petersen, William. 1975. *Population.* New York: Macmillan.

———. 1979. *Malthus.* Cambridge, MA: Harvard University Press.

Petulla, Joseph M. 1977. *American Environmental History: The Exploitation and Conservation of Natural Resources.* San Francisco, CA: Boyd & Fraser Publishing Co.

———. 1980. *American Environmentalism: Values, Tactics, Priorities.* College Station: Texas A&M University Press.

Pfiffner, James P. 1993. "Presidential policy-making and the Gulf War." Pp. 3–23 in Marcia Lynn Whicker, James P. Pfiffner, and Raymond A. Moore (eds.) *The Presidency and the Persian Gulf War.* Westport, CT: Praeger.

Phillips, James. 1992. "Rethinking U.S. policy in the Middle East." *Background Paper* no. 891. Washington, DC: The Heritage Foundation.

Pichardo, Nelson A. 1997. "New social movements: A critical review." *Annual Review of Sociology* 23:411–30.

Pigram, John J. 1992. "Alternative tourism: Tourism and sustainable resource management." In V.L. Smith and W.R. Eadington (eds.) *Tourism Alternatives: Potentials and Problems in the Development of Tourism.* Philadelphia: University of Pennsylvania Press.

Pimentel, David, and Marcia Pimentel. 1979. *Food, Energy, and Society.* New York: John Wiley.

Piotrow, Phyllis Tilson. 1973. *World Population Crisis: The United States Response.* New York: Praeger.

Plucknett, D. et al. 1987. *Gene Banks and the World's Food.* Princeton: Princeton University Press.

Popper, Frank J. 1992. "The great LULU trading game." *Planning* 58:15–17.

———. 1994. "Finding treasures in TOADS." *Planning* 60:25–28

Porter, G., and J.W. Brown. 1996. *Global Environmental Politics.* Boulder, CO: Westview Press.

Postel, Sandra. 1989. "Halting land degradation," In L.R. Brown (ed.) *State of the World* 1989. New York: W.W. Norton & Co.

———. 1990. "Water for agriculture: facing the limits." *Worldwatch Paper* no. 93. Washington, DC: Worldwatch Institute.

Potter, Harry R. 1996. "Precursors of ecopopulism." Presented at the American Sociological Association Meeting, August 16–20, New York.

———. 1997. "Precursors of the environmental movement II: Scientists and the legislation of the 1960s." Presented at the American Sociological Association, August 9–13, Toronto, Ontario, CAN.

———. 1999. "Public awareness of environmental issues before Earth Day, 1970." Presented at the North Central Sociological Association Meeting, April 15–18, Troy, Michigan.

Raikes, P. 1988. *Modernizing Hunger: Famine, Food Surplus, and Farm Policy in the EEC and Africa.* London: Catholic Institute for International Affairs.

Redclift, Michael. 1987. *Sustainable Development: Exploring the Contradictions.* London: Routledge.

———. 1996. *Wasted: Counting the Costs of Global Consumption.* London: Earthscan.

Redclift, Michael, and Graham Woodgate. 1997. "Sustainability and Social Construction." Pp. 55–70 in M. Redclift and G. Woodgate (eds.) *The International Handbook of Environmental Sociology.* Northhampton, MA: Edward Elgar.

Redclift, M., and T. Benton, (eds.) 1994. *Social Theory and the Global Environment.* London: Routledge.

Redefining Progress website. *www.rprogress.org.*

Revkin, Andrew. 1990. *The Buning Season: Murder of Chico Mendes.* Boston: Houghton Mifflin.

Rich, Bruce. 1994. *Mortgaging the Earth: The World Bank, Environmental Impoverishment, and the Crisis of Development.* Boston: Beacon Press.

Richardson, Dick. 1997. "The Politics of Sustainable Development." Pp. 43–60 in S. Baker et al., (eds.) *The Politics of Sustainable Development.* London: Routledge.

Ritchie, Mark. 1996. "Control of trade by multinationals: impact of the Uruguay Round of GATT on sustainable food security." *Development.* no. 4:40–44.

Roberts, J. Timmons, and Peter E. Grimes. 1997. "Carbon intensity and economic development 1962–1991: A brief exploration of the environmental Kuznets curve." *World Development* 25, no. 2:181–87.

Rosenbaum, W.A. 1973. *The Politics of Environmental Concern.* New York: Praeger.

———.2000. "Escaping the 'battered agency syndrome': EPA's gamble with regulatory reinvention." Pp. 165–89 in N.J. Vig and M.E. Kraft (eds.) *Environmental Policy.* 4th ed. Washington, DC: CQ Press.

Rothman, Franklin Daniel, and Pamela E. Oliver. 1999. "From local to global: The anti-dam movement in southern Brazil, 1979–1992." *Mobilization* 4(1):41–57.

Rowell, Andrew. 1996. *Green Backlash: Global Subversion of the Environmental Movement.* London: Routledge.

Rucht, Dieter. 1989. "Environmental Movement Organizations in West Germany and France Structure and Interorganizational Relations." *International Social Movement Research* 2:61–94.

Rudel, Thomas K. 1989. "Population, development, and tropical deforestation: A cross-national study." *Rural Sociology* 54:327–38.

———. 2000. "Organizing for sustainable development: Conservation organizations and the struggle to protect tropical rain forests in Esmeraldas, Ecuador." *Ambio* 29(2):78–82.

Rudel, Thomas K., with B. Horowitz. 1993. *Tropical Deforestation: Small Farmers and Land Clearing in the Ecuadorian Amazon.* New York: Columbia University Press.

Sachs, Wolfgang. 1992. "One world." Pp. 102–15 in Wolfgang Sachs (ed.) *The Development Dictionary.* London: Zed Books.

———. (ed.). 1993. *Global Ecology: A New Arena of Political Conflict.* London: Zed Books.

———.1997. "Sustainable development." Pp. 71–82 in M. Redclift and G. Woodgate (eds.) *The International Handbook of Environmental Sociology.* Northampton, MA: Edward Elgar.

———. 1999. *Planet Dialectics.* London: Zed Books.

Sachs, W., R. Loske, and M. Linz. 1998. *Greening the North.* London: Zed Books.

Sadik, Nafis. 1990. *The State of World Population, 1990: Choices for a New Century.* New York: United Nations Fund for Population Action.

Sale, Kirkpatrick. 1993. "The U.S. green movement today." *The Nation* 257: 92–96.

Sampson, Anthony. 1975. *The Seven Sisters: The Great Oil Companies and the World They Shaped.* New York: Viking Press.

Sanchez, P. 1976. *Properties and Management of Soils in the Tropics.* New York: Wiley-Interscience.

Sandbach, Francis. 1978. "The rise and fall of the *Limits to Growth* debate." *Social Studies of Science* 8:495–520.

Sanderson, Steven. 1989. "Mexican agricultural policy in the shadow of the U.S. farm crisis." Pp. 205–33 in David Goodman and Michael Redclift (eds.) *The International Farm Crisis.* New York: St. Martin's Press.

Scarce, Rik. 1990. *Eco-Warriors: Understanding the Radical Environmental Movement.* Chicago: The Noble Press, Inc.

Schmid, Carol L. 1987. "The green movement in West Germany: Resource mobilization and institutionalization." *Journal of Political and Military Sociology* 15:33–37.

Schmidheiny, Stephan (with the Business Council for Sustainable Development). 1992. *Changing Course: A Global Business Perspective on Development and the Environment.* Cambridge: The MIT Press.

Schnaiberg, Allan. 1972. "Environmental sociology and the division of labor." Unpublished manuscript, Department of Sociology, Northwestern University, Evanston, IL.

———. 1975. "Social syntheses of the societal-environmental dialectic: The role of distributional impacts." *Social Science Quarterly* 56:5–20.

———. 1980. *The Environment: From Surplus to Scarcity.* New York: Oxford University Press.

———. 1983. "Redistributive goals versus distributional politics: Social equity limits in environmental and appropriate technology movements." *Sociological Inquiry* 53:200–19.

———. 1994. "The political economy of environmental problems and policies: Consciousness, conflict, and control capacity." Pp. 23–64 in Lee Freese (ed.) *Advances in Human Ecology,* vol. 3. Greenwich, CT: JAI Press.

———. 1997. "Sustainable development and the treadmill of production." Pp. 72–88 in S. Baker et al., (eds.) *The Politics of Sustainable Development.* London: Routledge.

Schnaiberg, Allan, and Kenneth A. Gould. 1994. *Environment and Society: The Enduring Conflict.* New York: St. Martin's Press.

Schneider, S. 1989. *Global Warming: Entering the Greenhouse Century.* San Francisco: Sierra Club Books.

Schnore, Leo R. 1958. "Social morphology and human ecology." *American Journal of Sociology* 63:620–34.

Schumacher, E.F. 1973. *Small Is Beautiful: Economics As If People Mattered.* New York: Perennial Library.

Schwarzweller, Harry K., and Thomas A. Lyson. 1995. "Introduction: Researching the sustainability of agriculture and rural communities." *Research in Rural Sociology and Development* 6:ix–xvii.

Scott, A. 1996. *Ideology and the New Social Movements.* Boston: Unwin Hyman.

Seabrook, Jeremy. 1996. "Development as colonialism: The ODA in India." *Race and Class* 37(4):13–29.

Seccombe, Wally. 1983. "Marxism and demography." *New Left Review* 137:22–47.

Selden, T., and D. Song. 1994. "Environmental quality and development: Is there a Kuznets curve for air pollution emissions?" *Journal of Environmental Economics and Management* 27:147–62.

Sharlin, Allan. 1986. "Urban-rural differences in fertility in Europe during the demographic transition." Pp. 234–60 in Ansley J. Coale and Susan Cotts Watkins (eds.) *The Decline of Fertility in Europe.* Princeton, NJ: Princeton University Press.

Shiva, Vandana. 1989. *Staying Alive: Women, Ecology and Development.* London: Zed Books.

———. 1991. "The green revolution in the Punjab." *The Ecologist* 21:57–61.

———. 1993. "The greening of the global reach." Pp. 206–18 in Wolfgang Sachs (ed.) *The Development Dictionary.* London: Zed Books.

———. 1996. "The seeds of our future." *Development* 4:14–21.

Shove, E. 1997. "Revealing the invisible: Sociology, energy, and the environment." Pp. 261–73 in M. Redclift and G. Woodgate (eds.) *The International Handbook of Environmental Sociology.* Northhampton, MA: Edward Elgar.

Shover, Neil, Donald A. Clelland, and John Lynxwiler. 1986. *Enforcement or Negotiation: Constructing a Regulatory Bureaucracy.* Albany, NY: SUNY Press.

Silva, Eduardo. 1994. "Contemporary environmental politics in Chile: The struggle over the comprehensive law." *Industrial & Environmental Crisis Quarterly* 8(4):323–43.

Simon, Julian L. 1982. *The Ultimate Resource.* Princeton, NJ: Princeton University Press.

———. 1996. *The Ultimate Resource 2.* Princeton, NJ: Princeton University Press.

Simon, Julian L., and Herman Kahn, (eds.) 1984. *The Resourceful Earth.* New York: Basil Blackwell.

Skowronek, W. 1990. *Building a New American State.* New York: Cambridge University Press.

Sly, David F. 1972. "Migration and the ecological complex." *American Sociological Review* 37:615–28.

Sly, David F., and Jeff Tayman. 1977. "Ecological approach to migration reexamined." *American Sociological Review* 42:783–95.

Smelser, Neil J. 1962. *Theory of Collective Behavior.* New York: The Free Press.

Smil, Vaclav. 1991. "Population growth and nitrogen: an exploration of a critical existential link." *Population and Development Review* 17:569–601.

———. 1994. "How many people can the earth feed?" *Population and Development Review* 20:255–92.

Smith and Hawken website. *www.smith-hawken.com.*

Smith, Jackie. 1997. "Characteristics of the Modern Transnational Social Movement Sector." Pp. 42–58 in Jackie Smith, Charles Chatfield, and Ron Pacnucco (eds.) *Transnational Social Movements and Global Politics.* Syracuse, NY: Syracuse University Press.

Snow, David A., and Robert D. Benford. 1992. "Ideology, Frame Resonance and Participant Mobilization." in B. Klandermans, H. Kriesi, and S. Tarrow (eds.) *From Structure to Action: Social Movement Participation Across Cultures.* Greenwich, CT: JAI.

Soloway, Richard A. 1982. *Birth Control and the Population Question.* Chapel Hill: University of North Carolina Press.

Sonnenfeld, David A. 1998. "From brown to green? Late industrialization, social conflict and adoption of environmental technologies in Thailand's pulp industry." *Organization & Environment* 11(1):59–87.

———. 2000. "Social movements and ecological modernization: The transformation of pulp and paper manufacturing." Working Paper 00–6. Institute of International Studies, Berkeley: University of California.

———.2000. "Contradictions of ecological modernisation: Pulp and paper manufacturing in South-East Asia." *Ecological Modernization Around the World.* Arthur P.J. Mol and David A. Sonenfeld (eds.) London: Frank Cassy Publishers.

Spaargaren G. 1996. *The Ecological Modernization of Production and Consumption.* Wageningen, Netherlands: Ph.D. dissertation, Wageningen University.

Steinhart, Carol E., and John S. Steinhart. 1972. *Blowout: A Case Study of the Santa Barbara Oil Spill.* California: Duxbury Press.

Steinhart, John S., and Carol E. Steinhart. 1974. "Energy use in the U.S. food system." *Science* 184:307–16.

Stokes, C. Shannon. 1995. "Explaining the demographic transition: Institutional factors in fertility decline." *Rural Sociology* 60:1–22.

Stoncich, Susan C. 1989. "Dynamics of social processes and environmental destruction: A Central American case study." *Population and Development Review* 15:269–96.

Stone, Roger D., and Eve Hamilton. 1991. *Global Economics and the Environment: Toward Sustainable Rural Development in the Third World*. New York: Council on Foreign Relations Press.

Stren, Richard, Rodney White, and Joseph Whitney. 1992. *Sustainable Cities: Urbanization and the Environment in International Perspective*. Boulder, CO: Westview Press.

Sunderlin, William D. 1995. "Managerialism and the conceptual limits of sustainable development." *Society and Natural Resources* 8:481–92.

Sustainable Seattle website, 1995. *www.scn.org/sustainable/indicators/1995*.

Switzer, Jacqueline Vaughn. 1997. *Green Backlash: The History and Politics of Environmental Opposition in the U.S.* Boulder, CO: Lynne Reinner Publishers.

Szasz, Andrew. 1994. *EcoPopulism: Toxic Waste and the Movement for Environmental Justice*. Minneapolis: University of Minnesota Press.

Szasz, Andrew, and Michael Meuser. 1997. "Environmental inequalities: Literature review and proposals for new directions in research and theory." *Current Sociology* 45:99–120.

———. 2000. "Unintended, inexorable: The production of environmental inequalities in Santa Clara County, California." *American Behavioral Scientist* 43(4):602–32.

Szertzer, Simon. 1988. "The importance of social intervention in Britain's mortality decline c. 1850–1914: A reinterpretation of the role of public health." *Social History of Medicine* 1: 1998.

———. 1997. "Economic growth, disruption, deprivation, disease, and death: On the importance of the politics of public health for development." *Population and Development Review* 23:693–728.

Talbot, Lee M. 1986. "Demographic factors in resource depletion and environmental degradation in East African rangeland." *Population and Development Review* 12:441–51.

Tanzer, Michael. 1975. *The Energy Crisis, World Struggle for Power, and Wealth*. New York: Monthly Review.

———. 1991. "Oil and the Persian Gulf crisis." *Monthly Review* 42:12–16.

Taylor, Bron Raymond (ed). 1995. *Ecological Resistance Movements: The Global Emergence of Radical and Popular Environmentalism*. Albany: State University of New York Press.

Taylor, Bron, Heidi Hadsell, Lois Lorentzen, and Rik Scarce. 1993. "Grass-roots resistance: The emergence of popular environmental movements in less affluent countries." Pp. 69–89 in S. Kamieniecki (ed.) *Environmental Politics in the International Arena*. Albany: State University of New York Press.

Taylor, Dorceta E. 2000. "The rise of the environmental justice paradigm." *American Behavioral Scientist* 43(4):508–80.

Taylor, P.J., and F.H. Buttel. 1992. "How do we know we have global environmental problems? Science and the globalization of environmental discourse." *Geoforum* 23:405–16.

Tellegen, E., and M. Wolsink. 1998. *Society and Environment*. Amsterdam: Gordon and Breach.

Thornton, J. 2000. *Pandora's Poison*. Cambridge, MA: MIT Press.

Thrupp, Lori Ann, (ed.) 1996. "New partnerships for sustainable agriculture." Washington, DC: World Resources Institute Paper.

Tilly, Charles. 1978. *From Mobilization to Revolution*. New York: McGraw-Hill Publishing Co.

Tivy, J. 1990. *Agricultural Ecology*. Essex: Longman, Harlow.

Tokar, Brian. 1990. "Marketing the environment." *Zeta Magazine,* February 1990, Pp. 15–21.

———. 1997. *Earth for Sale: Reclaiming Ecology in the Age of Corporate Greenwash*. Boston: South End Press.

Turner, Jonathan H. 1998. *The Structure of Social Theory.* Belmont, CA: Wadsworth Publishing Co.

Turner, Jonathan H., Leonard Beghley, and Charles H. Powers. 1998. *The Emergence of Social Theory.* Belmont, CA: Wadsworth Publishing Co.

UNDP (United Nations Development Programme). 1992. *Human Development Report 1992*. New York: Oxford University Press.

———. 1998. *Human Development Report 1998*. New York: Oxford University Press.

UNEP (United Nations Environment Programme). 1992. "Committee of International Development Institutions on the Environment (CIDIE): Workshop on environmental and natural resource accounting: Summary record." *Environmental Economics Series,* Paper no. 3, Nairobi.

United Nations. 1992. *Agenda 21: Programme of Action for Sustainable Development*. Pp. 291–94 in *Statement of Forest Principles.* New York: United Nations Publication.

United Nations Children's Fund. 1992. *State of the World's Children.* New York: United Nations.

U.S. Bureau of the Census. 1975. *Historical Statistics of the United States: Colonial Times to 1970*. Washington, DC: Department of Commerce.

U.S. Bureau of the Census. 1997. *Statistical Abstract of the United States: 1997.* 117th edition. Washington, DC

U.S. General Accounting Office (USGAO). 1983. *Siting of Hazardous Waste Landfills and Their Correlation with the Racial and Socio-Economic Status of Surrounding Communities.* Washington, DC: GAS/RCED-83–168, June.

U.S. Office of Technology Assessment. 1986. *Technology and Structural Unemployment: Reemploying Displaced Workers.* Washington, DC: U.S. Government Printing Office.

USAID website, *www.info.usaid.gov/environment.*

U'wa, Berito Kuwar. 1999. "Banking on earth, light, water." *Yes! A Journal of Positive Futures.*

U'wa Defense Project. *www.solcommunications.com/usa.*

Van der Ryn, Sim, and Stuart Cowan. 1995. *Ecological Design*. Washington, DC: Island Press.

Vaughan, D., and T. Chang. 1993. "In situ conservation of rice genetic resources." *Economic Botany* 46:349–67.

Vig, N.J., and M.E. Kraft. 1984. *Environmental Policy in the 1980s: Reagan's New Agenda.* Washington, DC: CQ Press.

vonWeitzsacher, Ernst, Amory Lovins, and Hunter Lovins. 1997. *Factor Four: Doubling Wealth, Halving Resource Use.* London: Earthscan.

Wallerstein, Immanuel. 1974. *The Modern World-System I.* San Diego, CA: Academic Press, Inc.

Walley, Noah, and Bradley Whitehead. 1994. "It's not easy being green." *Harvard Business Review* 72:46–52.

Walsh, Edward. 1981. "Resource mobilization and citizen protest in communities around Three Mile Island." *Social Problems* 29:1–21.

Walsh, Edward J. 1986. "The role of target vulnerabilities in high-technology protest movements: the nuclear establishment at Three Mile Island." *The Eastern Sociological Society* 1:199–217.

Walsh, Edward, Rex Warland, and D. Clayton Smith. 1997. *Don't Burn It Here: Grassroots Challenges to Trash Incinerators.* University Park: Pennsylvania State University Press.

Walsh, J. 1991. "Preserving the options: food productivity and sustainability." Consultative Group on International Agricultural Research, *Issues in Agriculture* no. 2.

Wargo, John. 1996. *Our Children's Toxic Legacy: How Science and Policy Fail to Protect Us from Pesticides.* New Haven: Yale University Press.

Warming, J.E.B. 1909. *Oecology of Plants.* New York: Oxford University Press.

Warnock, John W. 1987. *The Politics of Hunger* Toronto, CAN: Methuen.

Warren, Dean. 1990. "Death and hype in Amazonia." *The Wall Street Journal,* July 12: Section A, p. 9.

Warren, D. Michael, L. Jan Slikkerveer, and David Brokensha, (eds.). 1995. *The Cultural Dimension of Development: Indigenous Knowledge Systems.* London: Intermediate Technology Publications.

Warren, Karen J. 1987. "Feminism and ecology: Making connections." *Environmental Ethics* 9:3–20.

———. 1993. "The power and the promise of ecological feminism." In Michael E. Zimmerman and J. Baird Callicott (eds.) *Environmental Philosophy: From Animal Rights to Radical Ecology.* Englewood Cliffs, NJ: Prentice Hall.

———. 1994. *Ecological Feminsim* London: Routledge.

Watkins, Susan Cotts. 1986. "Conclusion." Pp. 420–50 in Ansley J. Coale and Susan Cotts Watkins (eds.) *The Decline of Fertility in Europe.* Princeton, NJ: Princeton University Press.

———. 1991. *From Provinces into Nations: Demographic Integration in West Europe, 1870–1960.* Princeton, NJ: Princeton University Press.

Watson, Robert T. 1999. "Overview: New strategies, strengthened partnerships." *Earth Matters.* Washington, DC: World Bank.

Weale, A. 1992. *The New Politics of Pollution.* Manchester, England: Manchester University Press.

Weale, A., G. Pridham, A. Williams, and M. Porter. 1996. "Environmental administration in six European countries: Secular convergence or national distinctiveness?" *Public Administration* 74:255–74.

Webb, Patrick. 1995. "A time of plenty, a world of need: The role of food aid in 2020." *Brief* no. 10. Washington, DC: International Food Policy Research Institute.

Weber, Edward P. 2000. "A new vanguard for the environment: Grass-roots ecosystem management as a new environmental movement." *Society & Natural Resources* 13:237–59.

Weigert, Andrew J. 1994. "Lawns of weeds: Status in opposition to life." *The American Sociologist* 25:80–96.

Wessel, R. 1983. *Trading the Future: Farm Exports and the Concentration of Economic Power in our Food System.* San Francisco, CA: Institute for Food and Development Policy.

West, Patrick C., and Steven R. Brechin, (eds). 1991. *Resident Peoples and National Parks: Social Dilemmas and Strategies in International Conservation.* Tucson: The University of Arizona Press.

Wheat, Sue. 1994. "Taming tourism." *Geographical Magazine* 66(4):16–20.

White, Nadia. 1998. "Ecology group issues warning: Earth liberation front threatens further Vail action." *Boulder News,* 22 October.

Whyte, William F. 1943 *Street Corner Society: The Social Structure of an Italian Slum.* Chicago: University of Chicago Press.

Wikan, Unni. 1995. "Sustainable development in the mega-city." *Current Anthropology* 36(4):635–55.

Wilson, E.O. 1989. "Threats to biodiversity." *Scientific American* 271:108–16.

Wilson, William J. 1996. *When Work Disappears.* New York: Random House.

———. 1999. *The Bridge Over the Racial Divide: Rising Inequality and Coalition Politics.* Berkeley: University of California Press.

Wimberely, Dale W. 1991. "Transnational corporate investment and food consumption in the Third World: a cross-national analysis." *Rural Sociology* 56:406–31.

Wimberely, Dale W., and Rosario Bello. 1992. "Effects of foreign investment, exports, and economic growth on Third World food consumption." *Social Forces* 70:895–921.

Wood, R. E. 1986. *From Marshall Plan to Debt Crisis: Foreign Aid and Development Choices in the World Economy.* Berkeley: University of California Press.

The World Bank website. *www.worldbank.org.*

The World Bank. 1991. *The World Bank and the Environment: A Progress Report.* Washington, DC: The World Bank.

———. 1992a. *The World Bank and the Environment.* Washington, DC: The World Bank.

———. 1992b. *World Development Report 1992: Development and the Environment.* New York: Oxford University Press.

———. 1994. *Making Development Sustainable.* Washington, DC: The World Bank.

———. 1996. *Integrated Pest Management Strategy.* Washington, DC: The World Bank.

———. 1998. "Press release: Zimbabwe's parks get US$68 million." *www.worldbank.org/html.extdr.extme.1791.htm*

———. 1998. *World Development Indicators.* Washington, DC.

———. 1999. *World Development Indicators.* Washington, DC.

World Commission on Environment and Development (WCED). 1987. *Our Common Future.* New York: Oxford University Press.

World Debt Tables , vol. 1, 1996. Washington, DC: The World Bank.

World Energy Commission. 1993. *Energy for Tomorrow's World.* New York: St. Martin's Press.

World Resources Institute. 1986. *World Resources.* New York: Basic Books.

———. 1992. *World Resources 1992–93.* New York: Oxford University Press.

———. 1994. *World Resources 1994–1995: A Guide to the Global Environment.* New York: Oxford University Press.

———. 1996. *World Resources 1996–97.* New York: Oxford University Press.

Worster, Donald. 1979. *Nature's Economy: The Roots of Ecology.* New York: Anchor/Doubleday.

———. 1993. *The Wealth of Nature: Environmental History and the Ecological Imagination.* New York: Oxford University Press.

———. 1993. "The shaky ground of sustainability." Pp. 132–48 in Wolfgang Sachs (ed.) *Global Ecology: A New Arena of Conflict.* London: Zed Books.

Wrigley, E. A., and R. S. Schofield. 1981. *The Population History of England: 1541–1871.* London: Edward Arnold.

Yago, Glenn. 1980. "Corporate power and urban transportation: A comparison of public transit decline in the United States and Germany." Pp. 296–323 in Maurice Zeitlin (ed.) *Classes, Class Conflict, and the State.* Cambridge, MA: Winthrop Publishers, Inc.

Yeager, Peter. 1991. *The Limits of Law: The Public Regulation of Private Pollution.* New York: Cambridge University Press.

Yearley, S. 1996. *Sociology, Environmentalism, Globalization.* London: Sage.

Yearly, Steven. 1991. *The Green Case: A Sociology of Environmental Issues, Arguments and Politics.* London: Harper Collins Academic.

Yergin, Daniel. 1991. *The Prize: The Epic Quest for Oil, Money, and Power.* New York: Simon & Schuster.

Young, Michael, and Peter Willmont. 1957. *Family and Kinship in East London.* London: Routledge and K. Paul.

Zald, Mayer N., and John D. McCarthy. 1987. *Social Movements in an Organizational Society: Collected Essays.* New Brunswick, NJ: Transaction Publishers.

3M website. *www.3m.com.*

Index

D

E

G

H

M

N

O

P

R

S

X

Y

Z